Springer Complexity

Springer Complexity is an interdisciplinary program publishing the best research and academic-level teaching on both fundamental and applied aspects of complex systems—cutting across all traditional disciplines of the natural and life sciences, engineering, economics, medicine, neuroscience, social and computer science.

Complex Systems are systems that comprise many interacting parts with the ability to generate a new quality of macroscopic collective behavior the manifestations of which are the spontaneous formation of distinctive temporal, spatial or functional structures. Models of such systems can be successfully mapped onto quite diverse "real-life" situations like the climate, the coherent emission of light from lasers, chemical reaction–diffusion systems, biological cellular networks, the dynamics of stock markets and of the internet, earthquake statistics and prediction, freeway traffic, the human brain, or the formation of opinions in social systems, to name just some of the popular applications.

Although their scope and methodologies overlap somewhat, one can distinguish the following main concepts and tools: self-organization, nonlinear dynamics, synergetics, turbulence, dynamical systems, catastrophes, instabilities, stochastic processes, chaos, graphs and networks, cellular automata, adaptive systems, genetic algorithms and computational intelligence.

The three major book publication platforms of the Springer Complexity program are the monograph series "Understanding Complex Systems" focusing on the various applications of complexity, and the "Springer Series in Synergetics", which is devoted to the quantitative theoretical and methodological foundations, and the "SpringerBriefs in Complexity" which are concise and topical working reports, case-studies, surveys, essays and lecture notes of relevance to the field. In addition to the books in these two core series, the program also incorporates individual titles ranging from textbooks to major reference works.

Editorial and Programme Advisory Board

Henry Abarbanel, Institute for Nonlinear Science, University of California, San Diego, USA

Dan Braha, New England Complex Systems, Institute and University of Massachusetts, Dartmouth, USA

Péter Érdi, Center for Complex Systems Studies, Kalamazoo College, USA and Hungarian Academy of Sciences, Budapest, Hungary

Karl Friston, Institute of Cognitive Neuroscience, University College London, London, UK

Hermann Haken, Center of Synergetics, University of Stuttgart, Stuttgart, Germany

Viktor Jirsa, Centre National de la Recherche Scientifique (CNRS), Université de la Méditerranée, Marseille, France

Janusz Kacprzyk, System Research, Polish Academy of Sciences, Warsaw, Poland

Kunihiko Kaneko, Research Center for Complex Systems Biology, The University of Tokyo, Tokyo, Japan

Scott Kelso, Center for Complex Systems and Brain Sciences, Florida Atlantic University, Boca Raton, USA

Markus Kirkilionis, Mathematics Institute and Centre for Complex Systems, University of Warwick, Coventry, UK

Jürgen Kurths, Nonlinear Dynamics Group, University of Potsdam, Potsdam, Germany

Andrzej Nowak, Department of Psychology, Warsaw University, Poland

Hassan Qudrat-Ullah, School of Administrative Studies, York University, Canada

Linda Reichl, Center for Complex Quantum Systems, University of Texas, Austin, USA

Peter Schuster, Theoretical Chemistry and Structural Biology, University of Vienna, Vienna, Austria

Frank Schweitzer, System Design, ETH Zürich, Zürich, Switzerland

Didier Sornette, Entrepreneurial Risk, ETH Zürich, Zürich, Switzerland

Stefan Thurner, Section for Science of Complex Systems, Medical University of Vienna, Vienna, Austria

Understanding Complex Systems

Founding Editor: Scott Kelso

Future scientific and technological developments in many fields will necessarily depend upon coming to grips with complex systems. Such systems are complex in both their composition—typically many different kinds of components interacting simultaneously and nonlinearly with each other and their environments on multiple levels—and in the rich diversity of behavior of which they are capable.

The Springer Series in Understanding Complex Systems series (UCS) promotes new strategies and paradigms for understanding and realizing applications of complex systems research in a wide variety of fields and endeavors. UCS is explicitly transdisciplinary. It has three main goals: First, to elaborate the concepts, methods and tools of complex systems at all levels of description and in all scientific fields, especially newly emerging areas within the life, social, behavioral, economic, neuro- and cognitive sciences (and derivatives thereof); second, to encourage novel applications of these ideas in various fields of engineering and computation such as robotics, nano-technology and informatics; third, to provide a single forum within which commonalities and differences in the workings of complex systems may be discerned, hence leading to deeper insight and understanding.

UCS will publish monographs, lecture notes and selected edited contributions aimed at communicating new findings to a large multidisciplinary audience.

More information about this series at http://www.springer.com/series/5394

Alexander Mehler · Andy Lücking
Sven Banisch · Philippe Blanchard
Barbara Frank-Job
Editors

Towards a Theoretical Framework for Analyzing Complex Linguistic Networks

Springer

Editors
Alexander Mehler
Goethe-University Frankfurt am Main
Department of Computer Science
 and Mathematics
Frankfurt am Main
Germany

Andy Lücking
Goethe-University Frankfurt am Main
Department of Computer Science
 and Mathematics
Frankfurt am Main
Germany

Sven Banisch
Max Planck Institute for Mathematics
 in the Sciences
Inselstrasse 22
D-04103 Leipzig
Germany

Philippe Blanchard
Department of Physics
University of Bielefeld
Bielefeld
Germany

Barbara Frank-Job
Faculty of Linguistics & Literary Studies
University of Bielefeld
Bielefeld
Germany

ISSN 1860-0832   ISSN 1860-0840   (electronic)
Understanding Complex Systems
ISBN 978-3-662-47237-8   ISBN 978-3-662-47238-5   (eBook)
DOI 10.1007/978-3-662-47238-5

Library of Congress Control Number: 2015940024

Springer Heidelberg New York Dordrecht London
© Springer-Verlag Berlin Heidelberg 2016

This work is subject to copyright. All rights are reserved by the Publisher, whether the whole or part of the material is concerned, specifically the rights of translation, reprinting, reuse of illustrations, recitation, broadcasting, reproduction on microfilms or in any other physical way, and transmission or information storage and retrieval, electronic adaptation, computer software, or by similar or dissimilar methodology now known or hereafter developed.

The use of general descriptive names, registered names, trademarks, service marks, etc. in this publication does not imply, even in the absence of a specific statement, that such names are exempt from the relevant protective laws and regulations and therefore free for general use.

The publisher, the authors and the editors are safe to assume that the advice and information in this book are believed to be true and accurate at the date of publication. Neither the publisher nor the authors or the editors give a warranty, express or implied, with respect to the material contained herein or for any errors or omissions that may have been made.

Printed on acid-free paper

Springer-Verlag GmbH Berlin Heidelberg is part of Springer Science+Business Media
(www.springer.com)

Introduction

Alexander Mehler, Andy Lücking, Sven Banisch,
Philippe Blanchard, and Barbara Frank-Job

1 On the Content of This Book

Currently, we observe an advent of approaches to analyzing linguistic networks with the methods of stochastic physics and graph theory. Generally speaking, a linguistic network is represented by a graph whose vertices denote linguistic units (e.g., words, sentences, or textual units) and whose edges model linguistic (e.g. syntactic, semantic or pragmatic) relations of these units. The aim of models operating on such networks is to capture the synchronic, topological or evolutionary dynamics of linguistic systems, say, on the phonological, morphological, syntactic, semantic or pragmatic level. What these approaches have in common is that they model the structural or temporal dynamics of linguistic systems in order to test information-theoretical or linguistic hypotheses on the grounds of complex network theory. This is partly done in terms of a *strong network perspective* according to which the network approach is seen to be indispensable to test the focal hypotheses. Apparently, the area of language evolution provides a good test case for such an approach.

Alexander Mehler · Andy Lücking
Goethe-University Frankfurt am Main, Department of Computer Science and Mathematics, Frankfurt am Main, Germany
e-mail: {Mehler,Luecking}@em.uni-frankfurt.de

Sven Banisch
Max Planck Institute for Mathematics in the Sciences, Inselstrasse 22,
D-04103 Leipzig, Germany
e-mail: Sven.Banisch@mis.mpg.de

Philippe Blanchard
Faculty of Physics, Bielefeld University, Germany
e-mail: Blanchard@physik.uni-bielefeld.de

Barbara Frank-Job
Faculty of Linguistics & Literary Studies, University of Bielefeld, Bielefeld, Germany
e-mail: Barbara.Job@uni-bielefeld.de

Language evolution can be seen as a meso system that connects language as a macro system with the micro system of cognitive processes of language processing. Starting from such a unified approach to language structure, language change and processing, network approaches try to gain insights into the laws of linguistic information processing in communities of social agents.

In spite of the remarkable success regarding the development of expressive graph models of linguistic systems, these approaches are still in need of a unifying framework. To date, the models are connected by a common methodical stance based on complex network theory in addition to quantitative linguistics. Thus, we face a range of diverse network models that focus on laws of information processing without clarifying their synergetic interdependencies. This is partly due to the lack of shared standards of data modeling, of the interoperability of algorithmic graph models and of the sustainability of the underlying linguistic resources and corpora. Obviously, interdisciplinary research across the boarder of computer science, linguistics and stochastic physics may profit from the availability of such standards.

This book aims at making first steps into the direction of filling this gap. It presents theoretical and empirical results in support of a unifying approach to linguistic networks that may help to overcome bottleneck problems of this field of research. To this end, the book comprises recent research efforts in the area of linguistic networks. It brings together scientists with diverse backgrounds ranging from linguistics to text-technology, from computational humanities to statistical network theory. The book is organized, roughly, into six parts including semantic and syntactic networks, the interplay of language and cognition, the simulation of sociolinguistic dynamics and text-technological resources of network modeling. Special emphasis is put on critical articles and articles that review recent developments in the field. This includes the following fields of research:

- Resources of linguistic network analysis.
- Principles of linguistic network induction.
- Topological models of language structure.
- Models of language dynamics: evolution, diachrony, change.
- Unified models from stochastic physics.
- Network models from cognitive linguistics.
- Network models of phonological, lexical, syntactic, semantic or pragmatic systems.
- Network models of text systems in contrast to language systems.

Dealing with these and related topics, the aim of the book is to advocate and promote network models of linguistic systems that are both based on thorough mathematical models and substantiated in terms of linguistic interpretations. In this way, the book contributes first steps towards establishing a statistical network theory as a theoretical basis of linguistic network analysis across the boarder of the natural sciences and the humanities.

2 Overview of the Book

2.1 Part I: Cognition

Successful applications of network analysis with a particular focus on the interplay of language and cognition are reviewed in the chapter of Beckage and Colunga. Concentrating on semantic and phonological networks, it explores network features and their relation to human language performance including the application to cognitive impairment and atypical behavior.

The chapter by Vitevitch, Goldstein and Johnson combines network tools and data from a psycholinguistic experiment to explore speech perception errors with the aim to understand better what is perceived when a spoken word is misperceived. The experimental results of their phonological association task are evaluated in terms of path' on a network of phonological similarity.

The chapter by De Deyne, Verheyen and Storms compares semantic networks derived from text corpora with networks obtained through word association experiments by looking at macro- and mesoscopic properties of both types of graphs. While the analysis reveals structural similarities at the global level, significant differences between text and word association graphs emerge at a lower level of community structure or centrality. The chapter also presents a comparison with human relatedness judgments.

2.2 Part II: Topology

The chapter by Biemann, Krumov, Roos and Weihe presents a statistical analysis of the motif signatures of co-occurrence graphs including co-authorship networks, communication networks and linguistic co-occurrence graphs of natural and artificial languages. Based on the hypothesis that different word classes serve different functions in a language an analysis of co-occurrence graphs for different word classes (verbs vs. nouns vs. adjectives etc.) is performed which shows that especially verbs are distinguishable from other word classes by their motif signature – across different languages.

The chapter by Araújo and Banisch highlights the need to consider different ways of network induction in network-based analysis of language and reasons that induction and analysis are strongly interdependent tasks. Based on a framework comprising different abstraction levels along with levels of statistical analysis, the authors argue that the field of linguistic networks is challenged by the fact that an interpretation of topological indicators used in network analysis becomes the harder, the higher the abstraction level of the network.

The chapter by Masucci, Kalampokis, Eguíluz and Hernández-García presents an information-theoretic approach to derive a directed network of semantic flow between Wikipedia articles using a complete snapshot of the English Wikipedia. The authors show that the resulting semantic space is characterized by a scale-free behavior at different scales which implies a hierarchical organization of semantic spaces.

The chapter by Zweig confronts the physically-inspired context-free quest for universal structures with the need of contextual interpretations in sociology and in linguistics. Zweig questions the usefulness of network representations of word-adjacency relations, because most of the well-known topological indicators rely on a rather specific network process and they may therefore be misleading if this process is not known or not adequately modeled by the process underlying the method.

2.3 Part III: Syntax

The chapter by Čech, Mačutek and Liu presents a critical review of the application of complex network tools to the analysis of syntax and points out the main challenges for further research. Among many other things, the article discusses the impact of syntax on network properties, the preprocessing of data, and the application of network studies to language typology and acquisition.

A second chapter dealing with syntactic dependency networks is by Chen and Liu. Based on two syntactic dependency networks from different genres this chapter analyses the syntactic status of three function words in Chinese. The importance (the authors propose the notion of syntactic centrality) of the words is analyzed by independently removing them from the network and comparing their statistical characteristics before and after removal.

The chapter by Ferrer i Cancho challenges the existing theory of syntax by confronting the observation that syntactic dependencies between the words of a sentence rarely cross when drawn over a sentence with two null hypotheses for the expected number of crossings by chance. Relying on the trade-off between parsimony and explanatory power, the chapter argues that the minimization of syntactic dependency length (as a principle that derives from limited computational resources of the brain) can explain uncrossing dependencies and that this explanation is, from an economic point of view, preferable over explanations relying on grammar.

2.4 Part IV: Dynamics

The role of cultural transmission in language change across three generations is analyzed on the basis of an extended simulation model by Gong and Shuai. While transmission within the offspring generation and between the offspring and the parent generation fosters language change and leads, at the same time, to mutual understandability within generations and across consecutive generations, interaction between children and their grandparent's generation plays an important role in preserving mutual cross-generational understandability in the long run.

Another simulation study is presented by Baxter who complements his numerical results with analytical arguments. Drawing on an evolutionary approach to language change, the author looks in detail to the convergence behavior of the model on different social networks and with heterogeneous patterns of mutual influence that, taken together, may encode a variety of social structures.

Introduction

The chapter by Maity and Mukherjee presents a simulation study of the effect of inflexible individuals on the dynamics of the naming game and shows that rigid minorities lead to the emergence of dominant states in the population. The model is analyzed on a series of static networks of different complexity ranging from the complete graph to scale-free topologies and a dynamic network obtained from real-world time-varying face-to-face interaction data is also considered.

2.5 Part V: Resources

The requirements of a data format applicable to the wide range of linguistic network data are discussed in the chapter by Stührenberg, Diewald and Gleim. The authors analyze various existing graph formats in relation to their expressivity and support by common tools for network analysis and propose an extension of GraphML as a possibly complex data model of a graph which allows to quickly extract views for specific tasks, rather than extracting incoherent different views from raw data. It is noteworthy, that this chapter grew out of a working group that was constituted at the MLN conference.

The book concludes with the chapter by Mehler and Gleim who present the LN system, an online platform for the automatic generation of lexical networks from texts. It addresses two communities: on the one hand humanities scholars (e.g., historical semanticists) who aim at studying the change of language use as an indicator of social-semantic change. On the other hand, network theorists who are in need of null models for making linguistic networks comparable. The workflow of the LN system – using GraphML as an output standard for linguistic networks – is explained and exemplified.

Acknowledgements

The book collects selected and extended contributions to the workshop on *Modeling Linguistic Networks: from Language Structures to Communication Processes* held at Goethe University Frankfurt, December 10-11, 2012, as part of the research project *Linguistic Networks: Text Technological Representation, Computational Linguistic Synthesis and Physical Modeling* funded by the German Federal Ministry of Education (BMBF). Financial support by the BMBF is gratefully acknowledged.

Contents

Introduction V
Alexander Mehler, Andy Lücking, Sven Banisch, Philippe Blanchard, Barbara Frank-Job
    1   On the Content of This Book V
    2   Overview of the Book VII
        2.1   Part I: Cognition VII
        2.2   Part II: Topology VII
        2.3   Part III: Syntax VIII
        2.4   Part IV: Dynamics VIII
        2.5   Part V: Resources IX

Part I: Cognition

Language Networks as Models of Cognition: Understanding Cognition through Language 3
Nicole M. Beckage, Eliana Colunga
    1   Introduction 3
    2   Language as a Network 5
        2.1   Semantic Networks 5
        2.2   Phonological Networks 6
    3   Global Level Network Structure 7
        3.1   Small-World Structure 8
        3.2   Scale-Free Networks 10
    4   Human Performance in Relation to Network Structure 11
        4.1   Spreading Activation 11
        4.2   Frequency Effects 15
    5   Network Models within Linguistic Networks 16
        5.1   Acquisition 17
        5.2   Network Navigation 21
    6   Understanding Atypical Processes 23
    7   The Future of Language Networks 25
    References 26

Path-Length and the Misperception of Speech: Insights from Network Science and Psycholinguistics 29
Michael S. Vitevitch, Rutherford Goldstein, Elizabeth Johnson
1 Introduction 29
2 Network Analysis: What Can Be Perceived When Speech Is Misperceived? 31
3 Psycholinguistic Experiment: What Is Perceived When Speech Is Misperceived? 34
   3.1 Method 35
   3.2 Results 37
4 Conclusion 40
References 43

Structure and Organization of the Mental Lexicon: A Network Approach Derived from Syntactic Dependency Relations and Word Associations 47
Simon De Deyne, Steven Verheyen, Gert Storms
1 Introduction 47
   1.1 Macro-, Meso-, and Microscopic Properties of the Mental Lexicon 48
   1.2 Acquiring a Mental Lexicon through Language 50
   1.3 Chapter Outline 51
2 Constructing the Networks 53
   2.1 Mental Networks 53
   2.2 Language Networks 54
3 Exploring the Structure of Language and Mental Networks 56
   3.1 Macroscopic Structure 56
   3.2 Mesoscopic Structure 59
   3.3 Semantic Relatedness Evaluation 66
4 Discussion 70
   4.1 Relationship between Language and Word Associations 72
   4.2 Final Words 73
References 74

Part II: Topology

Network Motifs Are a Powerful Tool for Semantic Distinction 83
Chris Biemann, Lachezar Krumov, Stefanie Roos, Karsten Weihe
1 Introduction 84
2 Related Work 86
3 The Case Studies 87
   3.1 Co-occurrence Graphs from Natural Vs. Artificial Language 87
   3.2 Co-occurrence Graphs from Verbs Vs. Other Word Classes 94

		3.3	Peer-to-Peer Streaming Networks	99

        3.3   Peer-to-Peer Streaming Networks 99
        3.4   Co-Authorship Networks from Two Subdisciplines
              of Physics 100
        3.5   Mailing Networks 102
  4   Conclusions and Outlook 103
  References 103

Multidimensional Analysis of Linguistic Networks 107
Tanya Araújo, Sven Banisch
  1   Introduction 107
  2   Linguistic Networks Are Special 109
        2.1   Three Types of Networks 109
        2.2   Network Induction 112
  3   Three Levels of Statistical Analysis 114
        3.1   A Brief Note on Signal Processing on Graphs 115
        3.2   The Statistical Levels 115
        3.3   Stylized Facts in Network Analysis 116
        3.4   Levels in the Statistical Analysis of Networks 118
  4   On the Intelligibility of Statistical Indicators in Linguistic
       Networks 120
        4.1   Path-Based Measures 120
        4.2   Links and Flows, Structure and Function 121
        4.3   Types of Network Flow 122
        4.4   Flow in Linguistic Networks 122
  5   Examples 124
  6   Discussion 124
  7   Concluding Remarks 126
  References 127

Semantic Space as a Metapopulation System: Modelling the
Wikipedia Information Flow Network 133
A. Paolo Masucci, Alkiviadis Kalampokis, Víctor M. Eguíluz,
Emilio Hernández-García
  1   Introduction 133
  2   The Dataset 136
  3   Topology of the Semantic Space 136
  4   Modelling the Semantic Space 141
  5   Discussion 143
  Appendix 145
  References 148

Are Word-Adjacency Networks Networks? 153
Katharina Anna Zweig
  1   Introduction 153
        1.1   Perspectives of Network Analysis 154

2	Definitions		156
	2.1	Definition of Word-Adjacency Networks	156
3	Walk-Based Methods and Network Flows		157
	3.1	Models of Walks	159
4	Word-Adjacency Networks in the Literature		160
5	Summary		162
References			163

Part III: Syntax

Syntactic Complex Networks and Their Applications 167
Radek Čech, Ján Mačutek, Haitao Liu

1	Introduction	167
2	Basic Characteristics of Syntactic Networks	168
3	Early Development of Syntactic Complex Network Analysis	169
4	Role of Syntax in Syntactic Dependency Complex Networks	172
5	Preprocessing of Data for a Syntactic Complex Network Analysis – Pitfalls to be Avoided	177
6	Applications of Syntactic Complex Networks to Language Typology and Acquisition	179
	6.1 Language Typology	180
	6.2 Language Acquisition	181
7	Conclusion	182
References		183

Function Nodes in Chinese Syntactic Networks 187
Xinying Chen, Haitao Liu

1	Introduction	187
2	The Chinese Dependency Networks for This Study	189
3	Chinese Function Words	192
4	Chinese Function Words in the Language Networks	193
	4.1 Network Properties of Chinese Function Words	193
	4.2 Network Manipulation	196
5	Conclusion	198
References		199

Non-crossing Dependencies: Least Effort, Not Grammar 203
Ramon Ferrer-i-Cancho

1	Introduction	203
2	The Syntactic Dependency Structure of Sentences	207
3	The Null Hypothesis	208
4	Alternative Hypotheses	212
	4.1 A Principle of Minimization of Dependency Crossings	213
	4.2 A Principle of Minimization of Dependency Lengths	214

	4.3	The Relationship between Minimization of Crossings and Minimization of Dependency Lengths	216
5	A Stronger Null Hypothesis		219
	5.1	The Probability That Two Edges Cross	220
	5.2	The Expected Number of Edge Crossings	221
6	Another Stronger Null Hypothesis		224
7	Predictions, Testing and Selection		224
8	Discussion		227
	Appendix		229
	References		231

Part IV: Dynamics

Simulating the Effects of Cross-Generational Cultural Transmission on Language Change ... 237
Tao Gong, Lan Shuai

1	Introduction	237
2	Modified Acquisition Framework	240
3	Simulation Results	242
4	Discussions and Conclusions	244
	Appendix	248
	References	254

Social Networks and Beyond in Language Change ... 257
Gareth J. Baxter

1	Introduction		257
2	Utterance Selection Model of Language Change		258
3	Numerical Model		260
4	Analysis		264
5	Social Networks in the Neutral Model		267
6	Weighted Interactor Selection		268
	6.1	Asymmetry Independent of Network Structure	269
	6.2	Asymmetry Depends on Speakers Degree	272
7	Conclusions		274
	Appendix		275
	References		276

Emergence of Dominant Opinions in Presence of Rigid Individuals ... 279
Suman Kalyan Maity, Animesh Mukherjee

1	Introduction		279
2	Related Work		282
3	The Model Description		283
4	Results and Discussion		283
	4.1	The Mean-Field Case	283
	4.2	Scale-Free Networks	287

	5	Time-Varying Networks	290
		5.1 Dataset Description	291
		5.2 The Model Adaptation in the Time-Varying Setting	291
		5.3 Results and Discussion	292
	6	Conclusions and Future Works	294
		References	294

Part V: Resources

Considerations for a Linguistic Network Markup Language 299
Maik Stührenberg, Nils Diewald, Rüdiger Gleim

1	Introduction	299
2	Data Formats	299
	2.1 Data Models	300
	2.2 Data Structures	301
	2.3 Data Serialization	302
3	Existing Formats	304
	3.1 GML	305
	3.2 XGMML	306
	3.3 GraphXML	307
	3.4 GraphML	309
	3.5 GXL	311
	3.6 GrAF	313
	3.7 Summary	315
4	Network Tools	315
5	Proposal for a Linguistic Network Markup Language	318
	5.1 Extending GraphML by Redefinition	320
	5.2 Extending GraphML by XML Namespaces	322
	5.3 Example Instance	325
6	Conclusion	327
	References	327

Linguistic Networks – An Online Platform for Deriving Collocation Networks from Natural Language Texts 331
Alexander Mehler, Rüdiger Gleim

1	Introduction	331
2	On the Parameter Space of LN	334
3	The Software Architecture of LN	336
4	Summary	340
	References	340

List of Contributors

Tanya Araújo
ISEG (Lisboa School of Economics and Management) of the University of Lisbon and Research Unit on Complexity in Economics (UECE), Portugal
e-mail: tanya@iseg.utl.pt

Sven Banisch
Max Planck Institute for Mathematics in the Sciences, Leipzig, Germany
e-mail: Sven.Banisch@mis.mpg.de

Gareth Baxter
Department of Physics, University of Aveiro, Portugal
e-mail: gjbaxter@ua.pt

Nicole Beckage
University of Colorado Boulder, USA
e-mail: nicole.beckage@colorado.edu

Chris Biemann
Computer Science Department, Technische Universität Darmstadt, Germany
e-mail: biem@cs.tu-darmstadt.de

Radek Čech
Department of Czech Language, Faculty of Arts, University of Ostrava, Czech Republic
e-mail: cechradek@gmail.com

Xinying Chen
School of International Study, Xi'an Jiaotong University, Xi'an, China
e-mail: chenxinying@mail.xjtu.edu.cn

Eliana Colunga
University of Colorado Boulder, USA
e-mail: eliana.colunga@colorado.edu

Simon De Deyne
University of Adelaide, Australia
e-mail: simon.dedeyne@adelaide.edu.au

Nils Diewald
Institut für Deutsche Sprache, Mannheim, Germany
e-mail: diewald@ids-mannheim.de

Víctor M. Eguíluz
Instituto de Física Interdisciplinar y Sistemas Complejos IFISC,
Palma de Mallorca, Spain
e-mail: victor@ifisc.uib-csic.es

Ramon Ferrer i Cancho
Complexity and Quantitative Linguistics Lab, LARCA Research Group,
Departament de Ciències de la Computació, Universitat Politècnica de Catalunya
(UPC), Barcelona, Catalonia, Spain
e-mail: rferrericancho@lsi.upc.edu

Rüdiger Gleim
Text Technology Lab, Department of Computer Science and Mathematics,
Goethe-University Frankfurt, Frankfurt am Main, Germany
e-mail: gleim@em.uni-frankfurt.de

Rutherford Goldstein
University of Kansas, Lawrence, KS 66045 USA
e-mail: rgolds@ku.edu

Tao Gong
Haskins Laboratories, New Haven, CT, USA
e-mail: gtojty@gmail.com

Emilio Hernández-García
Instituto de Física Interdisciplinar y Sistemas Complejos IFISC,
Palma de Mallorca, Spain
e-mail: emilio@ifisc.uib-csic.es

Elizabeth Johnson
University of Kansas, Lawrence, KS 66045 USA
e-mail: libbyjohnson54@gmail.com

Alkiviadis Kalampokis
Dipartimento di Scienze e Tecnologie, Universitá degli Studi di Napoli
"Parthenope", Centro Direzionale di Napoli, Napoli, Italy
e-mail: alkis@marine.aegean.gr

Lachezar Krumov
Computer Science Department, Technische Universität Darmstadt, Germany
e-mail: lakkrumov17@yahoo.com

List of Contributors

Haitao Liu
Department of Linguistics, Zhejiang University, Hangzhou, CN-310058, China
e-mail: lhtzju@gmail.com

Ján Mačutek
Department of Applied Mathematics and Statistics, Comenius University,
Mlynská dolina, Bratislava, Slovakia
e-mail: jmacutek@yahoo.com

Suman Kalyan Maity
Dept. of Computer Science and Engineering, Indian Institute of Technology,
Kharagpur, India
e-mail: sumankalyan.maity@cse.iitkgp.ernet.in

Paolo Masucci
Centre for Advanced Spatial Analysis, University College of London,
London, UK
e-mail: paolo_masucci@yahoo.it

Alexander Mehler
Text Technology Lab, Department of Computer Science and Mathematics,
Goethe-University Frankfurt, Frankfurt am Main, Germany
e-mail: mehler@em.uni-frankfurt.de

Animesh Mukherjee
Dept. of Computer Science and Engineering, Indian Institute of Technology,
Kharagpur, India
e-mail: animeshm@cse.iitkgp.ernet.in

Stefanie Roos
Computer Science Department, Technische Universität Darmstadt, Germany
e-mail: stefanie.roos@tu-dresden.de

Lan Shuai
Haskins Laboratories, New Haven, CT, USA
e-mail: susan.shuai@gmail.com

Gert Storms
University of Leuven, Belgium
e-mail: gert.storms@ppw.kuleuven.be

Maik Stührenberg
Institut für Deutsche Sprache, Mannheim, Germany
e-mail: stuehrenberg@ids-mannheim.de

Steven Verheyen
University of Leuven, Belgium
e-mail: steven.verheyen@ppw.kuleuven.be

Michael Vitevich
University of Kansas, Lawrence, KS 66045 USA
e-mail: mvitevit@ku.edu

Karsten Weihe
Computer Science Department, Technische Universität Darmstadt, Germany
e-mail: weihe@cs.tu-darmstadt.de

Katharina Zweig
Department of Computer Science, TU Kaiserslautern, Germany
e-mail: zweig@cs.uni-kl.de

Part I
Cognition

Language Networks as Models of Cognition: Understanding Cognition through Language

Nicole M. Beckage and Eliana Colunga

Abstract. Language is inherently cognitive and distinctly human. Separating the object of language from the human mind that processes and creates language fails to capture the full language system. Linguistics traditionally has focused on the study of language as a static representation, removed from the human mind. Network analysis has traditionally been focused on the properties and structure that emerge from network representations. Both disciplines could gain from looking at language as a cognitive process. In contrast, psycholinguistic research has focused on the process of language without committing to a representation. However, by considering language networks as approximations of the cognitive system we can take the strength of each of these approaches to study human performance and cognition as related to language. This paper reviews research showcasing the contributions of network science to the study of language. Specifically, we focus on the interplay of cognition and language as captured by a network representation. To this end, we review different types of language network representations before considering the influence of global level network features. We continue by considering human performance in relation to network structure and conclude with theoretical network models that offer potential and testable explanations of cognitive and linguistic phenomena.

1 Introduction

Over the last 15 years network analysis has taken off as a rich and fruitful tool to study complex systems across many disciplines. The strength of this approach

---

Nicole M. Beckage
University of Colorado Boulder, Department of Computer Science
e-mail: nicole.beckage@colorado.edu

Eliana Colunga
University of Colorado Boulder, Department of Psychology and Neuroscience
e-mail: eliana.colunga@colorado.edu

lies in the fact that one can formally study a system of objects (nodes) and relations (edges). This is particularly important and relevant for studying language. This framework allows for the interactions of objects to be as important as the objects themselves–allowing the system, as well as the constituents of the system, to be studied. Language can be studied both as an object and as a process. Network analysis techniques allow us to consider both structure and process in turn. The network representation is both simple enough to be analyzed at the system level, and detailed enough to be examined at a more local level (constituents or constituent subgroups). The application of network analysis to language allows for the study of not only how language is structured, but the influence of this structure on human performance, and as we will argue, the cognitive processes that exploit and give rise to this structure.

In this paper we will review research demonstrating the usefulness of network representations for language. Importantly, we focus on the interplay of cognition and language, as captured by a network representation. Language is inherently cognitive and distinctly human. Separating the object of language from the human mind fails to capture the language system fully. Similarly, considering just the emergent properties of a language representation ignores the cognitive aspects of the representation and use of language. Linguistics, traditionally, has been focused on the study of language as a static representation, removed from the human mind. Network analysis has traditionally been focused on properties and structure that emerge from the network representation. Both disciplines could gain from looking at language as a cognitive process. In contrast, psycholinguistic research has focused on the process of language without committing to a representation. However, by considering language networks as approximations of the cognitive system, we can take the strength of each of these approaches to study human performance and cognition as related to language.

We review the successful demonstrations of applying network analyses to language and discuss the implications of placing these results in the context of cognition. We specifically focus on lexical networks in which the nodes are words, and edges are relations between the words. In the first section we introduce two main classes of network representations corresponding to two parts of language – semantics and phonology. We then consider network analysis techniques that have uncovered consistent structures in naturally occurring networks, such as biological networks or social networks. These structural features are also found in the domain of language. We review this literature and its implications for language and cognition.

Continuing our review of past work, we consider behavioral results as related to networks. Just as network analysis has had an influence on the study of language, psychological experiments have also defined aspects of human performance that must be captured in language representations. So we move from focusing on structural features to reviewing studies that relate human performance to network structure. From there we consider how network analysis and psycholinguistics may be able to offer a link between representation and process, explaining aspects of human language use and acquisition. Finally, we show how language networks, and their relation to process, may allow for the characterization of different (sub)populations of interest to cognitive scientists. We conclude by discussing claims made in this paper, as well as future research directions.

2 Language as a Network

Language is the product of humans and changes over time both at the level of the individual as well as at the level of the population. Work that discusses evolution of language across populations in a network framework has been well reviewed by Solé et al. (2010). Here we choose to focus on language at the level of the individual – where cognitive processes occur. We focus on network representations that are inherently cognitive, derived from individual language knowledge or human performance. Specifically, for the sake of capturing the influence of cognition on language networks, we focus on two broad classes of networks that have influenced linguistics and psychology– 1) semantic networks and 2) phonological networks.

2.1 Semantic Networks

The intuition behind semantic networks is that words should be connected if they are semantically related to other words. Further, the strength of the connection (edge weight) should represent the semantic similarity between two words. In practice, it is difficult to compute similarity. For one, words have multiple meanings and which meaning is used can be dependent on context. For example, the word *bone* could refer to the skeletal system or a treat for a dog. Also, semantic relations can capture more than shared meaning (*cup-mug*), referring to hierarchical relationships (*bird-robin*), inclusional relationships (*car-wheel*) and relationships between opposites (*hot-cold*). These semantic relationships are inherently of human creation; they are not necessarily directly measurable in the world, but are a product of human cognition. There is no way to approximate semantic similarity without human input. This human input can come from experiments, language databases or corpora analyses. We briefly cover two datasets that are commonly used to build up semantic networks–the Florida Free Association Norms and the McRae Feature Norms– before turning to other sources of semantic similarity data.

In the study that we refer to as the Florida Free Association Norms (FFA), Nelson et al. (1999) asked participants to respond to a target word with a word that was 'meaningfully related' or 'strongly associated' to the target in a free association task. All in all, over 6000 participants were asked to respond to over 5000 target words. The targets and their responses provide (directed) edges in the network representation. Often these edges are weighted such that the edge weight is proportional to the frequency of the response. In this way, associations were not only present or absent but weighted to approximate similarity.

The McRae Feature Norms are another normed experimental study that has come to be used as the underlying edge list of many semantic networks. In this experiment, participants were asked to list features of objects in an open-ended manner. This dataset contains multiple types of relations that have been categorized into mutually exclusive classes such as 1) perceptual (e.g. *has hands*) 2) functional (e.g. *can be eaten*) and 3) taxonomic (e.g. *is a vehicle*). Two words are said to be connected if they share the same feature. Edges can further be weighted by the number of shared features of the two corresponding words. Since these norms are based on open-ended responses, lack of shared features may not mean that two objects do not share a specific feature but rather that the feature in common is not salient enough to be generated as a response for at least one of the two objects.

Another source of semantic similarity can be language corpora. Here the edges can be co-occurrence of words or some other statistical method to compute similarity such as mutual information or vector representations. In all cases, there is an embodiment of human cognition in this network since human-produced language is used as the underlying input to the corpora. This is even further highlighted when networks are built from thesauri or databases like WordNet in which individuals explicitly state similarity between words.

In all these network representations, there is a common theme of relying on humans to provide the language input, if not directly, then at least the type and strength of associations between words. This alone suggests that we cannot consider semantic networks without considering cognitive processes. As we will see, many of these semantic networks are used to examine cognitive processes of language but before summarizing these results, we first explain phonological networks.

2.2 Phonological Networks

There have been important advances in our understanding of language and cognition through considering network representations constructed based on phonology. There are multiple ways of measuring phonological similarity such as behavioral measures of confusability, number of overlapping phonemes or how similar two spectrograms are. In this review we focus on a measure of similarity based on edit distance. In networks based on edit distance, an edge exists in the graph if one word can be transformed into another word by a series of phoneme changes. For example, given an edit distance of one, the word *kit* is connected to words like *pit* (substitution), *skit* (insertion) and *it* (deletion). The resulting graph is unweighted and undirected, but can be weighted if extended to an edit distance greater than 1. The cognitive implications of these phonological networks are as wide-ranging as for the semantic networks, as we see in work reviewed below. Networks built from an edit distance of one are the most commonly used networks, but it is a potential area of future research to consider other types of phonological network representations.

For most of the phonological networks studied so far, there seem to be very consistent structural properties in the network representation. We include them here because these features seem to be specific to phonological networks and do not extend to other language networks. One such feature is that there are a small number (30-40%) of nodes in the giant component. Other lexical islands (smaller components) or lexical hermits (isolates) reliably emerge in phonological network representations based on edit distance as well (Vitevitch 2008). Some hypotheses as to why this structure exists include the fact that longer words have fewer neighbors (for example words ending in *tion* in English are not connected to any other words because there is no single phoneme change that can lead to an English word not ending in *tion*) and that there are certain allowable phonemes and transitions between phonemes in a given language, constraining possible words. Gruenenfelder et al. even proposed that these types of structural features could emerge in a random vocabulary with minimal linguistic constraints (Gruenenfelder and Pisoni 2009). Another feature of the phonological network is the presence of assortative mixing by degree. Specifically there is a positive correlation between a node's degree and the degree of its neighbors, where words with many neighbors are connected to words that themselves have many neighbors. This property will come up later in informing how individuals retrieve acoustic-phonetic input.

As we have seen in both semantic and phonological networks, the network representation itself includes cognitive aspects. These aspects must be considered as they have implications for the interpretation of any results within language models using these networks. Further, these networks require constructions that indirectly take into account ways in which humans process and use language. We cannot utilize these networks (or variations of these networks) without considering the implicit assumptions of, and implications for, cognition. In the next section, we consider global level network structure that emerges consistently from these representations.

3 Global Level Network Structure

One strength of a network approach is that there are reliable and understood measures that can be considered when classifying networks. In many domains certain properties are considered key to network efficiency and growth. We choose to focus on a subset of these findings, specifically small-world structure and scale-free properties, as they have implications in other domains. Of particular interest is small-world structure because of the results of Watts and Strogatz that suggest that this type of organization strikes a balance between overly structured and completely unstructured (Watts and Strogatz 1998). We move from there onto scale-free networks, as this has been used as evidence for particular types of underlying growth processes that may be useful in explaining language acquisition (Barabási and Albert 1999; Steyvers and Tenenbaum 2005). Since the network of interest is one related to language, these claims need to be considered in light of cognitive implications and empirical evidence from linguistics as well as cognitive psychology.

3.1 Small-World Structure

It has been shown in a variety of areas that naturally occurring networks have similar structural properties that have collectively been called small-world structure. A network is said to have small-world structure if it is characterized by a high amount of local clustering but an average shortest path length (geodesic) similar to a random network, as defined by each pair of nodes having an equal probability of being connected. These properties have been found in many types of networks across many different fields including biological networks (Montoya and Solé 2002), information networks (Albert et al. 1999) and social networks (Milgram 1967), among many others. Watts and Strogatz studied these small-world properties in detail. In their paper, they began with a network initialized to a regular graph in which every node has the same degree and with nearest neighbors connected, forming a lattice. This network was then rewired by giving each edge a certain probability of randomly being reassigned. Even with retie probabilities as low as .01, path lengths near random were seen and a high amount of local clustering was maintained (Watts and Strogatz 1998). See Fig. 1 for a version of their experiment conducted on a language network.

There is small-world structure in every network of language considered in this review and possibly to date. Table 1 lists the graph statistics of the free-association norms (FFA), McRae feature norms, a co-occurrence network and the giant component of the English phonological network. For each of these networks we compute basic global network statistics and compare these original networks to random. We find that all of the observed networks show similar small-world structure, as defined and classified by Watts and Strogatz (1998). Regardless of network size and density, both the clustering coefficients and the average shortest path lengths lie in between those of a regular and a random graph. Figure 1 and Table 2 show visualizations and quantification of this analysis.

Table 1 Statistics associated with graphs reviewed in Section 2

	vertices	density	avg. deg	clustering	avg path
FFA norms	2392	.004	20	.10	4.19
McRae norms	332	.327	107	.585	1.69
Co-Occurrence	5000	.006	62	.184	2.15
Phono (giant)	6508	.001	18	.313	6.05

Small-world structure itself is not specific to language, which suggests that it may not be a property that emerges as a result of cognition. However, because of the ubiquity of small-world structure in language networks, it is possible that this structure can give us insight into human cognition. The fact that small-world structure occurs in many different language networks, at the very least, indicates that a cognitive system may take advantage of these structural properties. Assuming that this structure impacts cognition in some way, we can generate hypotheses and begin to ask questions about how small-world structure relates to language cognition. For example, small-world structure may allow for efficient navigation within semantic memory (Collins and Loftus 1975). Similarly, small-world structure may improve the robustness of a network such that catastrophic failure may be less likely to happen in a language system (Borge-Holthoefer et al. 2011). This type of intuition as to the efficiency of representation has been extended to a model of conversation. Conversation partners want to maximize the understandability of their speech but also want to put in minimal effort in communication. This trade-off between redundancy and effort could lead to observable small-world properties in the representation (Solé et al. 2010). In short, there are many reasons small-world structure might be present in a language network. If small-world structure is really linked to efficient and robust processing, this structure may suggest ways of investigating the cognitive processes that underlie the emergence of this structure.

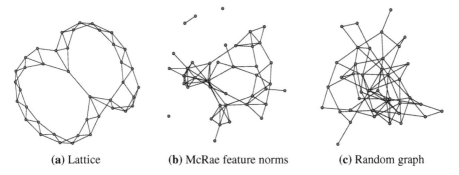

(a) Lattice (b) McRae feature norms (c) Random graph

Fig. 1 Networks with different structure. The first network (**a**) shows a highly structured, size-matched lattice-like graph, the second network (**b**) is a semantic network and the third network (**c**) is a size and density matched random network. Clustering decreases from left to right as does average shortest paths.

Table 2 Statistics associated with graphs in Fig. 1

	vertices	density	avg. deg	clustering	avg path
Lattice	40	.102	8	.446	4.21
McRae norms	40	.099	7.7	.439	2.87
Random graph	40	.103	8.1	.067	2.69

3.2 Scale-Free Networks

Another common structural feature of natural occurring networks is that the degree distribution is often well approximated by a power-law. This type of structure is evidenced by a few central hubs (words with many connections), but with most nodes having only a few connections. This network feature is important in the network literature not only because many naturally occurring networks show this property, but also because, as Barabasi and Albert suggest, this network structure may be evidence of a specific process known as preferential attachment (Barabási and Albert 1999). The basic idea is that nodes join a network by connecting to an already present node in the network. The attachment probability to a given node is proportional to the number of edges a node already has in the graph. This growth process results in a 'rich gets richer' effect where few nodes have many neighbors and most nodes have only a small number of neighbors.

Even though we reliably see power-law distributions in word frequency in language (e.g. Zipf 1949), there is no reason to assume that network representations would also give rise to this type of distribution. The results of analysis on language networks suggest that a degree distribution similar to a power-law is present in many different types of semantic networks, but not phonological networks. Steyvers and Tenenbaum (2005) considered three types of semantic networks: WordNet, Roget's thesaurus, and the FFA norms. They showed that all three of these types of semantic networks have degree distributions representing scale-free properties. The authors used this result as evidence that learning precedes in a fashion similar to preferential attachment. We will consider this finding in greater detail below when we consider process models of language. Phonological networks, however, are better fit by an exponential distribution as opposed to a power-law (Vitevitch 2008). Vitevitch suggests that this finding is explained by language specific constraints such as allowable phonemes and word length that may restrict the power-law formation. Vitevitch found this to be a confusing result, under the assumption that a process similar to preferential attachment likely leads to the observed phonological network as well (Vitevitch 2008).

The authors of the papers mentioned above suggest that scale-free networks and preferential attachment are inherently linked. They also indicate that finding scale-free structure is evidence of preferential attachment or, conversely, that the lack of this structure implies nothing like preferential growth can be happening. It is tempting to use structure as evidence of a specific process, especially when the process itself may be extremely difficult to study empirically. The idea that by identifying the structure of a representation, one can understand the underlying process is a powerful one, but it must be reigned in and validated against theoretical models and empirical findings. While the work of Barabási and Albert (1999) set forth to build a model of how a scale-free structure might emerge, it is not the only way such structure could emerge and thus finding a power-law distribution should not be taken as direct evidence of preferential attachment (Clauset et al. 2009). Nor should any structural feature be considered proof of a specific type of process. Instead structural features should be used as a way of narrowing down the types of processes that might be at play, and as a way of checking and verifying process models. In fact, we see this as a strength for the cognitive approach. If the structural features of language networks can be categorize and understood, testable hypotheses can be set forth, and can then be evaluated and refined with human data. The presence and emergence of network structure can be paralleled with human performance in tasks in which the network representation may provide a useful approximation to human language cognition. By keeping the representation of language in the mind of humans instead of as an entity independent of individuals (and independent of the cognitive processes that operate on the network) we can begin to study the whole language system. With this in mind we move on to consider experimental results that correspond, in one way or another, to a chosen network representation.

4 Human Performance in Relation to Network Structure

The global structural properties discussed above, and the hypotheses associated with them, are linked directly to the network representation. Even though we are often using human data to build up the network representation, it is a theoretical leap to assume that the network representation is itself the representation that humans store and use in their minds. If network representations are to be useful to language modeling, then this representation must capture some aspect of cognition. Many researchers have taken this approach to network analysis in language and have begun to look at the relationship between structural network properties and human performance in specific language tasks. We review some of this work in the remainder of this section.

4.1 Spreading Activation

4.1.1 Semantic Networks

As early as the late 1960's, Quillian (1967;1969) began to consider how semantic knowledge might be stored as a representation in computer memory. The resulting model is one in which words are stored with pointers to features and related words. Words are organized with increasing specificity and features that are related to multiple words are stored at the highest relevant level. For example *canary* has a pointer to *is yellow* and *bird*; features like *has feathers* are stored as pointers at the level of *bird* (see Fig. 2a). Collins and Quillian (1969) extended this work to semantic memory[1] in humans, testing the hypothesis that, if semantic knowledge is stored in such a fashion, the distance between the relevant concepts should affect retrieval

---

[1] In psychology semantic memory includes properties of language storage and retrieval. It is called semantic memory to contrast episodic memory, not to constrain the type of language properties it includes.

time. Specifically, it should take longer for individuals to retrieve feature information stored at super- or sub-ordinate classes than to retrieve information that is stored at the ordinate level. Participants had to evaluate sentences as true or false and their response times were recorded. The authors predicted that participants would take more time to process 'A canary has feathers' than 'A canary is yellow' since *has feathers* is a feature stored at the level of *bird* and thus the path of evaluation requires 2 steps (to bird and then to feathers) as opposed just one step for evaluation directly from *canary* to *is yellow*. The results showed that 1-step relations (such as a canary is yellow or a canary is a bird) took less time than 2-step relations (such as a canary has feathers) to evaluate. The change in response time between 1 and 2-step relations is the same as the change between 2 and 3 steps. This suggested to the original authors that the amount of time it takes between levels in their graph representation can be directly linked to cognition by means of response time, providing evidence that their constructed graph captures some cognitive aspect of semantic memory. In this way, Collins and Quillian provide the first link of geodesic distance to human search time in semantic memory, approximating the time it takes for humans to traverse an edge in their graph.

This work was further expanded and refined into a model known as the spreading activation model of semantic processing (Collins and Loftus 1975). This model relaxes the hierarchical assumption of the original network of Quillian. The network representation is similar to the semantic networks we defined above, with the strength of edges derived from the number of shared features (see Fig. 2b). Collins and Loftus successfully linked a variety of experimental results within semantic processing to a model of activation operating on this network. In this model, a concept is activated within the network; with time, this activation spreads in a decreasing fashion along accessible links, activating other nodes to varying levels. Activation of a single word is diffused throughout the network, taking into account both strength of connections and a decay of activation with distance. Activation spreads from the original, primed word to the primed word's neighbors. If those neighbors are themselves connected, activation will spread and build within the cluster. Thus, if the word *bus* is activated, *car* will also be activated since it is a neighbor of *bus*. Further *car* will receive activation from the neighbors it shares with *bus*, such as *truck* and *wheels*. This results in a high activation of *car* (among other words) from the originally primed word *bus* (see Fig. 2b, color scheme indicates activation if *fish* was the originally primed word on the Collins and Loftus network representation). This activation is assumed to be related to ease of retrieval. Among many other results, one thing Collins and Loftus showed was that participants are quicker to respond to 'name a *fruit* that is red' than 'name a *red* fruit.' Their explanation was that, when fruit was the first word said, activation began to spread within the category of fruit and activate all types of fruit. Whereas when red was said first, activation began to spread among all things red. The effect arises from the fact that the proportion of fruit that is red is higher than the proportion of red things that are fruit. The work of Collins and Loftus (1975) shows that human response times can be linked to processes, such as the spread of activation, that operate on a network.

Language Networks as Models of Cognition 13

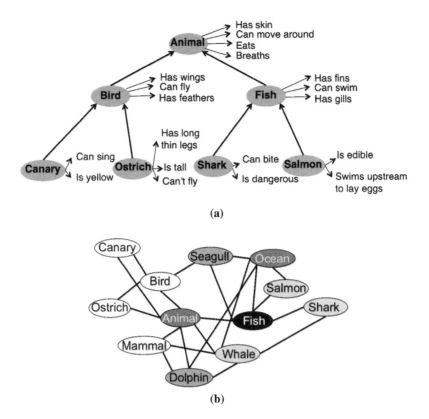

Fig. 2 Two semantic networks based on different types of relatedness. (**a**): Quillian's hierarchical semantic network. (**b**): Collins and Loftus' feature network based on approximate semantic similarity. The color coding represents how activation could spread and accumlate given initial priming or activation of the word *fish*. Darker color represents more activation. This is according to a definition of spreading activation that behaves similar to, and is approximated by, eigenvector centrality.

4.1.2 Phonological Networks

Phonological processing results can also be explained using the spreading activation model of semantic memory. This model has been useful in explaining and capturing retrieval times, confusability, false-memory and other types of lexical access results. However, the way in which this model has been applied has a slightly different history that we now review. The first exploration of the influence of network structure on phonology did not directly consider a network representation. Instead, psycholinguistic research considered the influence of phonologically similar words on spoken word recognition (e.g. Cluff and Luce 1990; Luce and Pisoni 1998). Luce and Pisoni saw linguistic structure and cognitive process as inextricably linked, so much so that their neighborhood activation model assumes that studying linguistic structure leads to understanding of lexical discrimination and vice versa. Luce and

Pisoni developed a measure known as neighborhood density to account for lexical retrieval times. This measure considers the degree[2] (number) and edge-weights (degree) of the neighborhood of a target word. The neighborhood density of a word is shown to be related to the confusability of that word in lexical tasks. Specifically, the authors looked at people's ability to recognize a word in the presence of noise, and found successful identification to be best predicted by a combination of the stimulus intelligibility, degree of target word (confusability of its neighbors), and frequency of the stimulus and its neighbors. The results showed that higher degree words (higher confusability with neighbors) are more often incorrectly recognized. Similar features were determined as important for detection of real vs. non-words. These experiments suggest that the degree and connectivity of a word in the graph (phonological neighborhood) influence recognition and retrieval of words. The results of their neighborhood activation model has been extended to account for many phonological effects on spoken word production, word recognition and other processes (for an overview, see Chan and Vitevitch 2010) and is even shown to relate to early word learning in children (Storkel 2004).

This idea that degree (or neighborhood density) can influence lexical access and retrieval was formalized in network terminology in a paper by Vitevitch (2008). Specifically, he considered the role of network structure as related to word learning and lexical retrieval, providing a link between the spreading activation model and a network representation of phonology. Vitevitch replicated Luce and Pisoni's work with small experimental variations, but recasting it in terms of network analysis. This reframing of lexical access as spreading activation led to new theoretical insights and suggested new research directions. With this framework, the effect of degree on lexical access can be readily explained and quantified. If the target word is activated and activation runs along phonological links, a word with many neighbors shares more of its activation with its neighbors than a word with few neighbors. For high degree target words, the difference between activation of the word and the activation of its neighbors is smaller than for low degree target words. Theoretically and experimentally, this makes target identification more difficult. This framework also suggests that other network features may influence lexical access. Such features were tested by Vitevitch and Chan using the English phonological network, showing that there is an influence of clustering coefficient (percent of neighbors who are themselves connected) on perceptual identification and lexical decision tasks (Chan and Vitevitch 2009). The results indicate that target words with lower clustering coefficients (fewer neighbors are neighbors of each other) are more accurately identified than words with higher clustering coefficients. In the case of higher clustering, activation is not only shared with the neighbors but, because of the cycles, this activation builds within the neighborhood. This reduces the relative activation of the target word, making identification more difficult for words in neighborhoods with high clustering. Here we see an example of how using network analysis to look at

---

[2] Because of the lack of a network formalization in the original paper, the terminology used in this early model of neighborhood activation can be confusing. To counter this, we include the standard network terminology in text and include the terminology of Luce and Pisoni in parentheses.

results in the domain of phonology provides a cognitive model that links phonological space to lexical access and retrieval. By extending the theory of spreading activation to phonology, we are able to refine models and test model predictions, given human performance in language related tasks.

4.2 Frequency Effects

A language measure that has been consistently shown to have an effect on human performance is word frequency. Words that are more frequent are recalled more quickly (Balota and Spieler 1999; Seidenberg and McClelland 1989), acquired earlier (Ellis and Morrison 1998) and are associated with more semantic relations (Steyvers and Tenenbaum 2005). However, network analysis has brought into question whether word frequency is the driving force of these findings or instead, a result of another process or phenomenon. Most lexical network representations do not directly encode word frequency, however there may be a strong relationship between frequency effects in language and lexical graph structure. In this section we review two network papers that look at the interaction of word frequency and human behavior. Together they show that network measures can account for frequency effects, sometimes better than frequency alone, challenging the idea that the underlying cause of performance differences is word frequency.

Griffiths, Steyvers and Firl use a semantic network representation to study word choice in a fluency task (such as 'name the first word that comes to mind starting with the letter A' or 'name as many animals as you can in 60 seconds' (Griffiths et al. 2007)). Earlier experimental results have suggested that the responses to fluency tasks are strongly biased by word frequency. The commonly offered explanation is that individuals have quicker access to words they hear more often (e.g. Balota and Spieler 1999; Seidenberg and McClelland 1989). Instead, Griffiths and colleagues consider network structure as an alternative explanation of human responses in this task. Specifically, they used a network-motivated measure known as PageRank (Brin and Page 1998). PageRank works similarly to spreading activation except that there is no single point of activation. Each word gets a baseline strength and this strength is shared with the node's neighbors. This is then repeated until a stable level of strength is attained. Once this stability is reached, the strength of each node can be seen as a property of the word (node) that indicates how central each node is to information flow or how 'important' the node is to the network. Nodes with a high number of neighbors receive some activation from their neighbors, but how much activation is passed on from the neighbor is dependent on how important the neighbors themselves are. Thus nodes with only a few neighbors can have higher activation than nodes with many neighbors, if the few neighbors are themselves important. The results of Griffiths et al. show that, when using the FFA norms, PageRank outperformed word frequency or FFA degree (called associate frequency in the paper) in predicting participant responses in a fluency task. In general, the findings suggest not only that PageRank corresponds to fluency data, but that a network representation, which inherently encodes the relational structure of language, is more capable of explaining human responses than raw word-level language statistics such as word frequency. This also shows how a network representation supports the creation of new hypotheses and provides a novel framework to reinterpret even robust effects such as the word frequency effect.

The experiment of Griffiths et al. (2007) suggests that human performance and behavior in a word fluency task may be better explained by the underlying graph structure than by word frequency alone. In some cases it might be possible to account for human performance without including frequency; in other cases, word frequency might interact with network structure. This is the case for work relating age-of-acquisition to a network representation. Steyvers and Tenenbaum considered the relationship between semantic connectivity, word frequency and age-of-acquisition as a way of understanding language development (Steyvers and Tenenbaum 2005). First, they found a strong correlation between high-frequency words and high-degree words in a semantic network. This is to say, words that occur more frequently in speech often have more associates or semantic connections (in the FFA norm network). Further, higher degree words are acquired earlier. Previous work has also shown an inverse relationship between age of acquisition and word frequency (Ellis and Morrison 1998). Considering these factors together, the authors showed an interaction of degree and word frequency on learning. The effect of degree on age of acquisition is strongest for words with high frequency. This is possibly because children do not only receive a learning boost for words that have many connections in the graph, but also are hearing these high degree words more often. This work shows how raw language measures such as frequency can be combined with relational language measures implicit in the network structure. When combining these two features we are better able to account for behavioral phenomena such as language learning and processing.

5 Network Models within Linguistic Networks

The idea that human performance can be related to network properties suggests that these network properties might provide evidence of, or have an influence on, human cognition as related to language. To fully understand the acquisition, evolution and even use of language, we need to consider the influence of structure in relation to human processing and, importantly, the process of language and cognition themselves. The idea that structure is evidence of process allows us to begin to build models of cognitive process, and to verify those models using observed network structure *and* measurable features of human behavior. We can extend the process models further by verifying that these process models propose logical (and testable) hypotheses of semantic or phonological processing and cognition in humans. Cognitive science and psychology cannot simply build models that reproduce the statistical properties and structures of interest. We must extend this further to build a model that seems cognitively plausible. Then we must design, test and refine our model in light of human data. This makes the interaction of network modeling and cognitive modeling particularly strong because we have two sources of knowledge to test and inform our theory–network theory and cognitive theory.

In organizing this section, we no longer distinguish between semantic and phonological networks. We do this here because, when process is considered, it becomes difficult to isolate the effect of semantics removed from phonology or phonology removed from semantics. For example, it is widely accepted that to access semantic meaning, the phonological word form must also be activated (Harley 1993) and thus semantic models must include phonological information and effects. We will also shortly see that in phonological research, semantic effects are needed to account for differences across languages (Stamer and Vitevitch 2011). In the following sections we focus on two areas of heavy research within network process modeling: language acquisition and semantic search/navigation.

5.1 Acquisition

One of the earliest attempts to model language acquisition adapted the Barabasi-Albert model of preferential attachment (BA model, Barabási and Albert 1999) to language acquisition. In the original BA model, nodes are added to a graph one at a time. Where they attach to the currently existing graph is proportional to the number of neighbors the attachment node already has. This produces an effect where words that have many neighbors will continue to gain more neighbors, giving rise to a scale-free network. However, while this model maintains scale-free structure, it cannot account for the high clustering found in language networks (Steyvers and Tenenbaum 2005; Vitevitch 2008). As discussed earlier, scale-free structure is present in a variety of different semantic networks but high local-level clustering is also reliably found (a feature of small-world networks (Steyvers and Tenenbaum 2005)). To achieve high clustering while still maintaining scale-free structure, Steyvers and Tenenbaum augmented the BA-model. Their small modification was that a newly added word also forms links to other neighboring nodes of the attachment node. With each new addition of a word to the existing graph, M links are added. First, the new word attaches to an already existing node i, then $M - 1$ edges are added to the neighbors of i. The bias for the new word to connect to already well-connected words allows scale-free structure to emerge (as in the BA model) and the $M - 1$ edges attaching to the neighbors of i maintains local clustering. Further, the authors suggest that the $M - 1$ remaining edges be added proportionally to the degree of the neighbor. Thus the preferential attachment idea is seen both in where the new word links to the existing graph, as well as in which neighbors the new node links to. The resulting network simulations showed similar network statistics and topology to the adult semantic networks (FFA norms, Roget's Thesaurus and WordNet). This not only fit much of the language data, it also provided a testable hypothesis on how structure in language might emerge during the process of acquisition.

While this work shows that it is possible to develop an end network that has similar properties to the language networks of interest, it does not directly consider whether the intermediate stages of the modeled network are similar to what we see in language acquisition. Hills and colleagues approached this aspect of the developmental question by modeling acquisition trajectories directly using the MacArthur-Bates Communicative Development Inventory (CDI, Dale and Fenson 1996). The CDI is a parent report checklist of about 700 words in which parents indicate the productive vocabulary of their child. For each word, the percentage of children at a given age (between 16 and 30 months) that can produce it is known. These norms (and individual checklists) can be used to study effects at the level of individual words and, possibly, individual learners. By combining these CDI vocabulary reports with network edge lists we can build developmental networks. This was first done by Hills and colleagues (Hills et al. 2009b; Hills et al. 2009a) when they considered the acquisition of the 130 earliest learned nouns of a 'normative child.' This normative vocabulary was built by considering a word to be part of the vocabulary once more than 50% of children, at a given age, were reported to produce that word. They used these developmental vocabularies as nodes in the graph and, in the original network modeling paper (Hills et al. 2009a), co-occurrences of words provided the graph edges. Considering each month to be a different network, they tested three developmental models of acquisition 1) preferential attachment[3], 2) preferential acquisition and 3) lure of the associates. In preferential attachment, words are more likely to enter the child's vocabulary the more connected a word is to the already known, well connected words. In preferential acquisition, words are more likely to be learned if they are well connected in the learning environment. As implemented, words are learned proportional to their in-degree in the adult (end) network. Lure of the associates suggests new words are learned if they have connections coming from already known words. This model lies between the two other models–it considers how much associative strength a word receives (in-degree) but conditioned on that the associative strength comes from already known words. See Fig. 3 for a graphical representation of these models.

In three related papers, the authors tested the performance of their models on the ability to account for the vocabulary trajectories of the 'normative child' (Hills et al. 2009b; Hills et al. 2009a; Hills et al. 2010). The results suggest that for many word classes, and specifically for nouns, preferential attachment fails to account for vocabulary growth. This is an important result because it is widely accepted that a scale-free structure is best accounted for by a model similar to preferential attachment (Barabási and Albert 1999). The fact that the authors took a developmental perspective and found that the preferential attachment model could not account for the intermediate observations suggests that there are other ways for scale-free structure to emerge. Further, the authors conclude that, of the three models tested, preferential acquisition is capable of modeling the developmental trajectory most accurately. From this, we can conclude that the learning environment, as captured by

---

[3] This is not the same as the other preferential attachment models because, unlike the BA-model, a new node can connect to more than one word and, unlike the modified BA-model (Steyvers and Tenenbaum 2005), the connections are not necessarily only to neighbors of the attachment word. In fact, the models of Hills et al. have the same underlying edge list such that when a word is added, the relations it forms are pre-defined by the end network.

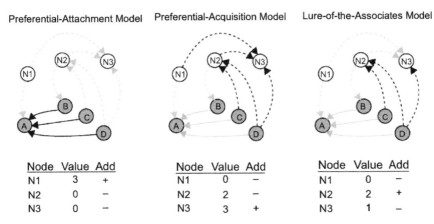

Fig. 3 Taken directly from Hills et al. 2009. The three growth models depicted in a simplified network. The network is the same in all three models, but the models assign different values to the unknown words. Gray shading and solid lines indicate nodes and links in the existing network (known words: A–D); no shading and dotted lines indicate nodes and links not yet incorporated into the known network (possible new nodes: N1, N2, and N3). Black lines indicate links relevant to the growth models, and gray lines indicate unimportant links. The "Add" column in each illustration indicates which node is favored for learning by the growth model in question; this is determined by the relative growth values of the possible new nodes. The growth values computed in this example are based on indegree for a directed network; arrow direction is important. For undirected networks, such as a feature network, arrow direction is not relevant. In the preferential-attachment model, the value of a new node is the average degree of the known nodes it would attach to (e.g., N1 is attached to A, which has an indegree of 3). In the preferential-acquisition model, the value of a new node is its degree in the learning environment – that is, the full network (e.g., N3 has an in-degree of 3, which includes one link from a known node and two links from unknown nodes). In the lure-of-the-associates model, the value of a node is its degree with respect to known words (e.g., N3 has a value of 1, based on its one connection from a known node).

the network, plays an important and non-negligible role in the process of language acquisition. Hills and colleagues characterize the role of the learning environment by emphasizing contextual diversity–words that occur in many different contexts have higher contextual diversity– as approximated by in-degree in the end network. Their findings indicate that words with higher contextual diversity are learned earlier. This relationship may be due to more quickly learned or more accurate word-meaning mappings from the varying presentation-contexts of a contextually diverse word. After considering the network models and the cognitive implications of these models, the authors also considered the network representation (Hills et al. 2009b). They found that certain representations were better able to account for language development than others. Specifically models based on the FFA norms accounted for growth more accurately than the McRae perceptual and functional features. This body of work suggests that researchers should consider not only the developmental model, but also the underlying language representation. The modeling results, indicating that preferential acquisition accounts for the developmental trajectories most accurately, suggest that we need to consider other growth processes than preferential attachment. Further, the developmental perspective offers us a domain where language change and acquisition can be observed, modeled and studied.

Other work on acquisition has focused more on the effects of phonological features and complexity on early learned words. Storkel considered the influence of phonological similarity on word acquisition. The main question was whether children build up dense phonological neighborhoods as they learn language. Her work suggests that words with higher degree[4] are learned earlier (Storkel 2004; 2009). She also found significant effects of length (with shorter words being learned earlier) and frequency (with more frequent words being learned earlier). Given that an interaction between word frequency and degree is found in semantic networks, and phonological networks show assortative mixing (Steyvers and Tenenbaum 2005; Vitevitch 2008), we might expect the same interaction in word learning. However, that is not the case–there is no interaction between degree and frequency in phonological representations. This suggests that the process of building up a language representation may be different at different times in the developmental process. We know that in adult semantic networks a relationship between degree and frequency exists (Steyvers and Tenenbaum 2005; Griffiths et al. 2007) but this may be a relationship that emerges only in mature language networks.

Vitevitch expands the results of Storkel and suggests how some basic stochastic growth processes could account for some of the non-trivial structural properties of the phonological network (Vitevitch 2008). Specifically, he considers the emergence of 1) small-world structure and 2) the global network structure as characterized by one giant component and many lexical islands and hermits. Using the psycholinguistic hypothesis of lexical restructuring in which children first learn very coarse representations and only gradually learn more detailed ones (Metsala and Walley 1998), edges can be added similarly to randomly choosing pairs of nodes to connect as an approximation of the early stages of the lexical restructuring hypothesis (Callaway et al. 2001). As the children's representation gains more detail, new edges are added even between already existing words in the graph. This results in the emergence of small-world structure over time. Further, an additional stochastic process in which new nodes are not required to attach to already existing nodes, can account for the rise of lexical islands and hermits that are found in the phonological graph. Such growth models give rise to network structure similar to the structure of the observed phonological network. While these models are not directly tested on developmental data, this exploration allows us to see how generic network process models can be used to model the emergent structural properties of a variety of graphs across domains. Such models can readily be adapted and used to test hypotheses of language processes. Importantly, this work shows how network models and linguistic theory can inform each other.

While a few methods of modeling language acquisition have been reviewed in this paper, these are still very rooted in network theory. Most of these models stem from the idea that degree influences the words that a child will learn next.

---

[4] Again we see differing notation. In the original work, degree is called density but refers to the number of phonological neighbors and not the density of the network or neighborhood.

Preferential attachment, and its variants, consider a word more likely to be learned if it links up to already existing, well connected words. Other variations consider the learning environment (the adult semantic network) or the interaction between productive vocabulary and the learning environment. In the case of modeling the phonological network, we see the application of random network models developed outside of the domain of language. While it is possible to take the theoretical claims of these models and map them to language processes, it would be interesting to ask the question in the reverse direction–if a linguistic process was acting on a network representation, how would it affect the network structure? We return to this idea in the discussion section and discuss some of the implications. Another notable feature of these acquisition models is that they mostly deal with semantic and phonological features of the words separately. It is much more likely that there is an interaction of phonology and semantics in early word learning. We return to this in the discussion section as well.

5.2 Network Navigation

Another rich area of research on modeling process within linguistic networks is in the domain of network navigation. We use this term very broadly to account for anything that might require search within, or reference to, lexical representations. We include tasks related to lexical retrieval search for a path between two words and other linguistic search tasks. We explain the specifics of each experiment and model in the context in which it is used. This area of research has been considered in more detail than we give it here in reviews such as those of Borge-Holthoefer and Arenas (2010) and Baronchelli et al. (2013).

A model of network navigation that was discussed above is that of spreading activation (Collins and Quillian 1969; Collins and Loftus 1975). As reviewed above, the degree of a word affects retrieval time and age of acquisition (Vitevitch 2008; Storkel 2004). Further, we also reviewed that the spread of activation does not stop after transferring to the nearest neighbors, instead the neighbors' neighbors also receive activation that can result in competition between neighbors and the target word. Behaviorally, this results in slower retrieval time and decreased performance (Chan and Vitevitch 2009). Considering this model as an approximation of a cognitive process allows the extension of this model to other languages as well. This model, when extended beyond English, captures differences across languages. In Spanish, we see the opposite effect of high clustering than in English–words are more quickly retrieved if the target word has higher clustering (Stamer and Vitevitch 2011). This is explained by the fact that semantic similarity varies across languages. For example, in English, many of the phonological neighbors of a word are not semantically related. However, in languages like Spanish, a much higher proportion of neighbors are semantically related. Thus, the activation of these semantically related neighbors helps, rather than hinders, recognition of the target word (Arbesman et al. 2010a; Arbesman et al. 2010b; Stamer and Vitevitch 2011). This highlights the fact that when we begin to consider network models as process models, we need to extend our views not only beyond English but also beyond phonology in isolation of semantics, or semantics removed from phonology.

The spreading activation model of semantic memory offers a cohesive framework to explain which options are considered. The framework, however, does not help us understand how individuals search through potential options. In an attempt to look directly at search in memory, Beckage and colleagues conducted a behavioral experiment to understand how individuals search within semantic memory (Beckage et al. 2012). Participants were asked to navigate within a semantic network from an arbitrary start word to arbitrary end word. To traverse the network, at each step, participants chose a word out of a set of words related to the current word. Figure 4 shows an example problem with a selection of participant paths. The question of interest is what information are people using to navigate the graph and complete the task. Overall, participants were successful at navigating to the end word 73% of the time and 22% of paths were optimal in length. Random walks based on the unweighted FFA network failed to model human performance. Even when associative strength of semantic relatedness is added, these random walks still have lower success rates than participants (Beckage et al. 2012). This difference in performance suggests that participants have access to some global level information, such as a general location of the goal word in semantic space, which is not available to random walks. Ongoing work shows that a dynamic programming algorithm more accurately models human performance. This algorithm works by selecting the option that minimizes approximate distance to the goal word. The ability of this model to capture human performance suggests that participants must have access to information related to the location of the goal word in semantic space. This is an example of how a process model can offer insight into the information available to humans during language processing.

A similar task was also conducted in orthographic/phonological space. In this task, participants were given a three-letter start word and a three-letter end word. They were then asked to build a path between the two words by changing one letter at each time, such that each newly proposed step was to an existing word (Sudarshan Iyengar et al. 2012). The authors compared performance in this task to landmark navigation and human wayfinding (Moore and Golledge 1967), showing that participants initially began by searching for words that shared a letter in common with the goal word. Over time however, participants began to navigate to specific intermediate words that then gave them access to many other words. These hubs offered a 'landmark' in the semantic space that participants could use to orient themselves before continuing their search for the goal word. Interestingly, these navigational hub words varied across participants but participants tended to use the same hub words across trials. Participants also went through stages of familiarity with the network and the task just as they do when navigating in the real world. They update their strategy and use the structure of the network differently at different points in the task. This work introduces a new perspective yet unseen in this review – psychological theory informing network modeling. Here the authors extend a theory of wayfinding and navigation in physical space to that of lexical space. The fact that

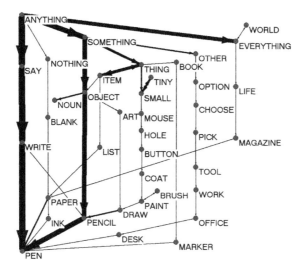

Fig. 4 Network based on successful paths in a semantic network navigation task. Paths are built from participant responses from the start word ANYTHING to the goal word PEN. Network neighbors of ANYTHING are SAY, NOTHING, SOMETHING, EVERYTHING. The thickness of the edges indicates the number of participants who made a given choice.

people go through similar stages as they become more familiar with the search space in both domains suggests that there are common principles of human cognition at work in both cases.

6 Understanding Atypical Processes

In the above sections, we saw that we can consider the process of language acquisition and navigation as opposed to just considering the relationship of structure to performance. If we assume that different processes are operating on network representations and, further, that the network structure is related to process, then we should be able to use differences in network structure to make inferences about differences in process. In the domain of cognitive science, this is an especially powerful tool as we consider the role of network analysis in distinguishing and categorizing atypical populations or individuals. In this section, we review work that attempts to characterize cognitive processes in atypical populations. Specifically, we review work looking at language degradation in Alzheimer patients and delayed language acquisition in a population of children called late talkers.

The work of Lerner et al. used network analysis to capture structural network differences in a category fluency task for elderly individuals, individuals showing mild cognitive impairment and those with early Alzheimer disease (Lerner et al. 2009).

This is of wide-ranging interest because Alzheimer patients show cognitive impairments in many predictable ways but there is no comprehensive model to account for this degradation. The original paper showed that it is possible to categorize these three groups based on summary statistics from a semantic network representation. Participants were asked to name as many animals as they could in 60 seconds and for each participant these responses provided the nodes in their graph. The responses from all participants were combined to construct the edge list, with an edge existing if two animals were named in succession by at least one participants (though see Goñi et al. (2011) for another way of building up these free-response networks). The results showed that the more pronounced the cognitive impairment, the higher the clustering and the lower the average shortest path. This suggests that participants with cognitive impairment may have more difficulty jumping between areas of the network (e.g. making the move from *pets* to *jungle animals*) accounting for less network traversal, and higher clustering of responses than individuals with no cognitive impairment. This idea was further extended, though with a slightly different task, by a model of semantic degradation (Borge-Holthoefer et al. 2011). In this model, degradation happens throughout the graph with more distant and weaker associates being lost more quickly. There is also a renormalization phase in which the remaining edges are strengthened, increasing clustering and decreasing path length of predicted responses. These modeling results can categorize the performance of individuals and also offer a potential explanation of a documented cognitive impairment in which certain relations become more salient and other relations disappear entirely (Borge-Holthoefer et al. 2011; Lerner et al. 2009; Goñi et al. 2011)

If we can model the decay of vocabulary structure, can we gain insight into language acquisition through the structural differences in developing vocabularies? Ke and Yao (2008) first looked at network structure differences in children acquiring language. They used a corpus of child directed speech with samples of child-parent interactions from 12 consecutive months (Manchester corpus from CHILDES, Theakston et al. 2001). The authors constructed child-based or caretaker-base co-occurrence networks, looking at both cumulative networks as well as networks built from a subset of the 12 month period. They found that, with structural measures such as network size and average degree, children could not be distinguished as showing typical or delayed language development. While this suggests that there may be no structural differences in the networks of late and typical talkers, they were able to show that the network structure changed and emerged over time across children. Beckage and colleagues asked a similar question on a larger sample of children, attempting to distinguish late and typical talkers on vocabulary structure alone (Beckage et al. 2011). Late talkers were defined to be children in the lower 25th percentile according to the CDI norms. By comparing a given child's network to a randomly generated bag-of-words model (random acquisition model), differences between these talker groups were indeed found. Typical talkers showed network structure that was indistinguishable from size-matched random acquisition networks. However, late talkers showed network structure that was statistically different than these random networks–and in a direction the authors believe to be detrimental for language acquisition. Children classified as late talkers showed higher geodesic distance and lower clustering than expected by random and thus differed from their typically developing peers. These structural differences correlated with the CDI-based classification, confirming that, not only are vocabularies smaller for these late talkers, but there are fundamental differences in the words and relations among the words for the vocabularies of these children.

The idea that network process is reflected in structure is not new (Barabási and Albert 1999) but utilizing these structural differences to distinguish different processes, and thus different population of learners or language users, is a concept that is especially important in psychology. If we are able to understand differences in network structure across populations, we may be able to build more accurate and specific models. These structural differences have potential use at distinguishing individuals in a population of concern earlier and with a higher diagnostic success. In the future, it may be possible to know whether an individual is showing signs of early cognitive impairment based on network structure before there is any other evidence of cognitive impairment. Similarly, it may be possible for language disorders to be diagnosed earlier based on emerging network structure. In the case of language acquisition, we may be able to analyze the structure of a child's vocabulary as opposed to simply considering the vocabulary size and waiting for crucial milestones to be missed. Along these same lines, by classifying different types of processes, we may be able to offer not only evidence for categorization, but also potential interventions. If structure and process are really as connected as they seem, there is a chance that correcting structural deficits may result in the correction of the underlying cognitive process. And even before classification and intervention can be achieved, comparison studies have the ability to challenge and inform our current understanding of linguistic and psychological performance in language.

7 The Future of Language Networks

In this review we attempted to highlight the role of cognition in understanding language, specifically in the context of network models. We first considered the history and evolution of lexical network representations before focusing on the classification of language topology such as small-world structure and power-law degree distributions. We then showed how network features such as distance between words or degree of a word are correlated with human performance on tasks related to language processing. From there we attempted to highlight an important theoretical shift from simply correlating human performance and network structure to considering language as a system that involves cognition. By modeling the process of acquisition and navigation, we can begin to answer questions about the role of cognition in language. The assumption that different language processes underlie cognitive differences can be used to investigate atypical cognition.

We provide evidence that human cognition cannot be excluded from the study of lexical networks. We further extend this claim to suggest that considering cognition and cognitive representations provide a theoretical framework that one can use to study language as a network. Importantly, with this framework, processes of growing a language network, and navigating on this network, become of crucial importance. Modeling language as a process involving cognition lends itself to a wide range of future research. One such area is to consider the intersection of phonology and semantics on acquisition. We documented results in both independently, but a network representation freely extends to include both of these language representations in one framework. For example, we can begin to consider navigation in semantic memory, capturing the influence of phonology on semantics in a new way. Another area of future research would be to understand how the cognitive system is able to take advantage of the structural features of a network. The very fact that small world structure is found in almost every emergent network suggests there may be some useful properties of 'small-worldness' that the cognitive system may be able to exploit.

To the extent that we can consider network representations of language as an approximation of semantic and phonological information in the brain, we can also begin to uncover the underlying representation of language in the mind. This allows us to ask new types of questions such as how an infant language network, and the constraints that come with it, develops into an adult language network. If we can model language acquisition as the growth of a network, we may begin to understand acquisition of language knowledge. We can also begin to consider individual differences in language acquisition and usage. We have already shown that different processes result in different language networks and that these networks can help distinguish populations of interest. But this idea can be extended further to the level of the individual. If we can model individual differences in processes, we can begin to understand how cognition varies across individuals. In that sense, it may someday be possible to model individual acquisition trajectories or individual search behaviors.

References

Albert, R., Jeong, H., Barabási, A.L.: Internet: Diameter of the worldwide web. Nature 401(130) (1999)

Arbesman, S., Strogatz, S.H., Vitevitch, M.S.: Comparative Analysis of Networks of Phonologically Similar Words in English and Spanish. Entropy 12(3), 327–337 (2010a)

Arbesman, S., Strogatz, S.H., Vitevitch, M.S.: The structure of phonological networks across multiple languages. International Journal of Bifurcation and Chaos 20, 679–685 (2010b)

Balota, D., Spieler, D.H.: Word Frequency, Repetition, and Lexicality Effects in Word Recognition Tasks: Beyond Measures of Central Tendency. Journal of Experimental Psychology: General (1999)

Barabási, A.L., Albert, R.: Emergence of Scaling in Random Networks. Science 286(5439), 509–512 (1999)

Baronchelli, A., Ferrer-I-Cancho, R., Pastor-Satorras, R., Chater, N., Christiansen, M.H.: Networks in cognitive science. Trends in Cognitive Sciences 17(7), 348–360 (2013)

Beckage, N., Smith, L., Hills, T.: Small worlds and semantic network growth in typical and late talkers. PloS One 6(5), e19348 (2011)

Beckage, N., Steyvers, M., Butts, C.T.: Route choice in individuals– semantic network navigation. In: Miyake, N., Peebles, D., Cooper, R.D. (eds.) Proceedings of the 34th Annual Conference of the Cognitive Science Society, pp. 108–113. Cognitive Science Society, Austin (2012)

Borge-Holthoefer, J., Arenas, A.: Semantic Networks: Structure and Dynamics. Entropy 12(5), 1264–1302 (2010)

Borge-Holthoefer, J., Moreno, Y., Arenas, A.: Modeling abnormal priming in Alzheimer's patients with a free association network. PloS One 6(8), e22651 (2011)

Brin, S., Page, L.: The anatomy of a large-scale hypertextual Web search engine. Computer Networks and ISDN Systems 30, 107–117 (1998)

Callaway, D.S., Hopcroft, J.E., Kleinberg, J.M., Newman, M.E.J., Strogatz, S.H.: Are randomly grown graphs really random? Physical Review E: Statistical, Nonlinear, and Soft Matter Physics 64, 041902 (2001)

Chan, K.Y., Vitevitch, M.S.: Network structure influences speech production. Cognitive Science 34(4), 685–697 (2010)

Chan, K.Y., Vitevitch, M.S.: The Influence of the Phonological Neighborhood Clustering-Coefficient on Spoken Word Recognition. Journal of Experimental Psychology: Human Perception and Performance 35(6), 1934–1949 (2009)

Clauset, A., Shalizi, C.R., Newman, M.E.J.: Power-Law Distributions in Empirical Data. SIAM Review 51(4), 661–703 (2009)

Cluff, M.S., Luce, P.A.: Similarity neighborhoods of spoken two-syllable words: Retroactive effects on multiple activation. Journal of Experimental Psychology: Human Perception and Performance 16, 551–563 (1990)

Collins, A.M., Loftus, E.F.: A Spreading-Activation Theory of Semantic Processing. Psychological Review 82(6), 407–428 (1975)

Collins, A.M., Quillian, M.R.: Retrieval Time from Semantic Memory. Journal of Verbal Learning and Verbal Behavior 247, 240–247 (1969)

Dale, P.S., Fenson, L.: Lexical development norms for young children. Behavior Research Methods, Instruments and Computers 28, 125–127 (1996)

Ellis, A.W., Morrison, C.M.: Real age of acquisition effects in lexical retrieval. Journal of Experimental Psychology: Learning Memory and Cognition 24, 515–523 (1998)

Goñi, J., Arrondo, G., Sepulcre, J., Martincorena, I., de Mendizábal, N.V., Corominas-Murtra, B., Bejarano, B., Ardanza-Trevijano, S., Peraita, H., Wall, D.P., Villoslada, P.: The semantic organization of the animal category: evidence from semantic verbal fluency and network theory. Cognitive Processing 12(2), 183–196 (2011)

Griffiths, T.L., Steyvers, M., Firl, A.: Google and the mind: predicting fluency with PageRank. Psychological Science 18(12), 1069–1076 (2007)

Gruenenfelder, T.M., Pisoni, D.B.: The Lexical Restructuring Hypothesis and Graph Theoretic Analyses of Networks Based on Random Lexicons. Journal of Speech, Language and Hearing Research 52(3), 596–609 (2009)

Harley, T.A.: Phonological activation of semantic competitors during lexical acces in speech production. Language and Cognitive Processes 8, 291–309 (1993)

Hills, T.T., Maouene, J., Riordan, B., Smith, L.B.: The Associative Structure of Language: Contextual Diversity in Early Word Learning. Journal of Memory and Language 63(3), 259–273 (2010)

Hills, T.T., Maouene, M., Maouene, J., Sheya, A., Smith, L.: Categorical structure among shared features in networks of earlylearned nouns. Cognition 112(3), 381–396 (2009a)

Hills, T.T., Maouene, M., Maouene, J., Sheya, A., Smith, L.: Longitudinal analysis of early semantic networks: preferential attachment or preferential acquisition? Psychological Science 20(6), 729–739 (2009b)

Ke, J., Yao, Y.: Analyzing language development from a network approach. Journal of Quantitative Linguistics, 1–22 (2008)

Lerner, A.J., Ogrocki, P.K., Thomas, P.J.: Network graph analysis of category fluency testing. Cognitive and Behavioral Neurology 22(1), 45–52 (2009)

Luce, P.A., Pisoni, D.B.: Recognizing Spoken Words: The Neighborhood Activation Model. Ear and Hearing 19(1), 1–36 (1998)

Metsala, J.L., Walley, A.C.: Spoken vocabulary growth and segmental restructuring of lexical representations: Precursors to phonemic awareness and early reading ability. In: Word Recognition in Beginning Literacy, vol. 4, pp. 89–120. Erlbaum, Mahwah (1998)

Milgram, S.: The small-world problem. Psychology Today 2, 60–67 (1967)

Montoya, J.M., Solé, R.V.: Small world patterns in food webs. Journal of Theoretical Biology 214, 405–412 (2002)

Moore, G.T., Golledge, R.G.: Environmental knowing. Hutchinson and Ross, Strodsberg (1967)

Nelson, D.L., McEvoy, C.L., Schrieber, T.A.: The University of South Florida Word Association Norms (1999)

Quillian, M.R.: Word concepts: A theory and simulation of some basic semantic capabilities. Behavioral Science 12, 410–430 (1967)

Quillian, M.R.: The teachable language comprehender: a simulation program and theory of language. Communications of the ACM (1969)

Seidenberg, M.S., McClelland, J.L.: A Distributed, Developmental Model of Word Recognition and Naming. Psychological Review (1989)

Solé, R.V., Corominas-Murtra, B., Valverde, S., Steels, L.: Language networks: Their structure, function, and evolution. Complexity 22, 1–9 (2010)

Stamer, M.K., Vitevitch, M.S.: Phonological similarity influences word learning in adults learning Spanish as a foreign language. Bilingualism: Language and Cognition 15(03), 490–502 (2011)

Steyvers, M., Tenenbaum, J.B.: The Large-Scale Structure of Semantic Networks: Statistical Analyses and a Model of Semantic Growth. Cognitive Science, 3–27 (2005)

Storkel, H.L.: Developmental differences in the effects of phonological, lexical and semantic variables on word learning by infants. Journal of Child Language 36, 291–321 (2009)

Storkel, H.L.: Do children acquire dense neighborhoods? An investigation of similarity neighborhoods in lexical acquisition. Applied Psycholinguistics 25(2), 201–221 (2004)

Sudarshan Iyengar, S.R., Veni Madhavan, C.E., Zweig, K.A., Natarajan, A.: Understanding human navigation using network analysis. Topics in Cognitive Science 4(1), 121–134 (2012)

Theakston, A.L., Lieven, E.V.M., Pine, J.M., Rowland, C.F.: The role of performance limitations in acquisition of verb-argument structure: an alternative account. Journal of Child Language 28, 127–152 (2001)

Vitevitch, M.S.: What Can Graph Theory Tell Us About Word Learning and Lexical Retrieval. Journal of Speech, Language and Hearing Research 51, 408–422 (2008)

Watts, D.J., Strogatz, S.H.: Collective dynamics of 'small-world' networks. Nature 393(6684), 440–442 (1998)

Zipf, G.K.: Human Behavior and the Principle of Least Effort. An Introduction to Human Ecology. Addison Wesley, Cambridge (1949)

Path-Length and the Misperception of Speech: Insights from Network Science and Psycholinguistics

Michael S. Vitevitch, Rutherford Goldstein, and Elizabeth Johnson

Abstract. Using the analytical methods of network science we examined what could be retrieved from the lexicon when a spoken word is misperceived. To simulate misperceptions in the laboratory, we used a variant of the semantic associates task—the phonological associate task—in which participants heard an English word and responded with the first word that came to mind that sounded like the word they heard, to examine what people actually do retrieve from the lexicon when a spoken word is misperceived. Most responses were 1 link away from the stimulus word in the lexical network. Distant neighbors (words >1 link) were provided more often as responses when the stimulus word had low rather than high degree. Finally, even very distant neighbors tended to be connected to the stimulus word by a path in the lexical network. These findings have implications for the processing of spoken words, and highlight the valuable insights that can be obtained by combining the analytic tools of network science with the experimental tasks of psycholinguistics.

1 Introduction

Network analysis has been used to examine semantic (De Deyne and Storms 2008; Hills et al. 2009; Kenett et al. 2011) and phonological (Arbesman et al. 2010; Carlson et al. 2011; Sonderegger 2011) relationships among words in a variety of languages. Although analyses of the structure of networks formed by the semantic and phonological relationships among words have provided unique insights into these languages, it is important to also examine how that observed structure influences language processing (Borge-Holthoefer and Arenas 2010). In the present chapter

---

Michael S. Vitevitch · Rutherford Goldstein · Elizabeth Johnson
University of Kansas, Lawrence, KS 66045 USA
e-mail: {mvitevit,rgolds}@ku.edu,libbyjohnson54@gmail.com

we used the analytic tools from network analyses and data from a psycholinguistic experiment to explore (1) the structure that exists in the network formed when words serve as vertices (or nodes) and edges (or links) connect words that sound similar to each other (i.e., they are phonologically related) (Vitevitch 2008), and (2) how that structure might influence what is perceived when listeners misperceive a spoken word.

Analysis of speech production errors, such as slips of the tongue, malapropisms, and tip-of-the-tongue experiences, has played an important role in increasing our understanding of the processes involved in speech production (Brown and McNeill 1966; Fay and Cutler 1977; Fromkin 1971). Curiously, however, there has been considerably less research examining speech perception errors, such as mondegreens and slips of the ear (Bond 1999). Instead, most research on speech perception and spoken word recognition has used laboratory-based tasks to examine how certain characteristics of words—such as the frequency with which the word occurs in the language—influence the speed and accuracy that a word can be successfully recognized, with little attention paid to the errors that occurred. The dearth of research on speech perception errors is unfortunate because analyses of such errors have the potential to inform and constrain models of speech perception and spoken word recognition just as similar analyses of speech production errors have informed and constrained models of speech production

There are several models of speech perception and spoken word recognition that have existed for several decades (Luce et al. 2000; Luce and Pisoni 1998; Marslen-Wilson 1987; McClelland and Elman 1986; Norris 1994). However, (to our knowledge) none of these models has been used to predict what will be perceived when a spoken word is misperceived. Given the basic assumptions of these models—multiple word-forms that resemble the acoustic-phonetic input are activated and then compete with each other for recognition—what was perceived when a misperception occurred might have appeared so obvious as to not require any further comment: one of the other partially-activated competitors will be perceived if the word that was actually spoken is not, for some reason, correctly perceived. This simple assumption raises an interesting question, however: what do the partially activated lexical competitors look like?

Of the studies that have examined speech perception errors, most have examined collections of actual perception errors, so-called slips of the ear, as in Bond (1999). Corpus analyses have much ecological validity and have increased our understanding of the spoken word recognition system, but concerns are often raised about the reliability of such data due to the possible influence of perceptual biases in the initial collection of the errors.

In the present study, rather than analyze a corpus of perceptual errors to examine the partially activated lexical competitors that might be erroneously perceived in a slip of the ear, we instead used techniques from network science and a laboratory-based psycholinguistic task. The techniques of network science enabled us to determine the range of lexical competitors that could be perceived as activation interesting questions about how such items might be retrieved from the lexicon, but those questions are beyond the scope of the present investigation.

Distance between nodes was assessed in terms of the smallest number of links between the two selected nodes. Recall that in the context of the phonological network, a link corresponds to a single phoneme change (i.e., an addition, deletion, or substitution) between adjacent nodes.

Figure 2 shows the Hop Plot for the 6,508 words in the giant component of the English phonological network examined in Vitevitch (2008). The x-axis is the number of links in the shortest path connecting two nodes. The y-axis is the cumulative percentage of node pairs that are at most d links from each other. Thus, a distance of 1 indicates the percentage of node pairs (i, j) that are reachable by 1-hop, or a distance of 1 link. The longest shortest-path between two nodes in the giant component consisted of 29 links, and exists between the words *connect* and *rehearsal*. The path from the word *connect* to *rehearsal* included the following words (each differing from immediately adjacent words by a single phoneme): *connect, collect, elect, affect, effect, infect, insect, inset, insert, inert, inurn, epergne, spurn, spin, sin, sieve, live, liver, lever, leva, leaven, heaven, haven, raven, riven, rivet, revert, reverse, rehearse, rehearsal.*

The Hop Plot shows that, on average, .14% of the words in the giant component (or 9.1 of 6,508 words) were reachable by going 1-link away from a given word. Thus, if one were to randomly select a word in the lexicon and change 1 phoneme in that word, one would have, on average, fewer than a dozen competitors (i.e., 9.1 words). If activation were to diffuse through the network to a distance of 2-hops away from a given word, one would activate, on average, 1.33% of the words in the giant component (86.6 of 6,508 words), increasing by an order of magnitude the number of competitors. The number of "phonologically similar" competitors continues to increase dramatically as the distance between words increases. At a distance of 3-hops, 7.9% of the words in the giant component were reachable (or 514.1 words of the 6,508 words in the giant component), and at a distance of 4-hops, 25.2% of the words in the giant component were reachable (or 1,640 words of the 6,508 words in the giant component).

The rapidly increasing number of nodes that can be reached as distance slowly increases visually illustrates one aspect of the small-world phenomenon (Albert et al. 1999; Kleinberg 2000; Watts and Strogatz 1998): despite the large number of items in the system, a large system like the phonological network can nevertheless be traversed quickly. However, the same small-world characteristic that contributes to efficient navigation in a network—being able to reach a large number of nodes very quickly—may lead to detrimental effects when trying to quickly and correctly perceive a spoken word (Luce and Pisoni 1998), or when trying to recover from the misperception of a spoken word. Restricting processing in some way to candidates that are 1 or 2 hops away from a given node may keep the number of competitors to a reasonable number, and may facilitate recovery when misperceptions do occur.

The "double-edged" nature of the path-length between two nodes should not be surprising given the "double-edged" nature of other network characteristics. For example, nodes with many connections, or a high degree (a.k.a. hubs), contribute

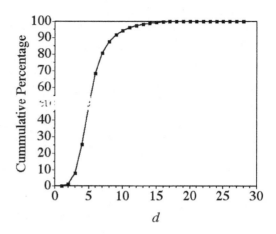

Fig. 2 The Hop Plot for the 6,508 words in the giant component of the English phonological network analyzed in Vitevitch (2008). The *x*-axis is d, the distance (i.e., number of links) of the shortest path connecting two nodes. The *y*-axis is the cumulative percentage of node pairs that are at most d links from each other.

to the stability of scale-free networks when the system is randomly attacked, but can also be the "Achilles heel" of the system when attacks target the hubs (Albert et al. 2000). Also compare the effects of clustering coefficient on speech perception (Chan and Vitevitch 2009) and production (Chan and Vitevitch 2010) to the effects of clustering coefficient on certain memory processes (Vitevitch et al. 2012) for another example of the "double-edged" nature of certain network characteristics.

3 Psycholinguistic Experiment: What Is Perceived When Speech Is Misperceived?

To further examine what might be perceived when a spoken word is misperceived, we used a variant of a well-known psycholinguistic task, the semantic associate task (Nelson et al. 1998), as a laboratory analogue of what happens when one experiences a slip of the ear. In a naturally occurring slip of the ear, a listener hears a word, but does not perceive the word that was uttered. Instead the listener perceives a word that is phonologically similar to the uttered word (Bond 1999).

In our variant of the semantic associate task—the phonological associate task—we presented an English word over a pair of headphones to participants, and asked them to respond with the first word that came to mind that *sounded like* the word they heard (see also Luce and Large (2001)). Note that each participant was allowed to define what "sounded like" meant. By leaving open the meaning of "sounded like," we were able to explore the parameters that listeners may use (implicitly) to define phonological similarity. The responses would also allow us to examine other characteristics of the words that listeners might perceive when they misperceive a word.

Admittedly, this task is contrived, and lacks the ecological validity found in the analysis of naturally occurring slips of the ear. However, the laboratory context of this task enables us to carefully control certain variables, and manipulate others to better explore what might be perceived when a spoken word is misperceived. Such precise control over relevant variables is not possible in analyses of naturally occurring slips of the ear, where one is at the mercy of the material reported in the corpus. For example, an analysis of slips of the ear showed that the words in the corpus that had been misperceived tended to have a higher degree (i.e., more phonological neighbors) than words in general (Vitevitch 2002b). Although this finding was consistent with predictions derived from models of spoken word recognition regarding the difficulty of recognizing spoken words with many phonological neighbors (Luce and Pisoni 1998), the prevalence of naturally occurring slips of ear in words with high degree limits our ability to understand what happens when one misperceives a word with low degree (i.e., few phonological neighbors). Given the prevalence of words with low degree in the network it is important to examine these words as well. Using a laboratory-based psycholinguistic task as in the present experiment, therefore enabled us to examine both types of words, those with high and those with low degree (i.e., many and few phonological neighbors), thereby giving us a more complete understanding of misperceived words than would be possible from an analysis of a corpus of slips of the ear. Finally, our use of the present task gave us the opportunity to obtain usable responses for every stimulus word rather than limiting our analysis to the smaller number of errors that might be obtained in a perceptual identification task, for example.

In psycholinguistic experiments in which a variable—like degree, or the number of phonological neighbors of a word—is manipulated, a prediction regarding performance as a function of that variable is often advanced. In the present case we are using a psycholinguistic task in a more exploratory manner—we wished to simply better understand what might be perceived when a spoken word is misperceived—so we will not advance any specific hypotheses regarding performance. Our inclusion of words with both high and low degree (i.e., many and few phonological neighbors) allows us to explore this question more completely, despite the typical distribution of such items in corpora of naturally occurring slips of the ear (i.e., predominantly words with high degree), and the typical distribution of such items in the lexicon itself (i.e., predominantly words with low degree).

3.1 Method

The same participants, materials, and procedure used in Experiment 2 of Vitevitch et al. (2014) were used in the present investigation. The responses from that experiment were analyzed in a different way in the present investigation. For the convenience of the reader we provide some details regarding the participants, materials, and procedure.

Fourteen native English-speaking students enrolled at the University of Kansas gave their written consent to participate in the present experiment. None of the participants reported a history of speech or hearing disorders.

The materials consisted of 100 English monosyllabic words containing three phonemes in a consonant-vowel-consonant syllable structure. A male native speaker of American English (the first author) produced all of the stimuli by speaking at a normal speaking rate and loudness in an IAC sound attenuated booth into a high-quality microphone. The pronunciation of each word was verified for correctness.

The words differed in degree/neighborhood density, but were similar with regards to a number of other lexical characteristics that are known to influence language processing. *Degree/Neighborhood density* refers to the number of words that sound similar to the stimulus word based on the addition, deletion or substitution of a single phoneme in that word (Luce and Pisoni 1998). A word like *cat*, which has many neighbors (e.g., *at, bat, mat, rat, scat, pat, sat, vat, cab, cad, calf, cash, cap, can, cot, kit, cut, coat*), has high degree and (in Psycholinguistic terms) is said to have a *dense phonological neighborhood*, whereas a word, like *dog*, that has few neighbors (e.g., *dig, dug, dot, fog*) has low degree and (in Psycholinguistic terms) is said to have a sparse phonological neighborhood (N.B., each word has additional neighbors, but only a few were listed for illustrative purposes). Half of the stimuli had high degree/dense phonological neighborhoods (*mean* = 27.7 neighbors, *sd* = 1.6), and the remaining stimuli had low degree/sparse phonological neighborhoods (*mean* = 14.9 neighbors, *sd* = 1.5; $F(1, 98) = 1648.62, p < .0001$).

Although the stimuli differed in degree/neighborhood density, they were comparable with regards to the following characteristics. *Subjective familiarity ratings* of the words, measured on a seven-point scale, were obtained from Nusbaum et al. (1984). Words with high degree/dense neighborhoods had a mean familiarity value of 6.87 (*sd* = .22) and words with low degree/sparse neighborhoods had a mean familiarity value of 6.82 (*sd* = .28, $F(1, 98) = 1.50, p = .22$). The mean familiarity value for the words in the two groups indicates that all of the words were highly familiar.

The mean *frequency of occurrence* in the language ($\log_{10}$ of the raw values from Kučera and Francis (1967)) was 1.03 (*sd* = .58) for the words with high degree/dense neighborhoods, and 1.00 (sd = .58) for the words with low degree/sparse neighborhoods ($F(1, 98) = .08, p = .77$).

Neighborhood frequency is the mean word frequency of the neighbors of the target word. Words with high degree/dense neighborhoods had a mean log neighborhood frequency value of 2.03 (*sd* = .24), and words with low degree/sparse neighborhoods had a mean log neighborhood frequency value of 1.94 (*sd* = .25; $F(1, 98) = 2.99, p = .09$).

Phonotactic probability refers to how often a certain segment occurs in a certain position in a word (*positional segment frequency*) and to how often two adjacent segments occur next to each other in a certain position (*biphone frequency*; as in Vitevitch and Luce (2005)). The mean positional segment frequency for words with high degree/dense neighborhoods was .147 (*sd* = .02) and for words with low degree/sparse neighborhoods was .140 (*sd* = .02, $F(1, 98) = 2.11, p = .15$). The mean biphone frequency for words with high degree/dense neighborhoods was .007 (*sd* = .003) and for words with low degree/sparse neighborhoods was .007 (*sd* = .003, F

$(1, 98) = .009$, $p = .93$). These values were obtained from a web-based phonotactic probability calculator (Vitevitch and Luce 2004).

Each participant was seated in front of a computer that controlled the presentation of stimuli and the collection of responses. In each trial, the word "READY" appeared on the computer screen for 500 ms. Participants then heard one of the randomly selected stimulus words through a set of headphones at a comfortable listening level. Each stimulus was presented only once. Participants were asked to type in the first English word that came to mind that "sounded like" the word that they heard over the headphones. The participants could use as much time as they needed to respond. Participants were able to see their responses on the computer screen when they were typing and could make corrections to their responses before they hit the RETURN key, which initiated the next trial. Although different effects might be found when a closed-response-set rather than an open-response-set is used, there does not appear to be any difference in performance depending on whether responses are spoken versus typed in tasks like that used in the present experiment (Clopper et al. 2006).

3.2 Results

Misspelled words and typographical errors in the responses were corrected to form English words according to the following criteria: (1) transposition of adjacent letters in the word was corrected, and (2) the addition of a single letter in the word was removed if the letter was within one key of the target letter on the keyboard. Of the 1400 responses, 4.56% were misspellings or typographical errors that could not be resolved into English words according to the criteria above, were semantically but *not* phonologically related to the stimulus, or were repetitions of the stimulus word. These responses could not be analyzed, leaving 1336 responses for examination.

Of the responses that we could analyze, 1125 (84.21% of the 1336 responses) were 1 link away from the stimulus word. That is, the responses differed from the stimulus word by one phoneme. We found 181 responses (13.54% of the 1336 responses) that were 2 links away from the stimulus word (i.e., differing from the stimulus word by two phonemes), 28 responses (2.1% of the 1336 responses) that were 3 links away from the stimulus word, 1 response (.07% of the 1336 responses) that was 6 links away from the stimulus word, and 1 response (.07% of the 1336 responses) that was 8 links away from the stimulus word. Thus, when asked to produce a word that "sounded like" a given word, listeners overwhelmingly selected a word that was a short path-length in the network of phonological word-forms away from the stimulus word, and only occasionally selected words at longer path-lengths from the stimulus, giving us additional insight into the criteria that typical users of language (rather than trained language scientists) employ to define phonological similarity.

We further examined in several ways the 1125 responses that differed from the stimulus word by 1 link as a function of degree/neighborhood density of the stimulus word. Our first analysis of these words examined how many different words were given in response to a stimulus word. That is, when presented with *cat*, did everyone give *hat* as the response, or was there some variety in the words that "sounded like" the word *cat*?

The 14 participants gave a mean of 9.16 different words ($sd = 1.74$) in response to words with a high degree/dense phonological neighborhood, and 8.66 different words ($sd = 2.03$) in response to words with a low degree/sparse phonological neighborhood. The analysis of the path-length between words indicated that there was remarkable consensus among participants regarding what "sounded like" the stimulus word: a word that was 1 link away from the stimulus word. Despite that agreement, the present analysis suggests that participants did not converge on the same path in the lexical network. Participants instead indicated that a variety of words in the phonological neighborhood (regardless of whether it was a dense or a sparse neighborhood) "sounded like" the stimulus word.

It is striking that the number of different words that participants indicated "sounded like" the stimulus word approximates the value of 9.1 obtained in the Hop Plot in Fig. 2 for the average number of words that could be reached by 1 hop in the network of phonological word-forms. Future research could explore whether the recurrence of this value is simply a coincidence, or is indicative of some sort of cognitive constraint on language processing, such as the well-known constraint in short-term memory of 7 plus or minus 2 chunks (Miller 1956). One way to distinguish between these two possibilities is to increase the number of respondents in this task. If, with additional participants, we obtain even more variety in the number of different words that "sounded like" the stimulus word, then we can rule out the possibility that the value of 9 is indicative of some sort of cognitive constraint on language processing. If that value is again observed, then additional investigation of some sort of cognitive constraint may be warranted.

A second analysis examined the percentage of responses that differed from the stimulus word by 1 link *as a function of degree/neighborhood density of the stimulus word*. For the stimulus words with high degree/dense phonological neighborhoods we found that 84.86% ($sd = 12.13$) of the responses given to these words were 1 link away from the stimulus word (meaning that 15.14% of the responses were more than 1 link away from the stimulus word), and for stimulus words with low degree/sparse phonological neighborhoods 76.43% ($sd = 16.78$) of the responses given to these words were 1 link away from the stimulus word (meaning that 23.57% of the responses were more than 1 link away from the stimulus word). This difference was statistically significant ($t(98) = 2.88$, $p < .01$), and on the one hand is not surprising. That is, words with low degree/sparse phonological neighborhoods have few words that "sounded like" the stimulus word that are 1 link away, so activation may diffuse across longer paths (i.e., two or more links) to activate a word that "sounded like" the stimulus.

On the other hand, however, the finding that a smaller proportion of stimulus words with low degree/sparse phonological neighborhoods had responses that were 1 link away from the stimulus word is peculiar, and raises additional questions. For example, consider this result in conjunction with the previous finding regarding the number of different words that participants indicated "sounded like" the stimulus word. If words with low degree/sparse phonological neighborhoods have few options to choose amongst for words that "sounded like" the stimulus word, then why was there variability in the number of different words that participants indicated "sounded like" the stimulus word? That is, why did participants give 2-hop neighbors as responses instead of simply producing the same 1-hop neighbors again and again (and therefore producing a smaller number of different types of words that "sounded like" the stimulus word for the words with low degree/sparse phonological neighborhoods)? This returns us to the provocative hypothesis that there may be some sort of cognitive constraint on language processing: during spoken word recognition a fixed number of candidates may be evaluated by the word recognition system. In the case of words with high degree/dense phonological neighborhoods, that fixed number of candidates may be reached (or exceeded) by 1-hop neighbors. Whereas in the case of words with low degree/sparse phonological neighborhoods, that fixed number of candidates may be reached only by considering more distant phonological neighbors (i.e., words more than 1-hop away). Additional analyses and psycholinguistic experiments may be warranted to examine further this speculative hypothesis.

Our next analysis examined the 211 responses that were more than 1 link away from the stimulus word. Given the insight provided by the Hop Plot, we again turned to the tools of network science, and examined the phonological network analyzed in Vitevitch (2008) to determine if a connected path of words existed between the stimulus word and the more distant responses. To illustrate (see Fig. 1), imagine spud was the stimulus, and the response was beach; one can get from *spud* to *beach* by going through the words *speed-speech-peach*, a path length of 4 links.

In 205 of the 211 cases (97.16%) there existed a path of words between the stimulus and the response. The 6 (2.84%) exceptions to this were (stimulus word → response): *lag* → *stagnant*, *niche* → *kitchen*, *poach* → *potion*, *poach* → *approach*, *noose* → *caboose*, and *bib* → *bibliography*. Note that the network analyzed in Vitevitch (2008) contained fewer than 20,000 words. If a larger network were analyzed—one that approached the higher estimates of vocabulary size offered by some (e.g., 216,000 words (Diller 1978))—it is possible that a path might be found between the stimulus and the response in these 6 cases as well.

Despite these 6 exceptions (less than .5% of the 1400 responses) the result of this analysis suggests that words that "sounded like" each other—even distantly related words—tend to connect to each other along a path of real words in the lexicon. The existence of lexical intermediaries observed in the present analysis raises some concerns about measures of word-form similarity that ignore such items, such as the Orthographic Levenshtein Distance-20 (OLD-20 (Yarkoni et al. 2008)), and the Phonological Levenshtein Distance-20 (PLD-20 (Suárez et al. 2011)). In OLD-20/PLD-20, Levenshtein distance is computed between a target word and all other words in the lexicon. OLD-20/PLD-20 is then the mean edit distance of the 20 closest neighbors. The computations of OLD-20/PLD-20 do not consider whether real-word intermediaries exist or not; the measure only considers the number of letter/phoneme changes (respectively) that are required to turn one word into another. However, the present findings show that distant phonological neighbors tend to be connected to a word via a path of real words, raising questions about the psychological validity of metrics such as OLD-20 and PLD-20 that do not consider the absence (or existence) of lexical intermediaries.

4 Conclusion

In the present chapter we used analytical tools from network science and experimental methods from psycholinguistics to examine a question about language processing that is less often examined: What is perceived when a spoken word is misperceived? A Hop Plot was used to assess the proportion of nodes that can be reached (on average) at a given distance, thereby providing us with information about the number of "phonologically similar" competitors one might expect to consider as activation diffuses across the network. This analysis revealed that a relatively small proportion of the network (.14% or 9.1 of 6,508 words) could be reached via 1 link. However, the proportion of words that could be reached increased dramatically as the number of links traversed increased.

With the information provided by our network analysis about how many candidates one might choose amongst when one misperceives a spoken word, we turned to the question of what those candidates actually look like, and examined that question with the phonological associate task, in which participants heard a word and responded with the first word that came to mind that "sounded like" that word. Although this task is admittedly somewhat artificial, it does mimic certain important aspects of the processes that are used "in the wild" to recover from the misperception of spoken words. Furthermore, the ability to carefully select certain words to use as stimuli enabled us to examine certain variables while controlling for other variables, which is something that cannot be done easily when analyzing a corpus of speech perception errors. Moreover, our ability to manipulate the variable of degree/neighborhood density allowed us to examine what happens when misperceptions occur in words with low degree/sparse neighborhoods; this is not possible in analyses of extent speech perception errors because such words rarely appear in such corpora (Vitevitch 2002b).

Several interesting results were observed in the phonological associate task: (1) most responses were 1 link away from the stimulus word, (2) responses that were more distant (>1 link away from the stimulus word) tended to occur more for words with low degree/sparse neighborhoods than for words with high degree/dense phonological neighborhoods, and (3) responses that were more distant tended to be connected to the stimulus word by a path of real words in the lexicon.

The observation that most responses were 1 link away from the stimulus word provides important insight into the criteria that listeners use to indicate that two words "sounded like" each other. Other logical and linguistically motivated possibilities exist, including responding with a longer word that contained the stimulus word (e.g., *cat* → *catalog*), or appending various morphemes to the stimulus word (e.g., *dog* → *doggedly*), but such alternative responses were quite rare in the present study.

The observation that distant responses (>1 link away from the stimulus word) tended to occur more for words with low degree/sparse neighborhoods than for words with high degree/dense phonological neighborhoods is also interesting, especially in light of the first observation. If most responses are 1 link away, one might expect that participants would have more consistency amongst themselves in identifying words that "sounded like" the stimulus words with low degree/sparse neighborhoods. That is, most participants should have provided the same word as a response to a given stimulus word instead of the wide variety of responses that was observed for each stimulus word. The fact that listeners instead went beyond the 1-hop neighbors even though there were still words to choose from—recall the mean number of neighbors for the stimuli with low degree/sparse phonological neighborhoods was 14.9 neighbors—is interesting, and opens up several new avenues for future research, including the hypothesis that a fixed number of candidates might be evaluated during spoken word recognition.

Another interesting avenue for future research is to examine the amount of time it takes to recover from the misperception of a spoken word. Unfortunately we did not measure the time to respond in the present study. Had we done so we could have compared the response times of the items that were 1-2 hops away from the stimulus to the response times of the items that were more than 2 hops away from the stimulus. Future experiments that compare a free-response condition in the phonological associate task to a condition with an imposed time-pressure to respond could shed light on the mechanisms that may be employed to recover from the misperception of a spoken word (De Deyne et al. 2012).

The present results also highlight the existence of lexical intermediates and the potential importance they may play in certain language-related processes. In the responses that were 2 or more hops away from the stimulus word, 97.16% of the responses had a path of extent words connecting the response to the stimulus. Recent work using a game called word-morph—in which participants were given a word, and asked to form a disparate word by changing one letter at a time—demonstrates that participants can exploit their knowledge of the paths between words to efficiently traverse large distances in a lexical network (Iyengar et al. 2012). For example, when asked to "morph" the word *car* into the word *shy* participants might have changed *car* into *cat-pat-pet-set-see-she* and finally into *shy*. Once participants in this task identified certain "landmark" words in the lexicon, the task of navigating from one word to another became trivial, enabling the participants to solve subsequent word-morph puzzles very quickly; solving times dropped from 10-18 minutes in the first 10 games, to about 2 minutes after playing 15 games, to about 30 seconds after playing 28 games. The results of the present study suggest that lexical intermediaries may also play a role in the misperception of a spoken word.

Another recent study further highlights the importance of intermediate lexical items (Geer and Luce 2012). In an auditory shadowing task and a lexical decision task distant neighbors (i.e., words 2 links away from the target) inhibited lexical intermediaries (i.e., words 1 link away from the target), thereby reducing the amount of inhibition that the target word receives from those intermediaries. Referring to Fig. 1, if *speech* is the target word, the word *spud* would inhibit the word *speed*, the word *beach* would inhibit *peach*, etc., thereby reducing the amount of inhibition that *speech* receives from *speed, peach*, etc. Said another way, the words that inhibit a target word are themselves inhibited by other words. Thus, the number of distant neighbors can influence retrieval of a target word by moderating the influence that near neighbors have on the target word (Geer and Luce 2012). The findings from the present study together with the findings from the word-morph game and the findings in Geer and Luce (2012) indicate that additional research on the role of lexical intermediaries on processing is warranted.

More broadly speaking, the present chapter illustrates how network science can be used to investigate questions related to complex *cognitive* systems, in addition to questions related to complex social, biological, or technological systems, areas typically analyzed by network scientists (Albert and Barabási 2002). Combining the power of laboratory-based experiments that are often used in the psychological sciences with the analytical tools and system-wide view of network science—as in the present chapter—holds much promise for advancing the psychological sciences into new areas of inquiry and for resolving ongoing debates. This approach has already increased our understanding of the brain (Sporns 2010), as well as the cognitive processes involved in human navigation (Iyengar et al. 2012), semantic memory (Hills et al. 2009; Marslen-Wilson 1987), and human collective behavior (Goldstone et al. 2008).

In the context of spoken language processing, the tools of network science have enabled us to measure the global as well as the local structure of words stored in the mental lexicon. Previous attempts to examine the structure of the lexicon have only focused on one level. Consider the work of Zipf (Zipf 1935), which found (among other things) a power-law relationship between the frequency with which a word occurs and its rank order. Consider other analyses (Baayen 1991; Baayen 2001; Frauenfelder et al. 1993; Landauer and Streeter 1973), which examined how certain lexical characteristics, such as word-frequency or phoneme frequencies, were related to other lexical characteristics, such as neighborhood density. Consider further the work on neighborhood spread (Vitevitch 2007), onset density (Vitevitch 2002a), and phonotactic probability (Vitevitch and Luce 2005). We see these and many other studies as attempts to measure some aspect of the structural relations among words in the lexicon with the statistical tools that were available at the time. Each of these attempts captured some aspect of that lexical structure, but only at one level of the system. Network science offers a more complete set of methodological tools that can be used to examine multiple levels of a system.

More important, network science offers a theoretical perspective that integrates the observations made at each level of the system. Previous observations of the structure of the lexicon were not only limited to one level of the system, but were often viewed as disparate findings instead of being cumulative, complementary, or somehow connected. That is, each of these previous findings provided yet another entry to the already long list of lexical variables that were known to influence processing in some way (Cutler 1981), instead of contributing to a cohesive description of the lexical system. We believe that the methods and theory of network science offer psychological scientists a unique and powerful framework to develop comprehensive models of cognitive processes and representations that can then be subjected to empirical tests. The present chapter serves as an example of how to combine the analytic tools of network science with the experimental tasks of psychology to examine (and raise) new questions about cognitive processing and representation.

References

Albert, R., Barabási, A.L.: Statistical mechanics of complex networks. Review of Modern Physics 74, 47–97 (2002)

Albert, R., Jeong, H., Barabási, A.L.: Diameter of the World-Wide-Web. Nature 401, 130–131 (1999)

Albert, R., Jeong, H., Barabási, A.L.: The Internet's Achilles' Heel: Error and attack tolerance of complex networks. Nature 406, 200 (2000)

Arbesman, S., Strogatz, S.H., Vitevitch, M.S.: The Structure of Phonological Networks Across Multiple Languages. International Journal of Bifurcation and Chaos 20, 679–685 (2010)

Baayen, R.H.: A stochastic process for word frequency distributions. In: Proceedings of the 29th Annual Meeting of the Association for Computational Linguistics (1991)

Baayen, R.H.: Word Frequency Distributions. Kluwer, Dordrecht (2001)

Bond, Z.S.: Slips of the Ear: Errors in the Perception of Casual Conversation. Academic Press (1999)

Borge-Holthoefer, J., Arenas, A.: Semantic Networks: Structure and Dynamics. Entropy 12, 1264–1302 (2010)

Brown, R., McNeill, D.: The "tip of the tongue" phenomenon. Journal of Verbal Learning and Verbal Behavior 5, 325–337 (1966)

Carlson, M.T., Bane, M., Sonderegger, M.: Global Properties of the Phonological Networks in Child and Child-Directed Speech. In: Danis, N., Mesh, K., Sung, H. (eds.) Proceedings of the 35th Annual Boston University Conference on Language Development. BUCLD, vol. 35, pp. 97–109 (2011)

Chan, K.Y., Vitevitch, M.S.: The Influence of the Phonological Neighborhood Clustering-Coefficient on Spoken Word Recognition. Journal of Experimental Psychology: Human Perception & Performance 35, 1934–1949 (2009)

Chan, K.Y., Vitevitch, M.S.: Network structure influences speech production. Cognitive Science 34, 685–697 (2010)

Clopper, C.G., Pisoni, D.B., Tierney, A.T.: Effects of Open-Set and Closed-Set Task Demands on Spoken Word Recognition. Journal of the American Academy of Audiology 17, 331–349 (2006)

Coltheart, M., Davelaar, E., Jonasson, J.T., Besner, D.: Access to the internal lexicon. In: Dornic, S. (ed.) Attention and Performance VI, pp. 535–556. Academic Press, New York (1977)

Cutler, A.: Making up materials is a confounded nuisance, or: Will we be able to run any psycholinguistic experiments at all in 1990? Cognition 10, 65–70 (1981)

De Deyne, S., Navarro, D.J., Perfors, A., Storms, G.: Strong structure in weak semantic similarity: A graph based account. In: Proceedings of the 34th Annual Conference of the Cognitive Science Society, Sapporo, Japan (2012)

De Deyne, S., Storms, G.: Word association: Network and Semantic properties. Behavior Research Methods 40, 213–231 (2008)

Diller, K.C.: The language teaching controversy. Newbury House, Rowley (1978)

Dorogovtsev, S.N., Mendes, J.F.F.: Language as an evolving word web. Proceedings of the Royal Society B: Biological Sciences 268, 2603–2606 (2001)

Fay, D., Cutler, A.: Malapropisms and the structure of the mental lexicon. Linguistic Inquiry 8, 505–520 (1977)

Frauenfelder, U.H., Baayen, R.H., Hellwig, F.M., Schreuder, R.: Neighborhood density and frequency across languages and modalities 32, 781–804 (1993)

Fromkin, V.A.: The Non-Anomalous Nature of Anomalous Utterances. Language 47, 27–52 (1971)

Geer, M., Luce, P.A.: Neighbors of Neighbors in SpokenWord Processing (2012) (unpublished manuscript submitted for publication)

Goldstone, R.L., Roberts, M.E., Gureckis, T.M.: Emergent processes in group behavior. Current Directions in Psychological Science 17, 10–15 (2008)

Greenberg, J.H., Jenkins, J.J.: Studies in the psychological correlates of the sound system of American English. Word 20, 157–177 (1964)

Hills, T.T., Maouene, M., Maouene, J., Sheya, A., Smith, L.: Longitudinal analysis of early semantic networks: Preferential attachment or preferential acquisition? Psychological Science 20, 729–739 (2009)

Iyengar, S.R.S., Madhavan, C.E.V., Zweig, K.A., Natarajan, A.: Understanding Human Navigation Using Network Analysis. Topics in Cognitive Science 4, 121–134 (2012)

Kenett, Y.N., Kenett, D.Y., Ben-Jacob, E., Faust, M.: Global and local features of semantic networks: Evidence from the Hebrew mental lexicon. PLoS One 6, e23912 (2011)

Kleinberg, J.M.: Navigation in a small world. Nature 406, 845 (2000)

Kuçera, H., Francis, W.N.: Computational Analysis of Present Day American English. Brown University Press, Providence (1967)

Landauer, T.K., Streeter, L.A.: Structural differences between common and rare words: Failure of equivalence assumptions for theories of word recognition. Journal of Verbal Learning & Verbal Behavior 12, 119–131 (1973)

Levenshtein, V.I.: Binary codes capable of correcting deletions, insertions and reversals. Soviet Physics Doklady 10, 707 (1966)

Luce, P.A., Goldinger, S.D., Auer Jr., E.T., Vitevitch, M.S.: Phonetic priming, neighborhood activation, and PARSYN. Perception and Psychophysics 62, 615–625 (2000)

Luce, P.A., Large, N.R.: Phonotactics, density and entropy in spoken word recognition. Language and Cognitive Processes 16, 565–581 (2001)

Luce, P.A., Pisoni, D.B.: Recognizing spoken words: The neighborhood activation model. Ear and Hearing 19, 1–36 (1998)

Marslen-Wilson, W.D.: Functional parallelism in spoken word-recognition. Cognition 25, 71–102 (1987)

McClelland, J.L., Elman, J.L.: The TRACE model of speech perception. Cognitive Psychology 18, 1–86 (1986)

Miller, G.A.: The magical number seven, plus or minus two: Some limits on our capacity for processing information. Psychological Review 63, 81–97 (1956)

Nelson, D.L., McEvoy, C.L., Schreiber, T.A.: The University of South Florida word association, rhyme, and word fragment norms (1998),
http://www.usf.edu/FreeAssociation

Norris, D.: Shortlist: A connectionist model of continuous speech recognition. Cognition 52, 189–234 (1994)

Nusbaum, H.C., Pisoni, D.B., Davis, C.K.: Sizing up the Hoosier Mental Lexicon: Measuring the familiarity of 20,000 words. Research on Speech Perception Progress Report 10, 357–376 (1984)

Sonderegger, M.: Applications of graph theory to an English rhyming corpus. Computer Speech and Language 25, 655–678 (2011)

Sporns, O.: Networks of the brain. MIT Press (2010)

Suárez, L., Tan, S.H., Yap, M.J., Goh, W.D.: Observing neighborhood effects without neighbors. Psychonomic Bulletin and Review 18, 605–611 (2011)

Vitevitch, M.S.: Influence of onset density on spoken-word recognition. Journal of Experimental Psychology: Human Perception and Performance 28, 270–278 (2002a)

Vitevitch, M.S.: Naturalistic and experimental analyses of word frequency and neighborhood density effects in slips of the ear. Language and Speech 45, 407–434 (2002b)

Vitevitch, M.S.: The spread of the phonological neighborhood influences spoken word recognition. Memory & Cognition 35, 166–175 (2007)

Vitevitch, M.S.: What can graph theory tell us about word learning and lexical retrieval? Journal of Speech Language Hearing Research 51, 408–422 (2008)

Vitevitch, M.S., Chan, K.Y., Goldstein, R.: Insights into failed lexical retrieval from network science. Cognitive Psychology 68, 1–32 (2014)

Vitevitch, M.S., Chan, K.Y., Roodenrys, S.: Complex network structure influences processing in long-term and short-term memory. Journal of Memory & Language 67, 30–44 (2012)

Vitevitch, M.S., Ercal, G., Adagarla, B.: Simulating retrieval from a highly clustered network: Implications for spoken word recognition. Frontiers in Psychology 2, 369 (2011), doi:10.3389/fpsyg.2011.00369

Vitevitch, M.S., Luce, P.A.: A web-based interface to calculate phonotactic probability for words and nonwords in English. Behavior Research Methods, Instruments, and Computers 36 (2004)

Vitevitch, M.S., Luce, P.A.: Increases in phonotactic probability facilitate spoken nonword repetition. Journal of Memory & Language 52, 193–204 (2005)

Watts, D.J., Strogatz, S.H.: Collective dynamics of 'small-world' networks. Nature 393, 409–410 (1998)

Yarkoni, T., Balota, D., Yap, M.J.: Moving beyond Coltheart's N: A new measure of orthographic similarity. Psychonomic Bulletin & Review 15, 971–979 (2008)

Zipf, G.K.: The Psychobiology of Language. Houghton-Mifflin (1935)

Structure and Organization of the Mental Lexicon: A Network Approach Derived from Syntactic Dependency Relations and Word Associations

Simon De Deyne, Steven Verheyen, and Gert Storms

Abstract. Semantic networks are often used to represent the meaning of a word in the mental lexicon. To construct a large-scale network for this lexicon, text corpora provide a convenient and rich resource. In this chapter the network properties of a text-based approach are evaluated and compared with a more direct way of assessing the mental content of the lexicon through word associations. This comparison indicates that both approaches highlight different properties specific to linguistic and mental representations. Both types of network are qualitatively different in terms of their global network structure and the content of the network communities. Moreover, behavioral data from relatedness judgments show that language networks do not capture these judgments as well as mental networks.

1 Introduction

In cognitive science semantic networks, in which words are connected with each other through a set of links, have been introduced over 50 years ago in the work of Collins and Quillian (1969) and Collins and Loftus (1975) and have remained an influential theoretical model of the mental lexicon ever since. Until very recently, this model has been employed mainly as an elusive metaphor and idealized theoretical construct, since sizable implementations of such a network were missing. This has changed through a combination of factors such as the availability of large corpora, increased computational resources, and accelerated advances in network theory.

In this chapter two approaches towards constructing a large-scale network model of the mental lexicon are compared that make use of novel corpora. A first one is based on word associations and a second one is based on linguistic representations derived from a syntactically annotated text corpus.

Simon De Deyne
University of Adelaide, Adelaide, Australia
e-mail: `simon.dedeyne@adelaide.edu.au`

Steven Verheyen · Gert Storms
Faculty of Psychology and Educational Sciences, University of Leuven,
Tiensestraat 102, Box 3711 B-3000 Leuven, Belgium
e-mail: `{steven.verheyen,gert.storms}@ppw.kuleuven.be`

While previous work has looked into both types of corpora (De Deyne and Storms 2008; Kenett et al. 2011; Motter et al. 2002; Steyvers and Tenenbaum 2005; Solé et al. 2010), the interpretation of the findings is complicated by the lack of control for factors such as the number of tokens or network size. Another reason why such a comparison has been lacking is the limited size of around 5,000 nodes of the frequently used word association networks based on the University of Florida dataset (Nelson et al. 2004). In this chapter, a new word association corpus based on over 12,000 cues and over 3 million responses will be described, which for the first time enables a comparison with a similar-sized network derived from text resources. Apart from comparability, the choice of these two types of corpora also allows for a comparison between a representation based on purely linguistic materials from text and a representation that accesses more mental properties present in the lexicon by looking at word associations. In other words, by matching quantitative properties regarding the size of the network, the comparison allows for the identification of qualitative differences between the two networks.

This chapter will compare the networks' structure at a global and intermediate level by capitalizing on the innovative contributions of network science as a unifying formal framework to examine the structure at different levels simultaneously.

1.1 Macro-, Meso-, and Microscopic Properties of the Mental Lexicon

The fundamental strength of the network account lies in the way it addresses the structure of the lexicon at the macroscopic, mesoscopic, and microscopic level simultaneously. The ability to do so is an important feat of network science, since studies of complex systems indicate that different functional patterns emerge depending on the level of analysis and complexity of the network.

The macroscopic or network level reflects the combined role of all the connections between the nodes of the network. In naturally occurring networks, this pattern of connections is often very distinct from comparable random networks, for instance in the case of small-world networks. Over the past years, studies have revealed a small-world structure in both linguistic and word association networks (Steyvers and Tenenbaum 2005; De Deyne and Storms 2008). In these small-world networks, regardless of the starting node, any other node can be reached in less than four steps on average. Moreover, in contrast to comparable random networks, the networks also contain a small number of highly connected nodes or hubs. Similarly, the interconnectivity among neighboring nodes indicated by the clustering coefficient, tends to be much larger in these networks than in comparable random networks.

The way a network is organized at the macroscopic level provides insight in its robustness against damage and efficiency of information dissipation (Bullmore and Sporns 2012). It also captures various dynamic properties such as the gradual growth (Steyvers and Tenenbaum 2005), abrupt emergence of new cognitive functions during development, as well as the degradation of these functions with aging or neurodegenerative illness (Baronchelli et al. 2013).

The mesoscopic or group level involves the properties of a considerable subset of nodes in the network. The structure at the mesoscopic level in the mental lexicon is informative of the meaning of words. This is achieved by computing the distance between a set of words through a set of direct and indirect paths connecting them. These distances allow us to identify closely knit regions in the network. In network science, this method is called community detection. It has been successfully applied in cognitive science to uncover the community structure of phonological networks (Vitevitch 2008), and to identify different word senses in small word association networks (Lancichinetti et al. 2011). Identifying the communities in the mental lexicon might reflect similarity in meaning on a variety of grounds. For instance, this could be a taxonomic structure with groupings for different types of animals like birds, mammals, or fish (Rosch 1973). Communities could also be thematic, where different members of a community occur in a specific script, like a restaurant community consisting of members such as *eating*, *bill*, *waiter*, and *dessert* (Schank and Abelson 1977). Perhaps the communities group together words in a manner reflecting the neuro-anatomic properties of the brain leading to a distinction between living kinds and artefacts (Warrington and Shallice 1984), abstract and concrete words (Crutch and Warrington 2005) or categories grounded by emotional responses (Niedenthal et al. 1999). These are just a few examples, and it is quite likely that the investigating of a large network of words might point towards a structure different from these.

Focusing on just a pair of nodes rather than a larger subset, the mesoscopic level is also informative about how related or close two nodes are and what types of paths exist between them. Since the early propositional network model by Collins and Quillian (1969), the closeness between a pair of nodes has been shown to predict the time to verify sentences like *a bird can fly* (Collins and Quillian 1969). To accommodate for a larger set of behavioral data, the theory was extended to include the notion of spreading activation (Collins and Loftus 1975), in which both direct and indirect paths contribute to the closeness of pairs of words. In network theory, spreading activation is often thought of as a stochastic random walk, resulting in a measure of relatedness that reflects both the number and the length of paths connecting two nodes in the network. Such a random walk model allows us to infer additional information beyond the direct connection between two nodes, which has been shown to improve predictions of human similarity judgments (Capitán et al. 2012; Van Dongen 2000), and the extraction of categorical relations between words (Borge-Holthoefer and Arenas 2010).

A quintessential example of the role of these connections is the study of word priming. In priming tasks, the processing of a word is enhanced when it is preceeded by a related word. In the case of associative priming this involves the presentation of a prime such as *dog* which facilitates processing of the word *bone*. In network terms, such facilitation might be explained by the presence of an associative link between these words. Closely related is mediated priming, whereby one word primes another because they are connected through a mediated link, as in the example of *stripes – tiger – lion*. This type of priming is of particular theoretical importance, as it allows testing the assumption of activation spreading throughout the network (Hutchison 2003) similar to the original proposals by Collins and Loftus (1975). A final type of priming that is often considered distinct from the two previous ones, is semantic priming. Here, an ensemble of shared features or links rather than a single connection determines whether or not priming occurs. From the provided examples it will be clear that a network account provides an elegant way to understand many of the documented priming effects. In this area as well, such an account has been mostly influential at a theoretical level, rather than has made use of a fully implemented model of the mental lexicon.

The microscopic or node level of analysis of the network focuses on how a single node is connected with the rest of the network. Examples are node centrality measures, such as the number of in- or outgoing links. These type of centrality measures have been studied quite extensively in psycholinguistics and explain why certain words are processed more efficiently than others (Nelson and McEvoy 2000; Chumbley 1986; de Groot 1989; Hutchison 2003). In this case network-derived measures provide a structural explanation for many lexical properties of words which have been demonstrated to facilitate word processing.

Structural explanations have been given for the effects of variables such as age-of-acquisition (Steyvers and Tenenbaum 2005) and word frequency (Monaco et al. 2007). An interesting example is the finding that highly imageable words such as *chicken* will be processed faster and more accurately across a range of tasks, including naming and lexical decision, compared to more abstract words such as *intuition*. Such a finding can be explained by looking at the set-size (i.e., summed in- and out-degree) of a word. Researchers believe concrete words have smaller associate sets than abstract ones (Galbraith and Underwood 1973; Schwanenflugel et al. 1992) while others believe that concrete words have more semantic properties than abstract words (de Groot 1989; Plaut and Shallice 1993). A network approach has the potential to tease these two explanations apart.

1.2 Acquiring a Mental Lexicon through Language

The rationale of the current approach, in which the mental lexicon is implemented as a network derived from language, is that this lexicon should reflect a repository of shared subjective meaning, allowing language users to communicate efficiently. It is shared under the assumption that with increasing proficiency a speaker acquires a lexicon that mimics the linguistic properties of his or her environment. It is efficient, assuming that it is organized in a non-trivial fashion to meet information retrieval demands. Represented as a network or graph, the mental lexicon consists of nodes corresponding to lexicalized concepts, and links between these nodes indicate lexico-semantic relationships between these nodes.

We believe the mental lexicon acquires meaning through the continuous exposure to words in context, following similar ideas by Wittgenstein (2001) and Firth (1968), where word meaning is equated to its use in language. This is also the idea that underlies many large-scale models which track the co-occurrence of words at the document level (e.g. Landauer and Dutnais 1997) or at the sentence level (e.g., Lund and Burgess 1996). However, as many studies have shown, humans do not merely encode the surface level-properties of a single sentence or a larger discourse unit. Instead, it is assumed that a mental model is constructed that conveys the crucial information of the utterance beyond the verbatim format and involves comprehension of the syntactic nature of its constituents (Dennis 2005; Kintsch and Mangalath 2011) and the integration of its meaning with prior knowledge (Kintsch 1998; Prior and Bentin 2003).

Indeed, in addition to learning about which words co-occur in language, knowledge about different parts-of-speech and syntactic constructions are likely to be used by humans to capture additional information about the meaning of an utterance (Dennis 2005; Kintsch and Mangalath 2011). In many languages word meaning and part-of-speech characteristics are highly correlated, which allows one to infer what the actions (verbs), entities (nouns) and properties of these entities (adjectives) are. Similarly, syntactic relationships between a subject and an object might reveal something about agency. Furthermore, various studies have shown that linguistic models that incorporate this information provide a better account of human relatedness judgments compared to *n*-gram models that do not (Heylen et al. 2008; Padó and Lapata 2007). Altogether, this suggests that a language network derived from a syntactically annotated text corpus will lead to a representations that capture some key properties of the mental lexicon.

One limitation of this linguistic approach is the fact that language is not merely representational, as it is used to convey a message between a speaker and a listener. Utterances comprise pragmatic factors as well. Compared to a text-based network, this is one of the main reasons to assume that a word association model is likely to encode mental representations differently, as they are considered to be free from pragmatics or the intent to communicate some organized discourse, and believed to be simply the expression of thought (Szalay and Deese 1978). Moreover, these associations do not necessarily reflect propositional information derived from the linguistic environment, but might reflect imagery, knowledge, beliefs, attitudes, and affect as well (De Deyne and Storms 2008; De Deyne et al. 2013a; Szalay and Deese 1978; Rensbergen et al. 2014; De Deyne and Storms 2008; Simmons et al. 2008). In other words, word associations tap directly into the semantic information of the mental lexicon.

1.3 Chapter Outline

The remainder of this chapter starts with an explanation of how a language network is derived from a syntactically annotated text corpus, and how a mental network is derived from a large corpus of word associations. The language network chapter refers to a syntactic language network, where nodes are words and where two nodes are connected through a syntactic dependency relationship such as the adjective *red* modifying the noun *car*. The mental network refers to a network where nodes are also words but the relationship between them is determined by how strongly a specific word is evoked by a cue word in a word association task. Compared to the language networks, these responses are not constrained by syntax but reflect mental constraints of what prominently comes to mind. Both networks aim to capture the mental lexicon in an unsupervised way. This contrasts with the original handcrafted Collins and Loftus network (Collins and Loftus 1975) or WordNet (Fellbaum 1998), where the representations are derived manually by expert linguists. It also differs from connectionist approaches (Rogers and McClelland 2004), where the set of nodes and types of relations is decided in advance and connection weights are estimated using supervised learning.

The focus will be on the macroscopic and mesoscopic levels of the networks, as these have only been recently introduced in the context of studying structure in the mental lexicon (Baronchelli et al. 2013). First, the macroscopic stucture of the networks will be addressed. It will provide a characterization of their global organization and explore the nature of network hubs.

Next, community detection will be used to explore which types of clusters of meaning are present in language and mental networks at a mesoscopic level. An inspection of these communities can reveal what the underlying structural principles are and how various parts of the network relate to each other. For instance, one possibility is that the hubs identified in the previous analysis are indicative of the important domains of knowledge represented in the network. Another possibility is that certain nodes in the network play a special role by connecting different clusters in the graph, for instance in the case of polysemous words. In both cases, communities of limited size might allow us to interpret hubs much easier in comparison with hubs identified at the macro-level. Community members can also provide us with some information about the nature of the organization of the network. According to the dominant view in psychology, concepts are organized in a hierarchical taxonomy of natural categories (Rosch 1973) on the basis of shared perceptual properties, whereas other views attribute a larger role to a structure based on thematic relations of the lexicon (Lin and Murphy 2001).

To test whether the communities correspond with a taxonomic organization, the classification performance for basic-level categories such as birds or fish, obtained from human behavioral data, will be used. This allows us to evaluate whether language and mental networks make similar distinctions and provides the opportunity to discuss alternative interpretations if such structure wouldn't be evident.

The final part of the mesoscopic analysis complements the classification study but uses a more direct way of assessing the underlying mesoscopic properties of the network. This is accomplished by using network-derived similarity measures to predict human relatedness judgments. Considering various levels of abstraction and different types of semantic relations (e.g., relations at the basic and domain level, and thematic relations) allows us to generalize the results beyond concrete basic level nouns, which have dominated the field of cognitive science for a long time (Medin et al. 2000). However, because the large-scale networks in this chapter are extremely sparse such an evaluation poses a specific challenge as a simple overlap measure for relatedness that only takes into account shared neighbors between words might not suffice. To address this issue a spreading activation mechanism similar to the one originally conceived by Collins and Loftus (1975) will be proposed. One way of implementing this is by using Markov random walks over the network, as these also take into account indirect paths that exist between a pair of nodes. Just like dimension reduction in high dimensional semantic spaces like Latent Semantic Analysis (Landauer 2007), the spreading activation mechanism introduces a mechanism to infer indirect links. This allows us to deal with the sparsity associated with linguistic representations and is assumed to lead to more reliable estimates of relatedness. This sections ends with a brief discussion of the role of this spreading activation for both language and mental networks in predicting different types of semantic relations.

2 Constructing the Networks

In the following section, the derivation of several networks based on word association and text corpora are given. Both types of networks are implemented as a unipartite localist network, where nodes correspond to words, and are connected through weighted directed edges with other nodes. To make the networks comparable, the set of words will be restricted to those that occur in both the text corpus and word association data.

2.1 Mental Networks

The mental network was derived from a large scale word association study conducted between 2003 and 2010 at the University of Leuven.[1] This study is described in detail in De Deyne et al. (2013a). In short, it involved a total of 71,380 native Dutch speakers. The association procedure differed from most large-scale studies (e.g. Kiss 1968; Nelson et al. 2004) because it used a continued response format, where each participant generated three different responses for each cue instead of one. This allows one to get a better approximation of weak links in the network (De Deyne et al. 2013a). This way, a total of 300 responses were obtained from 100 participants per cue, corresponding to 100 primary, 100 secondary, and 100 tertiary associations. In order to be able to compare the results with previous work based on a single response procedure, an additional network will be derived which only includes the primary responses.

---

[1] The word association project is ongoing. In 2014, the project contained at least 300 responses per cue for 16,000 Dutch cues and 8,000 English cues. The studies can be accessed from http://www.smallworldofwords.com.

Data Preprocessing and Network Construction

The word association data consisted of 3.77 million responses for a total of 12,581 different cues. About 0.20 million different response types were represented in the data. From these data, two weighted directed networks were derived. The first network G_{asso1}, is based on the primary responses, comparable to the common single response datasets (Nelson et al. 2004). The second network, $G_{asso123}$, includes the secondary and tertiary responses as well. Reducing the network from a bipartite representation to a unipartite representation involves the removal of responses that were not members of the set of cues. The removal of these responses did not affect the coverage in terms of token too much, as about 87% and 83% of the response tokens were retained in G_{asso1} and $G_{asso123}$. To allow a comparison with the language networks which will be explained in the next section, a total of 11,252 cues (94% of the original) were retained. With a total of 0.85 million response tokens G_{asso1} and 2.41 million tokens in $G_{asso123}$ it still represents a sizeable portion of nodes present in the original networks.

2.2 Language Networks

An advanced syntactic dependency parser was used to build a network from a small number of predefined syntactic relations (Heylen et al. 2008; Padó and Lapata 2007). This approach offers a number of advantages in comparison to simple n-gram models derived from raw text because it allows us to infer the part of speech of the words and the syntactic relation between the constituents of a sentence. Because many sentences exhibit a complex nested structure, a second advantage of this analysis is that it captures interesting relations between words even if they are not adjacent within an n-gram window.

Corpus

The corpus described in this chapter consists of a variety of language resources spanning three different registers (De Deyne et al. 2014): (1) text derived from Dutch articles in newspapers and magazines from the Twente Nieuws Corpus (Ordelman 2002) and the Leuven Newspaper Corpus (Heylen et al. 2008), (2) informal language retrieved from Internet web pages collected between 2005 and 2007 and the Dutch Wikipedia retrieved in 2008 (De Deyne 2008), and (3) spoken text from Dutch movie subtitles (Keuleers et al. 2010) and the Corpus of Spoken Dutch (Oostdijk 2000).

Each sentence in the corpus was parsed using Alpino, an advanced Dutch dependency parser (Bouma et al. 2000). Similar to Pereira et al. (1993) and Padó and Lapata (2007), two words were connected by a small number of predefined dependency paths. To reduce sparsity, part-of-speech tagged lemma forms provided by Alpino were used instead of word forms. In other words, plurals and inflections were all reduced to a more basic form. Next, all lemmas were counted and only adjectives, adverbs, nouns, and verbs occurring at least 60 times were retained. Applying

Structure and Organization of the Mental Lexicon 55

this cutoff removed very infrequent words and aided in keeping the computations manageable. The resulting corpus vocabulary consisted of 157 million tokens and 103,842 different lemmas; 82.7% were nouns, 12.6% adjectives, 4.5% verbs, and 0.2% adverbs.

Table 1 Overview of the syntactic dependency paths p and examples

Abbreviation	Full Path (p)	Example
ObjHd	$V \xleftarrow{\text{object of head}} N$	We <u>need</u> some more *coffee*.
HdMod	$N \xrightarrow{\text{modification}} A$	This is, excuse me, damn <u>good</u> *coffee*.
HdModObj	$N \xrightarrow{\text{modification}} NP \xrightarrow{\text{object of}} N$	Lucy takes a loud *sip* of <u>coffee</u>
SuObj	$N \xrightarrow{\text{subject of object}} N$	*Coffee* contains lots of <u>caffeine</u>.
SuHd	$N \xrightarrow{\text{subject of head}} V$	This *coffee* <u>tastes</u> delicious!
Cnj	$N \xleftarrow{\text{conjunction}} N$	Norma arrives with Cooper's *pie* and *coffee*.
SuPredc	$N \xrightarrow{\text{subject of predicative phrase}} N$	*Coffee* is a <u>drink</u>.
HdPredc	$V \xrightarrow{\text{predicative complement}} A$	This coffee *tastes* <u>delicious</u>!

Data Preprocessing and Network Construction

The syntactic relations coded as dependency paths, together with examples and the number of pairs for each of the eight paths are shown in Table 1. With the exception of the HdModObj pattern of length 2, all paths p had a length of 1. For each pattern a reverse path was created by transposing the path-dependent graph. For example, for pattern HdMod, the weight of a path for the adjective *good* and the noun *coffee* is derived from the transposed dependency matrix $G_{\text{HdMod}'}$. An example of the obtained dependencies based on the sum of the original and transposed paths for the word *coffee* is shown in Table 2. It illustrates how the most frequent relations uncovered by the syntactic dependencies are interpretable as corresponding to distinctions in terms of function, attributes, and related entities.

To allow a comparison with the mental networks, the network G_{lex} consisted only of words that also occurred in the mental lexicon G_{asso123} which resulted in a set of 11,252 cues. The total number of tokens in G_{lex} was 83.87 million, while G_{asso1} and G_{asso123} contained only a fraction of this amount of tokens (0.85 million and 2.41 million respectively).

To further improve comparability, a new network $G_{\text{lex}123}$ was derived to closely match the properties of $G_{\text{asso}123}$. This was accomplished by making two additional assumptions. First, apart from vocabulary size, the number of tokens in both networks was matched. This was achieved by sampling responses in a way that matched

Table 2 English translations of the 5 most frequent syntax dependencies derived for *coffee* in the G_{lex} network

ObjHd	HdMod	HdModObj	SuObj	SuHd	Cnj	SuPred	HdPred
drink	free	hand	visitor	serve	tea	coffee	ready
will	strong	man	person	offer	pastry	drink	cold
poor	fresh	taste	man	grow	tobacco	tea	free
get	fair	sugar	someone	drink	soda	water	delicious
sell	black	chance	company	cool	cookie	product	good

the out-strength (i.e. the total number of recorded association responses) of each cue in the mental network. In addition, because participants in the continued word association task were not able to provide the same associate twice, a sampling without replacement scheme was used.

3 Exploring the Structure of Language and Mental Networks

3.1 Macroscopic Structure

Previous studies have shown that a small-world structure is present in both language-derived networks and word association networks (De Deyne and Storms 2008; Solé et al. 2010; Steyvers and Tenenbaum 2005). In line with this work, such a structure should be present in all four networks derived in the previous section. By controlling the number of observations, the macroscopic network statistics of the language and mental networks can be directly compared. Moreover, since two different sampling regimes were applied, the effect of denser networks can be evaluated. Of particular interest is the clustering coefficient of the networks, as this measure provides an indication of the amount of structure present in the networks.

3.1.1 Network Statistics

For each of the four networks, the network statistics were calculated from the largest strongly connected component. The results are presented in Table 3. The largest difference between the two types of networks is based on their density D. In particular, the language network G_{lex} was over thirty times denser than G_{asso1}. The matched G_{lex123} had a higher density than the $G_{asso123}$ network, which indicates that language-based representations are more heterogeneous in terms of connected nodes even when the total number of responses is matched to those of $G_{asso123}$. Presumably this reflects the fact that by definition most relations in the language network are syntagmatic (i.e., fulfilling a different syntactic role, e.g., *captain–ship*), while in word associations paradigmatic responses (i.e., fulfilling a similar syntactic role, e.g., *captain–boss*) are more common (Cramer 1968).

Density also differed between the single and continued word association networks. As indicated by Table 3, the single response network G_{asso1} had a density of 0.22%. Including the secondary and tertiary responses increased the density considerably, to 0.64% for $G_{asso123}$. This confirms that the continued procedure draws on a more heterogeneous response set through the inclusion of weaker links that might go undetected in single response procedures (De Deyne et al. 2013b). Despite this increase, the density remains very small in comparison to G_{lex} and G_{lex123}. Related to the observed differences in density, Table 3 also shows how the continued response procedure increases the in-degree (k^{in}) and out-degree (k^{out}) substantially, from 24.3 for G_{asso1} to 71.5 for $G_{asso123}$. These values are again considerably smaller than the corresponding ones for the language networks G_{lex} and G_{lex123}, reflecting the same heterogeneous distribution of edges in the networks.

Table 3 Descriptive network statistics for each of the four graphs

	G_{asso1}		$G_{asso123}$		G_{lex}		G_{lex123}	
	M	SD	M	SD	M	SD	M	SD
D	0.0022		0.0064		0.0611		0.0091	
L	3.77	0.824	2.85	0.57	1.98	0.33	2.68	0.61
max(L)	10		7		5		7	
k^{in}	24.3	51.9	71.5	140.9	687.0	870.5	102.9	226.2
k^{out}	24.3	8.3	71.5	16.41	687.0	870.1	102.9	41.4
CC	0.0046	0.0036	0.0015	0.0006	0.0005	0.0010	0.0009	0.0027
CC_{rand}	0.0004	0.0000	0.0001	0.0000	0.0000	0.0000	0.0001	0.0000

All networks were characterized by small average paths L (ranging from 1.98 to 3.77 steps) and network diameters $max(L)$ ranging between 5 and 10. In comparison to a matched random network (see CC_{rand}), the clustering coefficient CC for weighted directed networks (see Fagiolo 2007) was considerably higher for the real networks indicating an extensive degree of organization. Moreover, combined with the average short paths lengths, such structure indicates a small-world organization and replicates earlier results for the language and mental networks (Steyvers and Tenenbaum 2005; Solé et al. 2010).[2]

3.1.2 Network Hubs

A second way to characterize the macroscopic structure of the network is by looking at the most central nodes or hubs in the network. For each of the four networks, the

---

[2] Note that the absolute values are lower than that of previous reports. This is a side-effect of using a weighted form of the clustering coefficient as defined by Fagiolo (2007).

ten most central nodes in terms of in-strength and PageRank with α set to .80 (Page et al. 1998) are listed in Table 4 and illustrated in Fig. 1. Using these measures to identify network hubs allows us to evaluate the qualitative nature of the most central words in the networks.

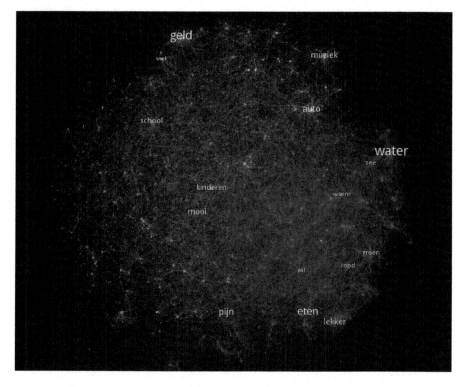

Fig. 1 Large-scale visualization of hubs and communities found in the $G_{asso123}$ network

If word associations are primarily based on associative learning from the linguistic environment, this should lead to hubs that closely match those in the language network. The hubs in the mental networks such as *water* (Dutch: *water*), *food* (*eten*), *money* (*geld*), *car* (*auto*), and *pain* (*pijn*) seem to reflect something about the basic human needs. The hubs in the language networks show some overlap with the mental networks' hubs, but tend to include more abstract words such as *year* (*jaar*), *new* (*nieuw*), *good* (*goed*), *human* (*mens*), *own* (*eigen*), *previous* (*vorig*), and *other* (*ander*).

Furthermore, despite the large differences in density, the hub nodes were quite similar in the mental networks and almost identical in the lexical graphs. The in-strength and PageRank measure of centrality capture slightly different patterns for the hub nodes, but were highly correlated overall, between .88 and .95. More than the type of centrality measure itself, the largest variability was due to the type of graph. In this case, only a moderate correlation existed between the centrality in mental and language networks (between .45 and .46 for in-strength and between .32 and .34 for PageRank) indicating a substantial difference in the identity of central nodes.

A final observation is that the hubs obtained here differ from those identified in previous reports. Where syntactic network hubs have been found to correspond to functional words, and semantic network hubs to polysemous words (Solé et al. 2010), the current results do not include functional hubs. This mainly reflects the fact that closed-form class words were excluded from the analysis as it would obscure any comparison between both types of graphs. In addition, hubs in both the language and mental networks cannot be considered polysemous in a classical sense, which likely reflects the fact that semantic networks reported in previous work (Solé et al. 2010), were based on linguistic expert knowledge derived from WordNet (Fellbaum 1998).

Table 4 Ten most central network hubs derived from in-strength and PageRank ($\alpha = .80$) centrality measures

In-strength				PageRank			
G_{asso1}	$G_{asso123}$	G_{lex}	G_{lex123}	G_{asso1}	$G_{asso123}$	G_{lex}	G_{lex123}
money	water	big	big	water	sun	big	big
water	money	human	human	warm	water	year	good
food	food	man	man	sun	warm	new	new
car	car	new	new	money	food	good	other
music	pain	good	good	green	money	other	year
pain	music	child	child	food	sea	human	human
child	pretty	other	other	car	pretty	man	man
school	school	woman	woman	fun	pain	previous	child
pretty	warm	year	year	sea	green	child	own
sea	sea	small	small	pretty	fun	own	woman

3.2 Mesoscopic Structure

The following analyses will compare clusters identified through community detection methods for language and mental networks. In particular, it will investigate the size and type of communities that can be derived from these graphs. Next, at the most detailed level of the community hierarchy, human data for basic-level categories will be used to explore to what degree these communities provide evidence for a hierarchical taxonomic structure of the kind proposed by Rosch and colleagues (Rosch 1973; Mervis and Rosch 1981) or supports alternative views based on thematic relations (Gentner and Kurtz 2005; Lin and Murphy 2001; Wisniewski and Bassok 1999). The last evaluation continues along these lines and uses human relatedness judgments to evaluate which relationships are best represented in the mental and language networks.

3.2.1 Community Clustering

To identify which clusters are represented at the mesoscopic level, the *Order Statistics Local Optimization Method* (OSLOM) community finding algorithm was applied (Lancichinetti et al. 2011). Using this method, communities (also called modules or clusters) can be identified by evaluating the likelihood that a found community can arise in a comparable random network (Lancichinetti et al. 2011). The proposal has a number of advantages in comparison to the many alternatives such as the Louvain method (Blondel et al. 2008). In particular, it operates on large, directed weighted graphs and allows for overlapping and hierarchical communities. Another advantage of OSLOM is that nodes that are not significantly associated with a community are not assigned. For each network, communities at different hierarchical levels were extracted.[3]

Hierarchical Organization and Interpretation of Communities

One of the interesting features of the OSLOM community procedure is that it identifies a hierarchical organisation by grouping smaller communities in larger ones by evaluating statistical evidence of such a structure to occur in random comparable networks. This allows us to investigate different levels of abstraction along the same lines of the hierarchical network as originally proposed by Collins and Quillian (1969) and taxonomy-based theories derived from the work of Rosch (1973). For G_{asso1} the hierarical structure had a depth of 4, while in G_{asso123} the depth was 5. The hierarchy was flatter for both language networks, with a depth of 3 in G_{lex} and a depth of 4 in G_{lex123}.

Starting at the highest level of the hierarchy, only a handful of communities were identified: 4 in G_{asso1}, 2 in and G_{asso123}. In the matched language networks the top level distinguished 2 communities in G_{lex} and 4 in G_{lex123}. In general, the large number of nodes in each community at the top level makes it difficult to interpret the meaning of these communities.

As the best community solution was found for G_{asso123} at the most detailed level (see Table 5 below), this network will be used to illustrate the structure of the communities at the higher levels of the hierarchy. To summarize the distinctions at the highest level, the most central words in each community were obtained by calculating the community specific in-strength. For each of the five levels of the hierarchy, the five most central items were computed and represented in Fig. 2.

---

[3] In contrast to the previous macroscopic analyses and similar to all subsequent analyses at the mesoscopic level, the weights in the networks were transformed using *positive pointwise mutual information* (PMI) weighting because of its good performance in the context of word co-occurrence models (Bullinaria and Levy 2007).

Structure and Organization of the Mental Lexicon

Table 5 Overview of community structure in the four networks at the lowest hierarchical level

	G_{asso1}	$G_{asso123}$	G_{lex}	G_{lex123}
# Communities	483	506	157	70
Average size	24	25	77	147
Standard dev. size	14	12	54	152
# Homeless nodes	1182	380	512	1721
# Overlapping nodes	3040	3624	2463	1509
Maximum overlap	8	5	5	15
Mean(p)	0.085	0.051	0.096	0.150

For illustration purposes, Dutch words that were synonymous in English (e.g., the Dutch words *fruit* and *vrucht*) were listed once in each community to convey a maximum of information.

At depth one, Fig. 2 shows the two distinct communities, with one of them containing highly central words with a negative connotation. To see whether this level distinguishes positive and negative words, a post-hoc test was set up using valence judgments for a large set of words from Moors et al. (2012). Ratings for a total of 3,642 non-overlapping words belonging to the two communities in the network were obtained. The difference in terms of valence was significant in an independent t-test ($t(3640) = 7.367$, $CI = [0.190, 0.327]$). This finding is in line with previous research that shows that valence is the most important dimension in semantic space (De Deyne et al. 2013; Samsonovic and Ascoli 2010) and proposals of emotion-based category structure (Niedenthal et al. 1999). However, a combination of factors might explain the observed high-level community structure and therefore strong conclusions might be preliminary.

From level 2 to 4, the interpretation of the communities becomes increasingly less abstract. For instance, level 2 shows that the "negative" community in level 1 also includes abstract words or words related to human culture (*knowledge, school, money, school, religion, time,...*) which is now differentiated from a pure negative community including community hubs like *negative, sadness* or *crossed*. The subdivisions of the "positive" community involve the central nodes *nature, music, sports,* and *food* which might be interpreted as covering sensorial information and natural kinds. At this level the communities point towards a distinction of concrete vs abstract words (Crutch and Warrington 2005) or natural kinds vs artifacts (Warrington and Shallice 1984) as structural principles of the lexicon. Clearly, such an interpretation is also suggestive, given the large size of the mental network communities and even larger size of the language network communities. More work is needed to confirm this result.

In order to be able to compare the different networks, the lowest level of the community structure provides us with the best chance of directly comparing results. An overview of the obtained community structure is shown in Table 5. In general, the average size of the communities was strongly related to the number of communities,

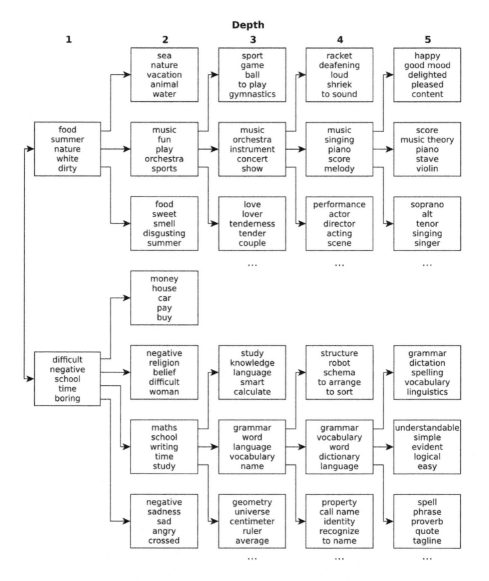

Fig. 2 Hierarchical tree visualization of communities in the $G_{asso123}$ network. Each community is indicated by five central members. At each depth beyond depth 2 a single example is shown of three descendant communities.

and the standard deviation for the community sizes in Table 5 was quite large. This is not surprising given that earlier studies show that in most networks the communities are not necessarily equal in size (Fortunato 2010).

Comparing the different networks, the most striking result is that both the number of communities, and the average significance p of the communities differ between the language and mental networks. The total number of communities was much smaller in the language networks than in the mental networks. The large difference between the two language networks (157 in G_{lex} vs 70 communities in G_{lex123}) can be explained by the difference in density between both graphs (see Table 3). The number of communities was quite similar in G_{asso1} and $G_{asso123}$, but the mean p-values of the identified communities indicate higher significance of identifying communities in the latter network when compared to a matched random network. The effect of increased density was also apparent for the language graphs, where in comparison with a random structure, the communities found in G_{lex} were more reliable, as the mean p was nearly half that of the sparser G_{lex123} network.

Similarly, there was a large difference in terms of the number of homeless nodes, with over three times more homeless nodes in the sparser networks (G_{asso1} and G_{lex123}). This could indicate that for these networks the density was simply too low to reliably assign nodes with either low in-strength and/or highly heterogeneous neighbors to a specific module. For example, in $G_{asso123}$ the in-degree for homeless nodes was on average 17, compared to 71 for the entire graph and the clustering coefficient was 0.0013 compared to 0.0015 (see Table 3).

At all hierarchical levels, nodes could be assigned to multiple communities and a large number of overlapping nodes were also present at the lowest level. As can be seen from Table 5, networks with many and highly significant communities also assigned more nodes to multiple communities which could indicate the ability to distinguish different senses for a specific word at this level. Moreover, in various cases words belonging to more than a single community corresponded to homonyms or words with related senses. For example, in $G_{asso123}$, the Dutch word *bank* which means bank or couch in English, belonged to both a community indicating finance and a community for furniture and sitting. Similarly, the word *language* was attributed to four different communities related to nationality, speech, language education, and communication. Again, the mental networks provided the clearest example of this, while the communities in the language-based networks were too coarse to uncover some of the polysemy or homonymy present in the mental networks.

3.2.2 Taxonomic Structure Evaluation

As mentioned in previous sections, there are many different ways in which the mental lexicon can be structured at the mesoscopic level and the previous exploratory approach indicates that various factors might contribute to the organization of the mental lexicon. However, one of the most influential ideas in psychological theories about knowledge representation is that of a hierarchical taxonomy, in which concepts are grouped in progressively larger categories (Collins and Quillian 1969; Rosch 1973; Murphy 2002). An example of such a hierarchy would be *living-thing <animal <bird <sparrow <house sparrow*. In this hierarchy, one particular level, the basic level, is of special significance as categories at this level capture the psychological structure of concepts that is maximal informative in communication. In this example the basic level category is that of birds, because, this level of description provides the best compromise between maximizing within-category similarity (birds tend to be quite similar to each other as they share many features) and

minimizing between-category similarity (birds tend to be dissimilar to fish) (Medin and Rips 2005).

Despite the large number of studies who have looked at hierarchical taxonomic structures for concepts and explanations of basic-level effects, most of them have limited themselves to concrete nouns (Medin et al. 2000). Moreover, as suggested by the community structure in the mental graphs and literature on a thematically or emotionally organized lexicon (Szalay and Deese 1978; Niedenthal et al. 1999; Samsonovic and Ascoli 2010), the omnipresence of hierarchical taxonomies might be partly due to a selection bias. The goal of this section is to evaluate whether the communities identified at the most detailed level support the idea of a hierarchical taxonomy with a special status for basic-level categories.

Data from an exemplar generation task were used to members of basic level categories. In this task, 100 participants generated as many exemplars they could think of for a list of six artifact categories and seven natural kinds categories (Ruts et al. 2004). The names of the categories and the number of exemplars obtained through this procedure are presented in the first two columns of Table 6.

If the communities in each network group together different types of birds, vehicles, fruits, and so on, this would indicate a taxonomic organization of semantic

Table 6 F-values and corresponding community sizes for 13 basic level categories consisting of human-generated category members

Category	Category size					F-values			
	Human	G_{asso1}	$G_{asso123}$	G_{lex}	G_{lex123}	G_{asso1}	$G_{asso123}$	G_{lex}	G_{lex123}
Fruit	40	93	50	142	106	0.54	0.47	0.20	0.52
Vegetables	35	42	58	132	105	0.47	0.50	0.31	0.46
Birds	53	58	63	63	55	0.61	0.53	0.64	0.63
Insects	39	53	34	83	109	0.67	0.46	0.49	0.43
Fish	37	46	48	44	53	0.55	0.57	0.47	0.53
Mammals	61	32	21	217	212	0.30	0.20	0.38	0.34
Reptiles	23	18	22	83	109	0.59	0.62	0.19	0.18
Mean	41	49	42	109	107	0.53	0.48	0.38	0.44
Clothing	46	77	70	98	536	0.36	0.35	0.28	0.15
Kitchen Utensils	71	33	18	63	58	0.29	0.20	0.30	0.25
Musical Instrum.	46	62	24	104	69	0.56	0.37	0.59	0.71
Tools	73	51	56	51	151	0.26	0.25	0.31	0.25
Vehicles	46	25	28	135	195	0.23	0.16	0.28	0.20
Weapons	46	33	25	51	151	0.30	0.37	0.27	0.17
Mean	55	47	37	84	193	0.33	0.28	0.34	0.29

memory. Table 6 shows the size of the best matching communities and the Jaccard index or F-measure for clustering performance based on precision and recall for each basic level category (Ball et al. 2011). A good solution would be found for a clustering with high precision and recall through a high number of true positives and a low number of true and false negatives. Starting with the category size, Table 6 shows that on average the best matching communities were of comparable size in G_{asso1} and slightly smaller (and thus more specific) in $G_{asso123}$. The sizes of the language network communities were larger than the number of generated exemplars by humans. This indicates that in these networks the communities are too general, which will affect their F-values.

For each of the four graphs, the F-values are generally not very high, which indicates that the communities obtained from the language and mental networks do not provide convincing evidence for a general and strict taxonomic organization. Notable exceptions for natural kinds categories were birds (all networks except $G_{asso123}$), insects (G_{asso1}) and reptiles ($G_{asso123}$). For artifacts, the only indication of a possible taxonomic structure was musical instruments for G_{lex123}.

Table 7 Top 5 false positives ordered by module in-strength for words belonging to the communities derived from $G_{asso123}$

Category	1	2	3	4	5
Fruit	fruit	juicy	pit	pick	summer
Vegetables	vegetable	healthy	puree	sausage	hotchpotch
Birds	bird	beak	nest	whistle	egg
Insects	insect	vermin	beast	crawl	animal
Fish	fish	fishing	rod	slippery	water
Mammals	rodent	gnaw	tail	pen	marten
Reptiles	reptile	scales	animal	tail	amphibian
Clothing	clothing	fashion	blouse	collar	zipper
Kitchen Utensils	cooking	kitchen	stove	cooker hood	burning
Musical Instruments	wind instrument	to blow	fanfare	orchestra	harmony
Tools	tool	carpenter	carpentry	wood	drill
Vehicles	speed	drive	vehicle	motor	circuit
Weapons	sharp	stab	blade	point	stake

On average, natural kinds resulted in higher F-values compared to artifacts. This result supports previous findings, showing that the inter-category structure of artifacts does not have a generally accepted delineation compared to the natural kind categories (Ceulemans and Storms 2010). A contributing factor for the higher F-values for natural kind categories in the mental networks, is that many people are less familiar with certain members of these categories, and predominantly generate taxonomic associates in response to these words. For example, in the case of *swallow* the dominant response was *bird*. This would also explain the better performance of G_{asso1} in this evaluation, as this network only contains the first responses, which frequently correspond to the category-label.

If the communities do not primarily consist of category coordinates, but also contain other words, one might question what factors other than taxonomic ones contribute to the structure found at the most detailed hierarchical level. To address this issue, the five most central false positives for each of the 13 categories were derived by looking at the community specific in-strength as was done for Fig. 2. The results in Table 7 are quite illuminating. First of all, for 8 out of 13 categories the most central item was the category label, which is in line with what can be considered a basic-level category in the literature (Ruts et al. 2004; Rosch 1973). However, this table also shows categories where the representation was too specific, for instance in the case of *rodent* or *wind instrument*, which is also confirmed by the category sizes in Table 6. One could argue that the inclusion of these category-labels might indicate that the F-values are actually underestimates of potential taxonomic structure. Furthermore, the human generated exemplars are not necessarily exhaustive (despite the fact that 100 participants generated exemplars for each category) or correct. For example, *marten* was wrongly identified as a false positive, which suggests this word might have been too infrequent to be captured by 100 participants. However, it is unlikely that this explanation suffices, as the other false positives clearly indicate that related properties, actions, and other thematic information are central. For example, in the case of fruit, other central community members were *juicy*, *pick*, and *summer*. Other examples at the most detailed level in Fig. 2 (e.g., *score*, *music theory*, *piano*, *stave*, *violin*) support this as well. Altogether, the absence of a basic-level taxonomy even for biological categories and the widespread thematic structure across nearly all communities for both the language and mental networks strongly suggest that multiple factors contribute to structure in the mental lexicon, and thematic relations are a major one of them.

3.3 Semantic Relatedness Evaluation

So far, the community detection approach provided some valuable insights about how the mental lexicon might be structured. However, the lack of well-defined small communities in the language networks did not allow us to fully evaluate and compare the language-based and word association-based network. A common direct way to compare these networks and see what kind of relationships they capture uses human relatedness judgments for pairs of words (e.g., Borge-Holthoefer and Arenas 2010; Capitán et al. 2012; Hughes and Ramage 2007). By manipulating the taxonomic and semantic relations between words, it is possible to precisely quantify to what extent each network captures various aspects of the mesoscopic structure. Three studies that were set up for this purpose are described below. In all three studies participants provided relatedness judgments for pairs of words using a 20-point Likert scale. The nature of the pairs differed between studies. They either captured relations at the basic level, at a more general domain level, or thematic relations that do not follow a classical taxonomy.

In a first study, similarity judgments for exemplars from concrete and abstract basic level categories, derived from De Deyne et al. (2008) and Verheyen et al. (2011) respectively, were used. The data consists of similarity judgments for all pairwise combinations of exemplars from 5 animal categories (*birds, fish, insects, mammals,* and *reptiles*), 6 artifact categories (*clothing, kitchen utensils, musical instruments, tools, vehicles,* and *weapons*) and 6 abstract categories (*art forms, crimes, diseases, emotions, media, sciences,* and *virtues*). Because the comparisons were performed at a basic-category level, they required an evaluation of nuanced and detailed properties (for instance, when comparing *hamster* and *mouse* or *kindness* and *helpfulness*).

In contrast to the information encoded at the basic category level, it is possible that the networks cover semantics at a wider range and capture a more course structure. According to this scenario, the networks would only capture a small amount of the variability of the relatedness structure within basic-level categories, but are well suited to distinguish between categories, at the domain level. This would mean that for instance natural kinds and artifacts can be distinguished at a high level in the hierarchy, perhaps at level 2 or 3 in Fig. 2. This might be especially true for the language networks. Since they tend to have broader clusters, they might adequately capture domain distinctions.

To test whether the networks differ in terms of how they capture domain differences apart, a second dataset was included. In this dataset, items from the 5 basic-level animal or 6 basic-level artifact categories introduced previously were paired, leading to pairs such as *butterfly* and *eagle* or *accordion* and *fridge*. If the networks are primarily sensitive to domain-level differences, this would lead to better predictions compared to basic-level categories. Since it is not feasible to present to participants all the pairwise combinations of the combined set of artifact or animal items, only five items from each of the artifact and animal categories were selected. Both items that were central to the category (e.g., *swallow* is a typical bird and thus a central member) and items that were not (e.g., *bat* is an atypical member of the mammals set, and is closely related to birds) were included.

As suggested by the findings on the network communities, it is quite likely that the lexicon reflects a thematic rather than taxonomic organization. If this is the case, this would suggest a high degree of agreement for human judgments of thematic pairs, compared to the basic-level pairs and domain-level pairs. In contrast to the previous pairs, thematic pairs can be closely related without necessarily belonging to a common category or domain. To test these hypotheses thematically related pairs, such as *boat* and *captain* and *rabbit* and *carrot* were used. The set of pairs included among others the items from the study by Miller and Charles (1991), a widely used benchmark test in computational linguistics.

For each of the three studies the number of pairs are listed in the first row of Table 8. An average of 17 participants provided relatedness judgments for a pair of words. The average judgments proved very reliable with Spearman-Brown split-half correlations ranging between .85 and .99. For details and stimuli see De Deyne et al. (2014).

Besides addressing how the networks predict judgments for distinct types of semantic relations, an important issue that remains is the role of sparsity in each graph. While all networks are extremely sparse, the network statistics in Table 3 indicate large differences in terms of network connectivity. The small out-degrees in the mental networks and the matched lexical network $G_{\text{lex}123}$ hint at potential limitations when overlap measures of similarity are used based on common neighbors. To investigate if indirect paths between nodes can contribute to model derived estimates of relatedness by reducing sparsity, a random-walk based measure for relatedness will be proposed.

3.3.1 Network Relatedness Measures

A widely used measure of similarity is the cosine measure. This distributional overlap measure captures the extent to which two nodes in the network share the same immediate neighbors. Two nodes that share no neighbors have a similarity of 0, and nodes that are linked to the exact same set of neighbors have similarity 1. Formally, it is defined as follows. Let $\mathbf{A}$ denote a weighted adjacency matrix, whose element a_{ij} contains a count of the number of times word j is given as an associate of word i in a word association task or the times it occurs in a syntactic dependency relationship. Each row in $\mathbf{A}$ is therefore a vector containing the associate / syntactic dependency frequencies for word i. The cosine measure of similarity is obtained by first normalizing each row so that all of these vectors are of length 1. This gives us a new matrix $\mathbf{G}$, where $g_{ij} = a_{ij}/(\sum_j a_{ij}^2)^{1/2}$, and the matrix of all pairwise similarities is now:

$$\mathbf{S} = \mathbf{G}\mathbf{G}^T \qquad (1)$$

The cosine measure defined in the previous section depends solely on the *local* structure of the graph: the similarity between two words is assessed by looking only at the words to which they are immediately linked. A different approach to similarity aims to take into account the overall structure of the entire network graph, and thus reflects a broader view of the relationship between two nodes. In this approach two nodes are similar if they share many direct or indirect paths. These paths are explored by a random walker, which stochastically follows local links in the network until the proportion of time it visits each node in the limit converges to a stationary distribution (Hughes and Ramage 2007).

Formally, this random walk corresponds to the regular equivalence measure by Leicht et al. (2006) and is specified by beginning with the weighted adjacency matrix $\mathbf{A}$. This time, however, we normalize the rows so that each one expresses a probability distribution over words. That is, we use the matrix $\mathbf{P}$ where $p_{ij} = a_{ij}/\sum_j a_{ij}$, and then calculate

$$\mathbf{G}' = (\mathbf{I} - \alpha \mathbf{P}^{-1}) \qquad (2)$$

where $\mathbf{I}$ is a diagonal identity matrix and the α parameter governs the decay in spread of activation by determining the relative contribution of short and longer paths. A path of length r is assigned a weight of α^r, so when $\alpha < 1$, longer paths get less weight than shorter ones.[4]

The resulting network G' can be thought of as a network of weighted paths. The similarity of two nodes in this network corresponds to the similarity of their stationary distributions. The value of α was fixed at 0.80 (similar to the α for PageRank used in previous sections). This represents a reasonable trade-off between some degree of decay and a non-trivial contribution of longer paths. As in the local relatedness measure above, a cosine measure can then be used to derive a pairwise similarity matrix $\mathbf{S}$ using these distributions. In contrast to the local relatedness measure, such random-walk based measure involves the entire network and is therefore sensitive to the global or macroscopic structure of the network.

3.3.2 Results

For each of the four graphs, relatedness measures were derived as defined above. The measures of relatedness were correlated with the human judgments after standardizing the measures for each category.

Table 8 Results of the similarity analyses for the four datasets (concrete, abstract, domain and thematic) and four graphs

		Basic Level		Domain Level	Thematic
		Concrete	Abstract		
	N	4437	694	1470	126
Local Overlap (Cosine)	G_{asso1}	.528	.505	.711	.567
	$G_{asso123}$	.593	.617	.786	.769
	G_{lex}	.338	.423	.623	.463
	G_{lex123}	.337	.319	.620	.481
Spreading Activation	G_{asso1}	.564	.616	.793	.802
	$G_{asso123}$	.590	.660	.824	.827
	G_{lex}	.370	.433	.718	.523
	G_{lex123}	.361	.375	.653	.504

The results for the local overlap measure presented in the top part of Table 8 show moderate to strong correlations between human judgments of relatedness and network-derived measures. One of the most striking patterns in Table 8 is the systematic difference between the amount of variability accounted for by the four graphs. Regardless of the dataset, the denser $G_{asso123}$ network shows substantial

---
[4] This approach is very similar to the PageRank measure ($\mathbf{X} = (\mathbf{I} - \alpha \mathbf{P}^{-1})\mathbf{1}$).

better agreement than all other graphs. Moreover, even the sparse mental network G_{asso1} outperforms the lexical networks in all cases. Since the G_{lex} network is almost 20 times denser than G_{lex123} (see Table 3), one would expect a better result for this denser graph, however, a significantly different correlation was only found for the basic-level abstract words $z = 2.25$, $p < .05$. A closer look at the different semantic relations indicates that the networks primarily capture the domain judgments, followed by thematic, and basic-level judgments. In line with the community clustering results, this confirms that the networks organize meaning in a thematic way but also include some taxonomic structure.

Next, the role of spreading activation in predicting human relatedness judgments was investigated. The results for the random walk-based spreading activation measure show a consistent improvement for all networks and datasets. The only exception were the results for the concrete words in $G_{asso123}$, where a slightly lower correlation was found. In this case, the setting of $\alpha = 0.80$ might have resulted in a detrimental contribution of longer paths. When α was systematically varied, the correlation improved to .601 for $\alpha = 0.6$, indicating that the optimal value of this parameter may depend on the type of relationships under consideration. However, in general, the correlation changes were very moderate across various parameter settings. Similar to the overlap measure, the large difference in density between G_{lex} and G_{lex123} did not systematically affect the performance in these language graphs, as only for the domain dataset the correlation values were significantly different $z = 3.33$, $p < .05$.

In conclusion, the use of human relatedness judgments to compare how different taxonomic and thematic relations are represented in the language and mental networks, resulted in findings similar to those from the community clustering of these networks described earlier. Language and mental networks capture primarily the domain level relations between words followed by the thematic relations. The mental networks also capture the basic-level conceptual structure, but the strength of this correlation was moderate. Regardless of the dataset, the mental networks provided a clearly better prediction of human judgments. Using longer indirect paths derived through a stochastic random walk led to systematic improvements in both types of networks, but did not alter the basic findings regarding the relationships captured by these networks.

4 Discussion

In this chapter, the main goal was to compare the macroscopic and mesoscopic properties of language and mental graphs, derived from text corpora and word associations, respectively. One of the key results was that representations systematically differ between both graphs. These differences in itself provide us with important pointers about what processes operate on the linguistic input humans are exposed to.

At a global, macroscopic level, the network-based approach unveiled a highly structured representation that is characterized by short average path-lengths and a significant degree of clustering in both language and mental graphs. This indicates that both graphs have a small-world structure. While there is some overlap between what constitutes a hub in the respective graphs, systematic differences between node centrality emerged. In mental graphs, a larger role for nodes that are presumably of psychological importance exists, while in the language networks hub nodes appear to be more abstract. The latter might reflect the frequency of words typical for language derived from newspaper and other written sources. Moreover, what constitutes a central hub in the mental network seems to be a universal property shared among multiple languages. For instance, for a similar ongoing word association project in English, the ten largest hubs in terms of in-strength in a network with 7,000 nodes corresponded to *money, food, water, love, work, car, music, time, happy,* and *green*.

Furthermore, the structure of the network argues against the view of the mental lexicon as exclusively and strictly taxonomically organised, where words are grouped in coherent semantic domains and categories. First of all, a substantial number of words were part of multiple communities, which argues against mutually exclusive categories. Second, while the representations can be described in a hierarchical clustered decomposition of the graph, most clusters or communities are characterized by thematic coherence rather than reflecting the type of structure that underlies thesauri, natural taxonomies, or WordNET.

The thematic structure was wide-spread, showing up in nearly all investigated communities at various depths of the hierarchy. The finding that many words from domains like animals, which traditionally are considered taxonomic, are thematically clustered at the lowest level of the hierarchy, corroborates the idea that the networks are organized along primarily thematic rather than categoric lines. In addition, evaluating the obtained structure in the language and mental networks through human relatedness judgments also confirmed the thematic nature of the networks as indicated by the large proportion of variance that was explained for thematic compared to basic-level judgments. This converges with recent evidence that highlights the role of thematic representations even in domains such as animals (Wisniewski and Bassok 1999; Lin and Murphy 2001; Gentner and Kurtz 2006) and the fact that a taxonomic organization of knowledge might be both heavily culturally defined (Lopez et al. 1997), a consequence of formal education (Sharp et al. 1979) or reflect different levels of expertise (Medin et al. 1997).

A number of explanations can account for why thematic structure was so central in both language and mental networks. One possible explanation is the wide coverage of all kinds of words in the network in terms of their abstractness, emotional connotation, and part of speech (verbs, adjectives, and nouns). By not restricting the type of words in the network, the risk of a selection bias towards concrete nouns (Medin et al. 2000) is reduced and the likelihood of identifying thematic relationships increases. In addition, it is quite likely that this reflects an inherent property of language, where most words are taxonomically related to only a small number of other words, but might occur in a variety of thematic settings. This is in line with previous findings showing that Zipf's law reflects the tendency to avoid excessive synonymy in semantic networks (Manin 2008). Clearly, many of these claims remain speculative, but given their potential implications for understanding the mental lexicon, it is hoped they will motivate future work.

One of the key features in many psychological network proposals is the idea of spreading activation. The current study showed that such a mechanism is of importance as it makes use of the network as a whole. The results show that by including not only direct paths that exist between two nodes (neighbors) but also indirect paths, leads to an improved ability to predict human judgments of relatedness. While this measure led to improvements in all networks, the current results also showed that the gain of indirect paths in predicting relatedness was modulated by the sparsity in the original graph, which is well exemplified by comparing the gains for G_{asso1} to those of $G_{asso123}$.

Similar to the spreading activation account at the mesoscopic level, access at the microscopic level might be governed by more than just the in-strength of a specific node. Measures such as eigen-centrality and PageRank make it conceptually clear that central nodes are those nodes which are easily reached among many possible paths in the graph. These measures are examples of recursive centrality measures, in which centrality is not only influenced by the neighbors of a node, but also takes into account the centrality of the neighbors themselves. This might result in similar benefits found for the spreading activation mechanism operating on sparse graphs. Support for this idea comes from recent studies showing that PageRank accounts for more variance than simple measures of in-strength (Griffiths et al. 2007) and detailed theoretical accounts that explain word frequency advantages in word recognition through higher level structural properties of the network (Monaco et al. 2007). Again, this illustrates the benefits of a network approach which simultaneously describes a macro-, meso- and microscopic level.

4.1 Relationship between Language and Word Associations

A number of studies have tried to predict word associations from text corpora (e.g. Griffiths and Steyvers 2003). While this prediction is often used as a yardstick to compare different text-based models, one of the striking patterns is the overall poor prediction. For instance, in a study by Griffiths and Steyvers (2003), the median rank for predicting the first word association in the University of Florida norms (Nelson et al. 2004) using a text-based topic model was 32. Prediction of the Dutch word association norms (which are considerably larger than the University of Florida norms) on the basis of $G_{asso123}$ resulted in a median rank of 129, and the correct prediction of the first associate in only 5.4% of the cases. Similar results were found when the overlap was calculated in terms of relatedness. Here both the association graphs were strongly correlated (.99), and so were the lexical graphs (.88). Crucially, between both types of graphs, the agreement was quite small: .14 for $G_{asso123}$ and G_{lex} and .11 for $G_{asso123}$ and G_{lex123}. Similar comparisons of microscopic measures of centrality showed only moderate correlations between language and word association graphs. The choice of network - language or mental - might thus lead to different conclusions about findings that show how semantic rich nodes (i.e., those with a high degree, or high clustering) are processed more efficiently in naming or word recognition (Buchanan et al. 2001; Pexman et al. 2003; Pexman et al. 2008).

The limited agreement between the networks and the systematic differences in how they account for specific types of words (especially concrete ones) provide further support to the idea that the association task does not rely on the same properties as common language production, but should rather be seen as tapping into the semantic information of the mental lexicon (Mollin 2009; McRae et al. 2011). This view resonates with the original ideas of Collins and Loftus (1975), in which the network depends both on semantic similarity and lexical co-occurrence in language, and other works that highlight the role of imagery and affect in the production of word associations (Szalay and Deese 1978). As mentioned in the introduction, the role of pragmatics in natural language explains why mentally central properties (e.g., the fact that bananas are yellow or apples are round) are very strong responses in word association data but much less prominently expressed in conventional written and spoken language. To some extent, this might also be the reason why the language networks did not fully capture the human judgments for concrete words (see Table 8). Of course, one could also argue that the language networks in this study are simply too limited due to the vocabulary size restrictions. It seems unlikely that this explanation can account for the entire set of findings. First of all, the results for G_{lex} and G_{lex123} showed that a sampled network based on only a fraction of the tokens produced comparable results for a number of domains. Second, a comparable study involving a language network consisting of a vocabulary of over 100,000 lemmas and the same human relatedness judgments, produced highly similar results (De Deyne et al. 2014). Naturally, this is not to say that additional data and pragmatics are irrelevant. In understanding a story, for instance, where representations that go beyond the word level are required, pragmatics are likely to play a more central role.

4.2 Final Words

In this chapter, a view of the mental lexicon, as a weighted directed graph, with words for nodes, has been advanced as a useful way to explore the structure and processing of word meaning. This account is limited in various ways and by no means complete. For instance, further studies are needed to investigate whether qualitatively different links could lead to a better model of the lexicon through differential weighting of different types of relations in the language network, either syntactic (conjunctions, modification of nouns, etc.) or semantic (hyperonymy, meronymy, etc.).

Similar to the language network, the connections in the word association network are presumably governed by a set of latent relations. In this area as well, the use of a multi-network representation with different weights for various types of relations is likely to explain additional properties of the data. On the basis of these relations, new studies might reveal distinct types of comparison processes as suggested by previous work on thematic and taxonomic comparisons (Wisniewski and Bassok 1999). In particular, a first type of process could be based on the integration of a word in a thematic context (e.g., *doctor* and *hospital*) while a second type might involve the alignment of shared properties between similar entities (e.g., *cat* and *tiger*). Presumably these processes might reflect a highly probable path in the former situation, while some kind of summation over a large number of different paths could be involved in the second process. Knowing something about the properties of nodes on a path (e.g., whether they refer to similar physical entities, a function, or thematic property) requires the derivation of a multi-network as mentioned earlier, and could inform us how such a differential comparison process takes place.

Of course, there are many other areas in which a network approach is likely to contribute in future studies of the lexicon, for instance by studying the development of the lexicon through dynamic networks (Beckage et al. 2010), the networks of individuals (Morais et al. 2013) or by comparing the networks of healthy individuals with clinical populations (Kenett et al. 2013). Presumably, better assumptions about how representations are extracted from the statistical regularities in the language environment will play an important role in these endeavors. In this respect, the application of a syntax-based dependency model represents a first, but certainly not the last step to build a more appropriate mental model of the lexicon. The close relation with empirical indices of mental organization such as human relatedness judgments, but potentially also online measures such as priming (Chumbley and Balota 1984) and word centrality (De Deyne et al. 2013a), suggests that a mental network derived from word associations represents a valuable alternative to model cognitive functions at various levels of abstraction offered through a network science framework.

Acknowledgements. My gratitude goes to Kris Heylen, Dirk Speelman, and Dirk Geeraerts for making the LeNC corpus available, and to Yves Peirsman for collaboration during the early stages of this project. This work was supported by Research Grant G.0436.13 from the Research Foundation - Flanders (FWO) to the first author and by the interdisciplinary research project IDO/07/002 awarded to Dirk Speelman, Dirk Geeraerts, and Gert Storms. Steven Verheyen is a postdoctoral fellow at the Research Foundation - Flanders. Comments may be sent to simon.dedeyne@adelaide.edu.au

References

Ball, B., Karrer, B., Newman, M.E.J.: Efficient and principled method for detecting communities in networks. Physical Review E 84(3), 036103 (2011)

Baronchelli, A., Ferrer-i-Cancho, R., Pastor-Satorras, R., Chater, N., Christiansen, M.H.: Networks in cognitive science. Trends in Cognitive Sciences 17(7), 348–360 (2013)

Beckage, N.M., Smith, L.B., Hills, T.: Semantic network connectivity is related to vocabulary growth rate in children. In: The Annual Meeting of The Cognitive Science Society (CogSci), pp. 2769–2774 (2010)

Blondel, V.D., Guillaume, J.-L., Lambiotte, R., Lefebvre, E.: Fast unfolding of communities in large networks. Journal of Statistical Mechanics: Theory and Experiment 2008(10), P10008+ (2008)

Borge-Holthoefer, J., Arenas, A.: Categorizing words through semantic memory navigation. The European Physical Journal B-Condensed Matter and Complex Systems 74(2), 265–270 (2010)

Bouma, G., van Noord, G., Malouf, R.: Alpino: Wide Coverage Computational Analysis of Dutch. In: Eleventh Meeting of Computational Linguistics in the Netherlands, CLIN, Tilburg, pp. 45–59 (2000)

Buchanan, L., Westbury, C., Burgess, C.: Characterizing the neighbourhood: Semantic neighbourhood effects in lexical decision and naming. Psychonomics Bulletin and Review 8, 531–544 (2001)

Bullinaria, J.A., Levy, J.P.: Extracting semantic representations from word co-occurrence statistics: A computational study. Behavior Research Methods 39, 510–526 (2007)

Bullmore, E., Sporns, O.: The economy of brain network organization. Nature Reviews Neuroscience 13(5), 336–349 (2012)

Capitán, J.A., Borge-Holthoefer, J., Gómez, S., Martinez-Romo, J., Araujo, L., Cuesta, J.A., Arenas, A.: Local-based semantic navigation on a networked representation of information. PloS One 7(8), e43694 (2012)

Ceulemans, E., Storms, G.: Detecting intra-and inter-categorical structure in semantic concepts using HICLAS. Acta Psychologica 133(3), 296–304 (2010)

Chumbley, J.I.: The roles of typicality, instance dominance, and category dominance in verifying category membership. Journal of Experimental Psychology: Learning, Memory, and Cognition 12, 257–267 (1986)

Chumbley, J.I., Balota, D.A.: A word's meaning affects the decision in lexical decision. Memory & Cognition 12, 590–606 (1984)

Collins, A.M., Loftus, E.F.: A spreading-activation theory of semantic processing. Psychological Review 82, 407–428 (1975)

Collins, A.M., Quillian, M.R.: Retrieval time from semantic memory. Journal of Verbal Learning and Verbal Behavior 9, 240–247 (1969)

Cramer, P.: Word Association. Academic Press, New York (1968)

Crutch, S.J., Warrington, E.K.: Abstract and concrete concepts have structurally different representational frameworks. Brain 128, 615–627 (2005)

De Deyne, S., Verheyen, S., Storms, G.: The role of corpus-size and syntax in deriving lexico-semantic representations for a wide range of concepts. Quarterly Journal of Experimental Psychology 26, 1–22 (2015), doi:10.1080/17470218.2014.994098

De Deyne, S., Voorspoels, W., Verheyen, S., Navarro, D.J., Storms, G.: Accounting for graded structure in adjective categories with valence-based opposition relationships. Language, Cognition and Cognitive Processes 29, 568–583 (2013)

De Deyne, S., Verheyen, S., Ameel, E., Vanpaemel, W., Dry, M., Voorspoels, W., Storms, G.: Exemplar by Feature Applicability Matrices and Other Dutch Normative Data for Semantic Concepts. Behavior Research Methods 40, 1030–1048 (2008)

De Deyne, S.: Proximity in Semantic Vector Space. Unpublished Doctoral Dissertation (2008)

De Deyne, S., Navarro, D.J., Storms, G.: Better explanations of lexical and semantic cognition using networks derived from continued rather than single word associations. Behavior Research Methods 45, 480–498 (2013a)

De Deyne, S., Navarro, D.J., Storms, G.: Associative strength and semantic activation in the mental lexicon: evidence from continued word associations. In: Knauff, M., Pauen, M., Sebanz, N., Wachsmuth, I. (eds.) Proceedings of the 33rd Annual Conference of the Cognitive Science Society, pp. 2142–2147. Cognitive Science Society, Austin (2013b)

De Deyne, S., Storms, G.: Word Associations: Network and Semantic properties. Behavior Research Methods 40, 213–231 (2008)

de Groot, A.M.B.: Representational Aspects of Word Imageability and Word Frequency as Assessed Through Word Association. Journal of Experimental Psychology: Learning, Memory, and Cognition 15, 824–845 (1989)

Dennis, S.: A Memory-Based Theory of Verbal Cognition. Cognitive Science 29(2), 145–193 (2005)

Fagiolo, G.: Clustering in Complex Directed Networks. Physical Review E 76, 026107 (2007)

Fellbaum, C.: WordNet: An electronic lexical Database. MIT Press, Cambridge (1998)

Firth, J.R.: Selected papers of JR Firth, 1952-59. Indiana University Press (1968)

Fortunato, S.: Community detection in graphs. Physics Reports 486(3), 75–174 (2010)

Galbraith, R.C., Underwood, B.J.: Perceived frequency of concrete and abstract words. Memory & Cognition 1, 56–60 (1973)

Gentner, D., Kurtz, K.: Relational categories. In: Ahn, W.K., Goldstone, R.L., Love, B.C., Markman, A.B., Wolff, P.W. (eds.) Categorization Inside and Outside the Lab, pp. 151–175. American Psychology Association, Washington, DC (2005)

Gentner, D., Kurtz, K.: Relations, objects, and the composition of analogies. Cognitive Science 30, 609–642 (2006)

Griffiths, T.L., Steyvers, M.: Prediction and Semantic Association. In: Advances in Neural Information Processing Systems, vol. 15, pp. 11–18. MIT Press, Cambridge (2003)

Griffiths, T.L., Steyvers, M., Firl, A.: Google and the Mind. Psychological Science 18, 1069–1076 (2007)

Heylen, K., Peirsman, Y., Geeraerts, D.: Automatic synonymy extraction. In: A Comparison of Syntactic Context Models. LOT Computational Linguistics in the Netherlands 2007, pp. 101–116 (2008)

Hughes, T., Ramage, D.: Lexical Semantic Relatedness with Random Graph Walks. In: Proceedings of the 2007 Joint Conference on Empirical Methods in Natural Language Processing and Computational Natural Language Learning (EMNLP-CoNLL), pp. 581–589. Association for Computational Linguistics, Prague (2007)

Hutchison, K.A.: Is semantic priming due to association strength or feature overlap? Psychonomic Bulletin and Review 10, 785–813 (2003)

Kenett, Y.N., Kenett, D.Y., Ben-Jacob, E., Faust, M.: Global and local features of semantic networks: Evidence from the Hebrew mental lexicon. PloS One 6(8), e23912 (2011)

Kenett, Y.N., Wechsler-Kashi, D., Kenett, D.Y., Schwartz, R.G., Ben-Jacob, E., Faust, M.: Semantic organization in children with cochlear implants: computational analysis of verbal fluency. Frontiers in Psychology 4 (2013)

Keuleers, E., Brysbaert, M., New, B.: SUBTLEX-NL: A new measure for Dutch word frequency based on film subtitles. Behavior Research Methods 42(3), 643–650 (2010)

Kintsch, W.: Comprehension: A paradigm for cognition. Cambridge University Press (1998)

Kintsch, W., Mangalath, P.: The construction of meaning. Topics in Cognitive Science 3(2), 346–370 (2011)

Kiss, G.R.: Words, Associations, and Networks. Journal of Verbal Learning and Verbal Behavior 7, 707–713 (1968)

Lancichinetti, A., Radicchi, F., Ramasco, J.J., Fortunato, S.: Finding statistically significant communities in networks. PloS One 6(4), e18961 (2011)

Landauer, T.K., Dumais, S.T.: A solution to Plato's Problem: The latent semantic analysis theory of acquisition, induction and representation of knowledge. Psychological Review 104, 211–240 (1997)

Landauer, T.K.: LSA as a theory of meaning. In: Landauer, T.K., McNamara, D.S., Dennis, S., Kintsch, W. (eds.) Handbook of Latent Semantic Analysis, pp. 3–35. Lawrence Erlbaum Associates, Mahwah (2007)

Leicht, E.A., Holme, P., Newman, M.E.J.: Vertex similarity in networks. Psychical Review E 73, 026120 (2006)

Lin, E.L., Murphy, G.L.: Thematic Relations in Adults' Concepts. Journal of Experimental Psychology: General 1, 3–28 (2001)

Lopez, A., Atran, S., Coley, J.D., Medin, D.L., Smith, E.E.: The tree of life: Universal and cultural features of folkbiological taxonomies and inductions. Cognitive Psychology 32(3), 251–295 (1997)

Lund, K., Burgess, C.: Producing high-dimensional semantic spaces from lexical co-occurrence. Behavior Research Methods, Instruments, and Computers 28, 203–208 (1996)

Manin, D.Y.: Zipf's law and avoidance of excessive synonymy. Cognitive Science 32(7), 1075–1098 (2008)

McRae, K., Khalkhali, S., Hare, M.: Semantic and associative relations: examining a tenuous dichotomy. In: Reyna, V.F., Chapman, S., Dougherty, M., Confrey, J. (eds.) The Adolescent Brain: Learning, Reasoning, and Decision Making, pp. 39–66. American Psychological Association, Washington, DC (2011), doi:10.1037/13493-002

Medin, D.L., Lynch, E.B., Coley, J.D., Atran, S.: Categorization and reasoning among tree experts: Do all roads lead to Rome? Cognitive Psychology 32(1), 49–96 (1997)

Medin, D.L., Lynch, E.B., Solomon, K.O.: Are there kinds of concepts? Annual Review of Psychology 51(1), 121–147 (2000)

Medin, D.L., Rips, L.J.: Concepts and categories: memory, meaning, and metaphysics. In: Holyoak, K., Morrison, R. (eds.) The Cambridge Handbook of Thinking and Reasoning, pp. 37–72. Cambridge University Press, Cambridge (2005)

Mervis, C.B., Rosch, E.: Categorization of natural objects. Annual Review of Psychology 32, 89–115 (1981)

Miller, G.A., Charles, W.G.: Contextual Correlates of Semantic Similarity. Language and Cognitive Processes 6, 1–28 (1991)

Mollin, S.: Combining corpus linguistics and psychological data on word co-occurrence: Corpus collocates versus word associations. Corpus Linguistics and Linguistic Theory 5, 175–200 (2009)

Monaco, J.D., Abbott, L.F., Kahana, M.J.: Lexico-semantic structure and the word-frequency effect in recognition memory. Learning & Memory 14, 204–213 (2007)

Moors, A., De Houwer, J., Hermans, D., Wanmaker, S., van Schie, K., Van Harmelen, A.-L., De Schryver, M., De Winne, J., Brysbaert, M.: Norms of valence, arousal, dominance, and age of acquisition for 4,300 Dutch words. Behavior Research Methods, 1–9 (2012)

Morais, A.S., Olsson, H., Schooler, L.J.: Mapping the structure of semantic memory. Cognitive Science 37(1), 125–145 (2013)

Motter, A.E., de Moura, A.P.S., Lai, Y.-C., Dasgupta, P.: Topology of the conceptual network of language. Physical Review E 6, 065102 (2002)

Murphy, G.L.: The big book of concepts. MIT Press, Cambridge (2002)

Nelson, D.L., McEvoy, C.L.: What is this thing called frequency? Memory & Cognition 28, 509–522 (2000)

Nelson, D.L., McEvoy, C.L., Schreiber, T.A.: The University of South Florida free association, rhyme, and word fragment norms. Behavior Research Methods, Instruments, and Computers 36, 402–407 (2004)

Niedenthal, P.M., Halberstadt, J.B., Innes-Ker, A.H.: Emotional response categorization. Psychological Review 106(2), 337 (1999)

Oostdijk, N.: The Spoken Dutch Corpus: Overview and first evaluation. In: Piperidis, S., Stainhaouer, G. (eds.) Proceedings of Second International Conference on Language Resources and Evaluation, vol. 2, pp. 887–894. ELRA, Paris (2000)

Ordelman, R.J.F.: Twente nieuws corpus (TWNC). Tech. rep. Parlevink Language Technology Group, University of Twente (2002)

Padó, S., Lapata, M.: Dependency-Based Construction of Semantic Space Models. Computational Linguistics 33(2), 161–199 (2007)

Page, L., Brin, S., Motwani, R., Winograd, T.: The pagerank citation ranking: Bringing order to the web. Tech. rep. Computer Science Department, Stanford University (1998)

Pereira, F., Tishby, N., Lee, L.: Distributional clustering of English words. In: Proceedings of the 31st Annual Meeting on Association for Computational Linguistics, pp. 183–190. Association for Computational Linguistics, Columbus (1993)

Pexman, P.M., Hargreaves, I.S., Siakaluk, P.D., Bodner, G.E., Pope, J.: There are many ways to be rich: Effects of three measures of semantic richness on visual word recognition. Psychonomic Bulletin & Review 15(1), 161–167 (2008)

Pexman, P.M., Holyk, G.G., Monfils, M.-H.: Number-of-features effects in semantic processing. Memory & Cognition 31, 842–855 (2003)

Plaut, D.C., Shallice, T.: Deep dyslexia: A case study of connectionist neuropsychology. Cognitive Neuropsychology 10, 377–500 (1993)

Prior, A., Bentin, S.: Incidental formation of episodic associations: the importance of sentential contex. Memory & Cognition 31, 306–316 (2003)

Van Rensbergen, B., De Deyne, S., Storms, G.: Cue-association correspondence on valence, dominance, and arousal. Manuscript submitted for publication (2014)

Rogers, T.T., McClelland, J.L.: Semantic cognition: A parallel distributed processing approach. MIT Press, Cambridge (2004)

Rosch, E.: Natural Categories. Cognitive Psychology 4, 328–350 (1973)

Ruts, W., De Deyne, S., Ameel, E., Vanpaemel, W., Verbeemen, T., Storms, G.: Dutch norm data for 13 semantic categories and 338 exemplars. Behaviour Research Methods, Instruments, and Computers 36, 506–515 (2004)

Samsonovic, A.V., Ascoli, G.A.: Principal semantic components of language and the measurement of meaning. PloS One 5(6), e10921 (2010)

Schank, R.C., Abelson, R.P.: Scripts, plans, goals, and understanding: An inquiry into human knowledge structures. Lawrence Erlbaum, Hillsdale (1977)

Schwanenflugel, P.J., Akin, C., Luh, W.M.: Context availability and the recall of abstract and concrete words. Memory & Cognition 20, 96–104 (1992)

Sharp, D., Cole, M., Lave, C., Ginsburg, H.P., Brown, A.L., French, L.A.: Education and cognitive development: The evidence from experimental research. In: Monographs of the society for Research in Child Development, pp. 1–112 (1979)

Simmons, W.K., Hamann, S.B., Harenski, C.N., Hu, X.P., Barsalou, L.W.: fMRI evidence for word association and situated simulation in conceptual processing. Journal of Physiology - Paris 102, 106–119 (2008)

Solé, R.V., Corominas-Murtra, B., Valverde, S., Steels, L.: Language networks: Their structure, function, and evolution. Complexity 15(6), 20–26 (2010)

Steyvers, M., Tenenbaum, J.B.: The Large-Scale Structure of Semantic Networks: Statistical Analyses and a Model of Semantic Growth. Cognitive Science 29, 41–78 (2005)

Szalay, L.B., Deese, J.: Subjective meaning and culture: An assessment through word associations. Lawrence Erlbaum, Hillsdale (1978)

Van Dongen, S.: Graph clustering by flow simulation. PhD thesis. University of Utrecht (2000)

Verheyen, S., Stukken, L., De Deyne, S., Dry, M.J., Storms, G.: The generalized polymorphous concept account of graded structure in abstract categories. Memory & Cognition 39, 1117–1132 (2011)

Vitevitch, M.S.: What can graph theory tell us about word learning and lexical retrieval? Journal of Speech, Language, and Hearing Research 51(2), 408–422 (2008)

Warrington, E.K., Shallice, T.: Category specific semantic impairments. Brain 107, 829–854 (1984)

Wisniewski, E.J., Bassok, M.: What Makes a Man Similar to a Tie? Cognitive Psychology 39, 208–238 (1999)

Wittgenstein, L.: Philosophical Investigations: 50th Anniversary Commemorative Edition. Blackwell Publishing, Incorporated, Oxford (2001)

Part II
Topology

Network Motifs Are a Powerful Tool for Semantic Distinction

Chris Biemann, Lachezar Krumov, Stefanie Roos, and Karsten Weihe

Abstract. *Motifs* are a general network analysis technique, which statistically relates network structure to epiphenomena on the network. This technique has been developed and brought to maturity in molecular biology, where it has been successfully applied to network-based chemical and biological dynamics of various types. Early on, the motif technique has been successfully applied outside biology as well – to social networks, electrical networks, and many more. Results by Milo et al. showed that the *motif signature* of a network varies from realm to realm to some extent but is significantly more homogenous within a realm. This observation has been the starting point of the thread of research presented in this paper. More specifically, we do not compare networks from different realms but focus on networks from a given realm. In several case studies on particular realms, we found that motif signatures suffice to distinguish certain classes of networks from each other. In this paper, we summarize our previous work, and present some new results. In particular, in Biemann et al. (2012), we found that natural and artificially generated language can be distinguished from each other through the motif signatures of the co-occurrence graphs. Based on that, we present work on co-occurrence graphs that are restricted to word classes. We found that the co-occurrence graphs of verbs (and other word classes used like predicates) exhibit strongly different motif signatures and can be distinguished by that. To demonstrate the general power of the approach, we present further original work on co-authorship networks, peer-to-peer streaming networks, and mailing networks.

Chris Biemann · Lachezar Krumov · Stefanie Roos · Karsten Weihe
Computer Science Department, Technische Universität Darmstadt,
Hochschulstr. 10, 64289 Darmstadt, Germany
e-mail: {biem,weihe}@cs.tu-darmstadt.de,
    lakkrumov17@yahoo.com, stefanie.roos@tu-dresden.de

1 Introduction

The Basis: Motif Signatures

By a *network*, we mean an undirected graph $G = (V, E)$, where, as usual, V is a finite set of *nodes* and E a set of *edges* linking nodes to each other. In this paper, we consider networks from several realms with thousands of nodes. We use the words *graph* and *network* synonymously.

A *motif* $M = (V_M, E_M)$ is a fixed graph and is typically very small, usually not more than a few nodes. An *occurrence* of a motif M in a network G is a subset $V' \subseteq V$ of size V_M such that the subgraph of V induced[1] by V' is isomorphic to M. For a fixed set of motifs, the *motif signature* of a network is a vector with one component for each motif, and the component associated with a motif contains the number of occurrences of this motif in the network. In order to compare networks of different size, we normalize this vector with the overall count of all motifs of fixed size, such that the sum of its entries is 1. In this way, values for single motifs correspond to the fraction of single motif counts from the overall count of all motifs in the network.

In this paper, the fixed sets of motifs are the one depicted in Fig. 1 and Fig. 2, respectively.[2] Note that motifs are defined over connected subgraphs, and do not contain each other: e.g. if four nodes are fully connected, they exhibit only the clique motif, and not e.g. any semi-clique motifs, as for a semi-clique, two of the four nodes must NOT be connected.

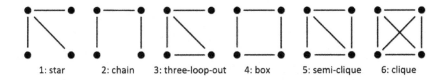

1: star   2: chain   3: three-loop-out   4: box   5: semi-clique   6: clique

Fig. 1 The undirected motifs of size 4 with their common names, which will be used in this paper as well.

---

[1] General definition of *induced subgraphs*: Let $G_1 = (V_1, E_1)$ and $G_2 = (V_2, E_2)$ such that $V_1 \leq V_2$, and let $V_2' \subseteq V_2$ such that $|V_2'| = |V_1|$. Then we say that G_1 is the *subgraph* of G_2 *induced* by V_2', if there is a bijection $\varphi : V_1 \to V_2'$ such that for all $v, w \in V_1$, it is $(v, w) \in E_1$ if and only if $(\varphi(v), \varphi(w)) \in E_2$. In contrast to the unique *induced* subgraph, an ordinary subgraph may contain fewer edges, that is, $(\varphi(v)\varphi(w)) \in E_2$ does *not* imply $(v, w) \in E_1$.

[2] Quite often in the literature, not all considered subgraphs are called *motifs*, but only the ones that occur significantly more frequently than in some *null model*. The ones that occur *less* frequently are then either called *motifs* as well, or they are called *anti-motifs*.

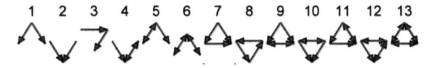

Fig. 2 The directed motifs of size 3. Motif 1 is called *V-Out*, motif 3 is the *3-Chain*, cf. Sect. 3.3.

Use of Motif Signatures

Generally speaking, every case study in this paper focuses on networks from a particular realm. Examples of realms are: co-occurrence networks, word similarity networks, lexical semantic networks, co-authorship networks, mailing networks, subnetworks of the web, peer-to-peer streaming networks, etc. Even within a realm, classes of networks are often difficult to distinguish from each other. For example, for the type of co-occurrence graphs, all graphs derived from English text form a subtype, and all graphs derived from language generated by a 3-gram model (Biemann et al. 2012) form another subtype.

Our general working hypothesis is that networks may be classified to some extent by their motif signatures. In this paper, we demonstrate that this hypothesis is indeed true in various domains.

Presented Studies

More specifically, we present five case studies, in which two subtypes of a network type can be quite accurately distinguished from each other through characteristic differences in the motif signatures. In the fourth case study, the situation is a bit more complex than in the first three ones. More specifically, it does not suffice to look at a well-chosen single motif. Instead, we need conjunctions of single-motif distinctions because here no single motif is distinctive for all instances. Finally, in the fifth case study, our attempt was not successful but still exhibits a promising tendency. The first case study has already been published, so we merely review it; the other four case studies are original work.

- Section 3.1: Co-occurrence graphs of text: artificially generated language vs. natural language (Biemann et al. 2012).
- Section 3.2: Co-occurrence graphs of word classes: verbs vs. other word classes.
- Section 3.3: Artificially generated peer-to-peer streaming networks from two different generators: Partial Streams vs. Node-disjoint Streams
- Section 3.4: Co-authorship networks from two different sub-disciplines of physics: condensed matter vs. astrophysics.
- Section 3.5: Mailing networks from two different contexts: the European Research Institute vs. Enron.

Before we present the studies, we briefly review the state of the art in the next section.

2 Related Work

Motif Analysis

Motif analysis has first been investigated in computational biology (Shen-Orr et al. 2002) and has since been applied to a variety of network types in biology and biochemistry (Alon 2007; Schreiber and Schwöbbermeyer 2010). The underlying insight is that biological and biochemical dynamics are statistically related to the occurrence of small *functional blocks*, which have specific structures. This insight is well captured by motif signatures, and in fact, many computational studies reveal significant relations. Due to this success, it did not take long time until this technique has been applied to networks from other domains. For example, Milo et al. (2002) and Milo et al. (2004) compare networks from biology, electrical engineering, natural language and computer science and find that the motif signatures from different domains are so different that they may serve as a "fingerprint" of the respective domain.

The idea of functional blocks applies in domains beyond biology and biochemistry as well, surprisingly, even in social networks. In Krumov et al. (2011), we analyze citation networks, which we model as undirected graphs on the authors. An edge indicates at least one joint publication. In a sense, the citation numbers of individual publications within an occurrence of a motif can be aggregated to a citation number of the entire occurrence. We consider four natural ways for aggregation. Roughly speaking, the main result of Krumov et al. (2011) is this: the average citation number of the *box motif* (#4 in Fig. 1), taken over all occurrences, is statistically significantly larger than expected. This effect occurs for all four ways of aggregation. A deeper look revealed that certain occurrences of the box motif explain this result: two "seniors," A and B, have jointly published, A has published with a "junior" C, B with a junior D, and C and D have joint publications as well, but neither A with D nor B with C. Among these occurrences, the ones that serve as "bridges" in the network in a certain sense are particularly responsible for the observed effect.

We further mention recent work that uses the concept of motifs for other purposes than network analysis. Krumov, Andreeva, et al. (2010) and Krumov, Schweizer, et al. (2010) develop an algorithm to optimize the structure of peer-to-peer networks based on local operations only. Each node manipulates the local structure in its vicinity in order to thrive the local motif signature towards the average local motif signature of an optimal network. We are not aware of any other usage of motifs for algorithmic purposes.

See Wong et al. (2012) for an overview of motif detection tools.

Graph Classification

A domain of graphs may decompose into classes of graphs. Often, for a given graph, the class to which it belongs cannot easily be identified. The general idea of *graph classification* is to use additional structural properties that are characteristic of the graph class. Chapter 11 of Aggarwal and Wang (2010) provides a good overview.

Using this terminology, the topic of the herewith presented paper and the five case studies can be concisely characterized as *graph classification using networks motifs*. Besides our own work, we are only aware of one other piece of scientific work of this type, which we briefly discuss next (Juszczyszyn and Kolaczek 2011).

Anomaly detection can be regarded as a special case of graph classification: determine whether or not the current state of a dynamically changing network exhibits an anomaly of a certain type at some point in time amounts to deciding whether the current snapshot of the network belongs to the class of abnormally shaped networks or not. We are aware of only one original result where motifs were used for anomaly detection. More precisely, the motif signature is observed to determine whether or not a communication network is under attack. See Juszczyszyn and Kolaczek (2011) for details.

In a broader sense, *frequent pattern mining* is the dual problem to graph classification using motifs: for some graph classification problem and some large set of patterns (motifs), find a small subset such that the frequencies (*i.e.* the motif signature) has maximal discriminative power. Jin et al. (2009) may serve as a entry point into the literature on frequent pattern mining.

3 The Case Studies

3.1 Co-occurrence Graphs from Natural Vs. Artificial Language

In this section, we review our work that has been previously published in Biemann et al. (2012) and complement it with a more detailed analysis. Though it is generally accepted that language models do not capture all aspects of real language, no adequate measures to quantify their shortcomings have previously been proposed. We use n-gram model generators to demonstrate that the differences between natural and generated language are indeed quantifiable with motif analysis based on the analysis of co-occurrence networks. The motif approach allows a deeper insight into those semantic properties of natural language that evidently cause these differences: polysemy and synonymy. With our method, it becomes possible for the first time to measure deficiencies of generative language models with regard to lexical semantics of natural language.

Our results are generated in a three-step process: First, the text needs to be selected, respectively generated, before the graphs can be derived from the texts according to a parameterizable strategy. In the last step, we evaluate our proposed metrics on these graphs.

For our experiments, we use corpora of different languages of one million sentences each, provided by LCC[3] (Biemann et al. 2007). We use the same corpus of real language for training the n-gram model and for comparison. For comparison between real and generated language, we generate text according to the same sentence length (number of tokens) distribution as found in the respective real language corpus, since we have found in preliminary experiments that co-occurrence network structure is dependent on the sentence length distribution. We have found in further experiments, that the general picture of results is stable for corpora of different sizes, starting from about ten thousand sentences, and insensitive to changing the parameters of graph construction (see below) in a wide range.

Text Generation with N-gram Models

For the scope of this work, we chose n-gram models, which are the standard workhorses of language modeling. A language model assigns a probability to a sequence of words, based on a probabilistic model of language. This can be used to pick the most probable/fluent amongst several alternatives, e.g. in a statistical machine translation system (Koehn 2010). An n-gram language model (cf. Manning and Schütze (1999)) over sequences of words is defined by a Markov chain of order $(n-1)$, where the probability of the next word only depends on the $(n-1)$ previous words, and the probability of a sentence is defined as $P(w_1...w_k) = \prod_{i=1..k} P(w_i|w_{i-1}..w_{i-n+1})$. We add special symbols, BoS and EoS, to indicate sentence beginning and end. Then we generate sentences word by word, starting from a sequence of $(n-1)$ BoS-symbols, according to the probability distribution over the vocabulary. As soon as the EoS symbol is generated, we generate the next sentence. Probabilities are initialized by training on the respective corpus of real text (see above) from the relative counts, i.e. $P(w_i|w_{i-1}..w_{i-n+1}) = count(w_i..w_{i-n+1})/count(w_{i-1}..w_{i-n+1})$. Despite their simplicity, n-gram models still excel in NLP applications (cf. Ramabhadran et al. (2012)).

Yet, shortcomings of n-gram models are obvious: no long-range relations are modeled explicitly, thus n-gram models produce locally readable but semantically incoherent text. This study is, to our knowledge, the first attempt to quantify this phenomenon.

Network Generation from Text

The nodes of the derived graphs correspond to the m most frequent words in the considered text. An edge from node A to B exists if the word corresponding to A *co-occurs*, i.e. occurs together in a well-defined context, with the word corresponding to B significantly often. Different kinds of co-occurrence contexts are considered, as well as significance thresholds and graph sizes m. We consider co-occurrence

---

[3] see http://corpora.informatik.uni-leipzig.de/

within a sentence (sequences as limited by *BoS* and *EoS*). Thus, for each corpus of text, composed of sentences, we can compute the co-occurrence graph by connecting word nodes with edges, if words co-occur. It is known (Biemann et al. 2004) that sentence-based co-occurrences, besides capturing collocations, often reflect semantic relations and capture topical dependencies between words. Since mere co-occurrence results in a large number of edges and very dense networks, we apply a significance test that measures the deviation of the actual co-occurrence frequency from the co-occurrence frequency that would have been observed if the two co-occurring words would be distributed independently. Here, we use the log-likelihood test (Dunning 1993) to prune the network: We only draw edges between word nodes, if the words co-occur with a significance value above a certain threshold. We have found in preliminary experiments that pruning non-significant edges moderately increases the number of chain motifs for all graphs. We did not find any differences when pruning with different levels of significance: no matter whether 10%, 50% or 80% of the non-significant or low-significant edges were pruned, the motif signatures are indistinguishable except for absolute counts. In the interest of the speed of the analysis, we apply a rather extensive pruning in our experiments.

Parameter Choice for Text and Network Generation

In our study, we used n-gram models with $n \in \{1,2,3,4\}$. The number of nodes m in the graph, corresponding to the most frequent words in the considered text, was set to be $5,000$, as to match the commonly assumed size of the core vocabulary of a language (Dorogovtsev and Mendes 2001). A number of thresholds for the significance value were tested. For brevity, we only present the results for a threshold of 10.83, which corresponds to a significance level of $p < 0.001$ and retains about 1/6th of edges of the co-occurrence graph. The co-occurrence graph was computed using the TinyCC corpus production engine (Quasthoff et al. 2006).

How Motifs Capture Natural Language Semantics

Recall the concept of *functional blocks* from Section 2. Next we show that, in our context here, the chain and the box motif are functional blocks in quite an analogous sense. Figure 3 shows the motif signatures of networks on a log-scale, depicting the motif signature for English networks of real and generated language for $n = 1,2,3,4$. Real language networks exhibit fewer star (#1) motifs and a higher amount of all other motifs. Differences for the chain (#2) and the box (#4) motifs are especially pronounced. The observed features can be found in all observed languages, as exemplary shown in Fig. 4 for six languages. In particular, the fraction of box motifs is reduced to less than 10%, 40% and 65% for 2-grams, 3-grams, and 4-grams, respectively, for all considered languages. As the absolute count of box motifs for real language graphs ranges in the 100,000s and the absolute counts of chains ranges in the millions in all experiments reported here, these differences can hardly be explained by random fluctuations.

Examining instances of these motifs more closely, we are able to link these differences to properties of natural language semantics.

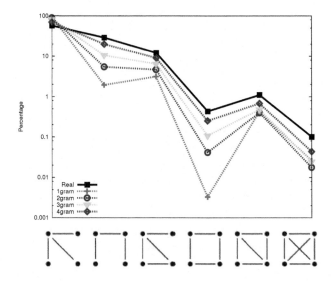

Fig. 3 Motif signatures of English Language for real and generated text networks based on sentence co-occurrence

Semantic polysemy refers to the phenomenon that a word, denoted as a string of characters, can have different denotations in different contexts, e.g. "board" as an assembly or a piece of wood. In real sentences, words are not co-occurring at random, but usually revolve around a certain topic. Thus, it is not likely to find the word "wood" in a sentence that talks about a "board of directors", and sentences about wooden planks usually do not contain the word "chairman". In co-occurrence networks, polysemy leads to chains: ambiguous words connect words that are not connected to each other, and act as a bridge between different topical word clusters. In a chain of length four, one more word from a topical cluster is observed, which does not connect to the polysemous word since it seems that their occurrences are deemed rather independent by the significance measure.

Enumerating the chain motif instances of the English real network, we exemplify this point with the following chains (the ambiguous word is emphasized in each line):

- total - km^2 - **square** - root
- Democrats - **Social** - Sciences - Arts
- Number - **One** - Formula - Championship
- Abraham - **Lincoln** - Nebraska - Iowa

Network Motifs Are a Powerful Tool for Semantic Distinction

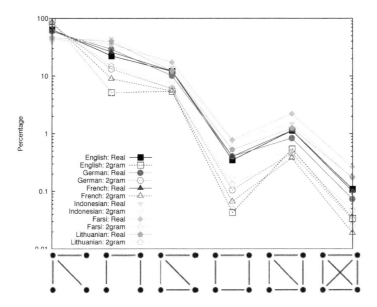

Fig. 4 Motif signatures of various languages for real and generated text networks based on sentence co-occurrence

N-gram models are oblivious to these sense distinctions. Thus, nothing prevents e.g. a 3-gram model from generating e.g. a sequence "Abraham Lincoln , Nebraska" with high likelihood, confusing the two senses of "Lincoln" as a last name and a city. In the co-occurrence network, this can result in a connection between "Abraham" and "Nebraska", which decreases the chain motif count. The remaining chains of n-gram networks, on the other hand, consist mostly of highly frequent words that occur next to each other, e.g. "slowly started on finals", "personal taste good advice". These are also present in the real language network. We observe a much smaller number of chains formed of words of lower frequencies in n-gram generated text. Note that it is neither the case that all polysemous words cause chains, nor do all words in the central positions of a chain exhibit lexical ambiguity – differences in chain motif counts rather quantitatively measure the amount of such polysemy than qualify as an instrument to find single instances.

Hence, the lower amount of chain motifs can be explained by the creation of links that are not present in real language. On the first glance, this should lead to a higher clustering, contradictory to the results for motifs #3,5,6. Although some of the 4-nodes sets that represent boxes or chains in real languages form (semi-)cliques in n-grams, instances of motif #3,5,6 in real languages are replaced by stars in n-gram graphs more frequently.

From these observations it becomes very clear that chain motif counts reflect polysemy. The lower n is chosen in the generating n-gram model, the smaller is the modelling context for ambiguities, resulting in lower chain motif counts.

Synonymy means that different words refer to the same concept. Two words are synonyms if they can be used interchangeably without changing the meaning, but there are also rather syntactic variants of words that refer to the same concept, such as nominalizations of adjectives or verb forms of different inflections.

In natural language, the principle of *parsimony* leads to the effect that the same concept is rarely referred to several times in the same sentence. In fact, synonyms usually do not co-occur, but they share a large number of significant co-occurrences – an observation that leads to the operationalization of the distributional hypothesis (Miller and Charles 1991). When two such concepts are mentioned together frequently since they belong to the same topic, this leads to box motifs, as the following examples from the English real language network illustrate:

- - Ancient - Greek - ancient - Greece -
- - winning - award - won - price -
- - Ph.D - his - doctorate - University -
- - said - interview - stated - " -
- - wrote - articles - published - poems -

We observe different kinds of word pairs for the same concept: synonyms like (award, price), same word stem within or across word classes like (winning, won) or (Greek, Greece), and artifacts of punctuation or spelling (ancient, Ancient) or (interview, "), Thus, box motifs capture a very loose notion of synonymy: "interview" and the double quote ", e.g. both refer to a (indirect or direct) speech act. Also, inhibiting concepts need not be synonyms – co-hyponyms seem to inhibit each other as well, as e.g. "articles" and "poems" above.

Again, n-gram models are not aware of concepts and references to them, so there is no mechanism that prevents the n-gram model from generating sentences that refer to the same concept several times or even use the same word repeatedly. This possibly results in a connection between those pairs, reducing the box motif count. Box motifs that can be found in both real and n-gram language are again resulting from local sequences of highly frequent words that are possibly circular, not necessarily from the same contexts. Examples include "desktop cover art background", "hall nearby on church" and "these will ask why".

Quantitative Assessment

To quantify our observations, we randomly sampled 100 box motifs and chain motifs from the English real language network. We excluded motifs containing one or more of the 100 most frequent words in order to concentrate our study on content words as opposed to words like prepositions, determiners and pronouns, and to exclude syntactic patterns of highly frequent words as mentioned above.

In the sample of 100 box motifs, we found the following:

- 40 instances where the opposite nodes in the motif are semantically related – sometimes as synonyms, but often as co-hyponyms or antonyms. While true synonyms are mostly found as capitalized or plural variants (e.g. "- board - member

- Board - members -", semantically close terms are more frequent, such as in " - Wednesday - night - Monday - morning - ".
- 31 instances where only one of the two opposing node pairs stand in such a semantic relation. Often, the other opposing nodes would be a compatible connection, but their connection was pruned by the significance threshold, e.g. as in "- Economists - polled - earnings - surveyed -", where "polled" and "surveyed" are almost synonymous, whereas "Economists" and "earnings" are not.
- 3 instances where the box motif was caused by ambiguous parts of (capitalized) names
- 13 instances where the box was caused by the significance threshold, as all nodes belong to the same topic, such as "- court - case - trial - date -".
- 13 instances that were unclear, such as "- runs - seven - lead - gave - "

In the sample of 100 chains, the distribution of classes was as follows:

- 26 instances with one ambiguous term, e.g. "slot - machines - voting - Early". Many of them were found to be parts of names, e.g. "First" in "Baptist - Church - First - Thomson" originating from some "First Baptist Church" and from "Thomson First Call".
- 6 instances where both middle nodes correspond to ambiguous terms, as in "Series - Division - II - War", where "Division" appears in a narrow sense of the "Baseball Division Series" and a wider sense with a Division number II, whereas II is part of "Word War II". All of these instances were parts of such named entities.
- 15 instances where the set of nodes contains synonyms or co-hyponyms. These would have fit the interpretation of a box pattern, except that one link was dropped because of the significance threshold, e.g. "Credit - card - credit - cards"
- 48 instances where the chain was caused by the significance threshold, i.e. all nodes are topically related
- 5 unclear instances

This confirms that while motif analysis cannot be used as an instrument to find synonymous resp. ambiguous words, the fraction of boxes and chains caused by these words is nevertheless sizeable enough to cause measurable differences in the motif signature.

Summary on Natural Vs. Artificial Language

These observations lead to the conclusion that synonymy of natural language leads to box motifs in sentence-based co-occurrence networks, and the difference in the box motif count quantifies the amount of capability of the language model to inhibit the generation of words that refer to concepts which already have been mentioned. N-gram models have this capability only for a very limited context, which again increases with higher n.

In this work, we have demonstrated that motifs in fact capture natural language semantics. Specifically, we were able to show that motifs detect deficiencies in n-gram models that stem from their obliviousness towards synonymy and polysemy.

An interesting fact to note here is that lexical meaning manifests itself in the *absence* of connections in the co-occurrence network.

Since the concept of motifs has recently been used for a constructive purpose (cf. Section 2), we anticipate that this change of perspective is also promising in the realm of language networks and may well guide the design of new, semantically more adequate, language models.

3.2 Co-occurrence Graphs from Verbs Vs. Other Word Classes

Now, we are interested in the difference of motif networks generated for different parts of speech (POS). Starting from the hypothesis that different word classes (such as verbs, common and proper nouns, adjectives, prepositions) serve different functions in language, we expect to be able to distinguish their respective networks. For experiments reported in this section, we use the same way of constructing sentence co-occurrence graphs from text as outlined in the previous section. Using a POS tagger (Schmid 1994) on real-language corpora of 1 million sentences, we can examine subnetworks from the co-occurrence network that are produced by only selecting the most frequent 3,000 nodes per open POS class: adjectives (ADJ), common nouns (NOUN), proper nouns (NAME) and verbs (VERB).

Figure 5 depicts the motif signatures for three languages. It becomes apparent that verbs show a pattern that is clearly distinct to the other open word classes.

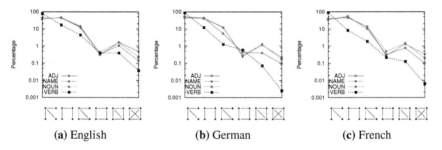

(a) English   (b) German   (c) French

Fig. 5 Motif signatures for Parts-of-Speech by a linguistically motivated POS-tagger for three languages

Verb networks are clustered much less (less cliques, semi-cliques, three-loopouts) than other word classes, and also form a smaller fraction of chains. This can be linguistically explained as follows: There are only few verbs per sentence, usually one verb in the main clause, and one verb per subordinate clause. While cliques are often formed by semantically similar words that occur together in enumerations (nouns) or co-modifiers (adjectives), there is no frequent grammatical construction that yields a co-occurrence of many verbs. The high fraction of star motifs can be explained by the co-occurrence of nearly all verbs with auxiliary or modal verbs, such as "have", "be", "shall", "can", "must" etc.

From this, we can conclude that motif analysis can distinguish sentence-based co-occurrence graphs of verbs from the graphs of other word classes, which is rooted in the fact that the function of verbs in natural language is very distinct from the function of the other open word classes. Note that if we combine word classes, e.g. verbs and adjectives, the distinctive features of verbs are lost, providing motif signatures close to the non-verb class.

Now, we attempt to use this knowledge to find the verbs for languages where we do not have a linguistically motivated POS tagger. In unsupervised natural language processing, knowing which word in a sentence has the function of the predicate (i.e. is a verb) greatly improves the induction of syntactic structures, see e.g. Søgaard (2012) and Bisk and Hockenmaier (2013).

Motivated by the distinctive motif signatures of verbs, we want to categorize verbs in an unsupervised manner as follows: First, an unsupervised POS tagger is applied generating word clusters from a corpus. Secondly, we categorize clusters into verb and non-verb clusters based on the motif signature and the transitivity[4] of their co-occurrence graphs.

The analyzed corpora consider a wide range of languages: Czech, Dutch, English, Estonian, French, German, Indonesian, Italian, Lithuanian, Portuguese, and Russian. Two unsupervised POS tagger are applied to generate word clusters: The first one (Biemann 2006) finds the number of tags automatically, whereas the approach by Clark (Clark 2003) requires setting the number of POS classes to a predefined number. Only clusters of at least 1000 words are included in the analysis because small cluster do not provide a sufficient sample of the motif signature.

After obtaining the word clusters, we first exemplary performed a manual categorization. Furthermore, two potential categorization methods are compared by their accuracy, recall, and precision. We define the accuracy as the fraction of clusters that have been accurately classified as verb or non-verb. The recall is the fraction verb clusters that were correctly classified, whereas the precision denotes the fraction of verb clusters within all clusters that have been classified as verbs.

The first algorithm only considers a single-scalar metric M, either one motif percentage or the transitivity. For each set of n clusters of the same corpus, the values $m_1, \ldots, m_n$ of M are sorted in ascending order. Then the two categories $c_1 = \{m_1, \ldots, m_t\}$ and $c_2 = \{m_{t+1}, \ldots, m_n\}$ are determined, so that $m_{t+1} - m_t$ is maximized. The verb cluster is chosen based on our observations: For the star motif the cluster with the higher values is labeled as verbs, otherwise the cluster with the lower values.

---

[4] Transitivity: Let $G = (V,E)$ be an undirected graph. A closed triangle is a set of three nodes such that all three possible edges do exist. On the other hand, a triple is any set of three nodes and two edges (in other words, a chain of two edges). The transitivity of $T(G)$ of G is three times the total number of closed triangles divided by the total number of triples, as defined by Newman et al. (2002).

The second approach is to apply a 2-means clustering to the word clusters with the goal of distinguishing verbs from non-verbs. The motif percentages as well as the transitivity are identified with points in an Euclidean space. Initially, two cluster centers are chosen uniformly at random from the hypercube defined by the lowest and highest values of each single-scalar metric. Then the standard k-means clustering algorithm (MacQueen 1967) is performed for a maximum of 100 rounds with $k = 2$, assigning each point to the closest cluster center and computing the new cluster centers. Finally, the cluster with the lower value for the chain motif is categorized as verbs. The experiment is repeated for 100 runs, and each word cluster is assigned the category it was assigned the majority of runs.

Given the clear difference between verbs and the other open word classes, we expect that verbs can be easily distinguished from other classes. However, the existence of additional classes and the distribution of verbs over various clusters can be expected to complicate the categorization, possibly reducing the accuracy of the algorithms.

We summarize the results achieved on clusters produced by the unsupervised POS-tagging by Biemann (Biemann 2006). Figure 6 shows three exemplary motif signatures for Portuguese, German, Italian, and Indonesian. The categorization for Portuguese is obvious. There is only one verb cluster, which is identified by the high fraction of stars as well as the low fractions of chains and cliques. For German, there are two obvious verb candidates are given by clusters 7 and 18, which indeed correspond to verb clusters. Furthermore, 12 is a verb cluster, which shows the characteristic high fraction of stars as well as a low fraction of chains, but does not have the distinctive features for cliques and semi-cliques. Such a clear distinction of the most dominant verb cluster can be observed in other Germanic languages such as Dutch and English, but also for Czech. Italian, too, exhibits clusters which are obvious verb candidates, namely 19 and 1. However, these are not actually verbs, but rather participles derived from verbs, e.g. *insoluti, risucchiati, uguali, accadute, affissi, determinanti*. The actual verb clusters 3 and 9 are less distinguished, only showing chain motif counts slightly below average. A similar, though less pronounced, relation can be observed in French.

Another indefinite picture is presented by the motif signatures in Indonesian. In contrast to most other languages, there are no obvious outliers. The verb clusters 14 and 15 show characteristic behavior for the semi-cliques and cliques. Furthermore, the values for chains are slightly lower than for the remaining languages. However, also 5 and 9, two clusters of nouns, could be identified as verbs by their high fraction of star motifs or the slightly lower than average values for chains and cliques.

The results from this manual categorization can be summarized as that there are features visible in verb clusters, however, these are not always as distinctive as when using a language-specific supervised POS-tagging. Moreover, these features can also be found in some non-verb classes. In particular, we also consider participles as verbs, since they have a similar function in the sentence, i.e. lexicalize a predicate. Furthermore, mixed classes are not considered verbs in our automatic classification due to the observation that verb characteristics are lost when mixed in with other classes.

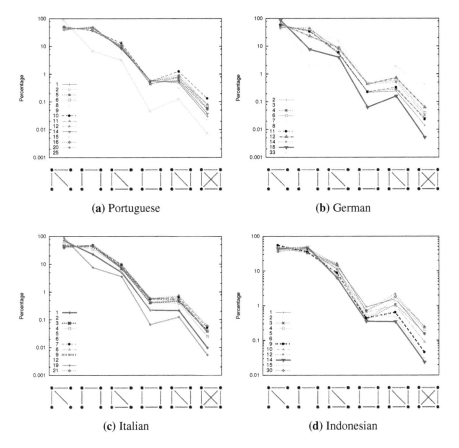

Fig. 6 Motif signatures of unsupervised POS-tagger

The performance results of our automatic algorithms differs for various languages. When only considering a single-scalar metric, there is no metric that performs well on all languages: For example, the transitivity is discriminative for Portuguese and Estonian, achieving an accuracy of 100%, but not so much for the others. Motif 1, 2 and 4 on the other hand produced almost perfect results for German, English, Dutch, Italian and Portuguese, but not for French, Estonian and Lithuanian. Motif 3 never achieved good results, which is indicated in the data by the fact that verbs do never show significantly higher or lower values.

As a consequence, the result of our first categorization is that threshold-based one-dimensional categorization is not sufficient, despite the clear differences that can be observed for some motifs in data from supervised POS taggers. The result could be expected by looking at motif signatures due to the fact that some verb clusters are clear outliers whereas some only show slight variations.

However, the results from this first experiment help us to identify potential subsets of metrics that can be used for the 2-means clustering. Besides considering all metrics for the categorization, we also considered only the motifs, all metrics but Motif 3, all motifs but motif 3, only the motifs 2 and 4, as well as the motifs 2,4,5 and 6.

The best results for all languages, presented in Table 1, are achieved when using Motif 2,4,5, and 6. Interestingly, adding the star motif did not increase the performance despite the observed high fraction of star motifs in verb clusters. Nevertheless, star motifs are mainly created by modal verbs, which are not contained in many verb clusters, leading to a negative impact for categorizing these clusters. An accuracy of 100% is achieved for Dutch and Portuguese. For English, German, and Lithuanian, there is only one verb cluster that is frequently assigned to the wrong category. When considering German, this is cluster number 12, which has already been noted during our manual categorization. Having a closer look on the individual runs of the algorithm, 12 is categorized as a verb cluster in about one third of the runs, whereas the verb clusters 18 and 7 are categorized as verb clusters roughly 80 % of the time. Moreover, no non-verb cluster is ever categorized as a verb cluster in case of German. Czech, French, and Italian exhibit two errors, which reduce either precision or recall drastically due to the low number of verb clusters. For Italian, the correctly identified verb clusters are actually participles, while the verb clusters are not found. Note that for Indonesian, Estonian, and Russian, both recall and precision are 0. This is due to the lacking distinctiveness of the verb features, the verb cluster is in general empty. As for Indonesian, the verb patterns in those languages do not seem to be distinctive enough to be detected by the clustering algorithm.

Table 1 Results for a categorization into verb vs. non-verb clusters based on Motif 2,4,5, and 6. Accuracy is the percentage of POS clusters that are classified correctly into these two classes; Precision is the percentage of classified verb clusters that are really verb clusters; Rcall is the percentage of true verb clusters found by the algorithm.

Language	Accuracy	Precision	Recall
Czech	0.8	0.3333	1.0
German	0.9091	1.0	0.6667
English	0.8333	0.6667	1.0
Estonian	0.8	0.0	0.0
French	0.6923	0.5	0.5
Indonesian	0.5833	0.0	0.0
Italian	0.8333	1.0	0.5
Lithuanian	0.9	0.6667	1.0
Dutch	1.0	1.0	1.0
Portuguese	1.0	1.0	1.0
Russian	0.8667	0.0	0.0

Applying the algorithm by Clark to the same data, we first noted that using a low number of classes (32) produces very mixed POS clusters that did not result in any distinctive outliers in the motif signature. When using a higher number of clusters (128), classes still are not separable into verbs and non-verbs, since verbs get scattered over many small clusters, which inhibits our analysis: motif analysis requires a certain number of nodes to produce stable results.

In summary, motif signatures for word clusters of verbs tend to have distinctive features such as a high fraction of star motifs and a low fraction of clustered motifs such as semi-cliques. Simple classification algorithms such as 2-means clustering detect the predominant verb clusters, but miss some of those that offer less distinct motif signatures. It might be possible to further improve the results by fine-tuning the algorithm and normalizing the values for each dimension. We hope that in this way, this approach extends also to the language families where our current setup fails. This study is to our knowledge the first partially successful attempt to assign linguistic word classes (here: verbs) to unsupervised POS classes.

3.3 Peer-to-Peer Streaming Networks

Peer-to-peer live-streaming networks are distributed communication architectures, where the participants connect directly to each other, not to central servers. The participants are represented by nodes. There is a directed edge from participant v to participant w if, and only if, v sends a streaming signal to w. We consider two types of generated streaming networks:

- The first type is the streaming topology of Strufe et al. (2006), abbreviated here as *BCBS*, which splits the stream into several partial streams (commonly named *stripes*) and then delivers these stripes along multiple paths in the network.
- The second type of streaming networks are a class of balanced topologies, which exhibit short, node-disjoint paths and are hence optimally resilient to network attacks and failures (Brinkmeier et al. 2009), abbreviated here as *Optimal*.

We generated 64 *BCBS* and 64 *Optimal* networks, each consisting of 250 nodes. For each network we count the number of the two three-node directed motifs, see Fig. 2. The first motif is the *V-Out*. It represents the case when the participant A streams a signal to two other participants B and C simultaneously, that is, $A \rightarrow B$ and $A \rightarrow C$. The second motif is the *3-Chain*. It represents the case when the participant B forwards to C the signal it has just received from A, that is, $A \rightarrow B \rightarrow C$.

For each network we calculate the percentage of *V-Out* and *3-Chain* motifs from the overall motif signature of the network. Thereafter, dividing the percentage of the corresponding motifs in bins and plotting the classes of networks gives a clear picture as displayed in Fig. 7, where the x-axis represents the bin size and the y-axis the number of networks in that bin.

Evidently, each of *V-Out* and *3-Chain* alone allows one to distinguish the *BCBS* and the *Optimal* networks from each other, although all networks belong to the same realm: live streaming networks. Thus, motif analysis can be used to analyze real-world networks and to compare them to model topologies, such as the two topologies discussed here.

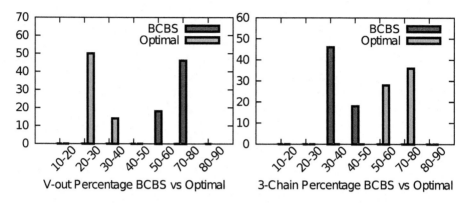

Fig. 7 Percentage of *V-Out* (left) and *3-Chain* (right) motifs for both types of live streaming networks. In both cases there is a clear distinction between the *BCBS* (red solid line) networks and the *Optimal* (green dashed line) network types.

3.4 Co-Authorship Networks from Two Subdisciplines of Physics

We define co-authorship networks as follows. Every node represents an author; two nodes are connected if, and only if, the authors have at least one joint publication. The edges are undirected.

We analyze two different subdisciplines of physics: condensed matter and astrophysics. The condensed matter collaboration network comprises 23,133 nodes / authors and 93,497 edges, whereas the astrophysics comprises 18,772 authors and 198,110 edges. Both data sets cover scientific publications over a period of 10 years and represent the complete history of the Condensed Matter Physics section and the Astro Physics section of the e-print arXiv up to April 2003, respectively. This data is available from the Stanford SNAP[5] Large Network Dataset Collection.

However, we do not need one network but a number of networks of either type to determine the extent to which networks of both types may be distinguished from each other. Therefore, we extract 10 connected subnetworks of 1,000 nodes from either network. In each of the 20 subnetworks, we count the number of the undirected motifs on four nodes, cf. Fig. 1.

The subnetworks are generated as follows. A random node is selected along with all of its neighbors. The neighbors of the neighbors are selected next, and so forth. The procedure stops once 1,000 nodes have been selected. Each edge that connects two selected nodes is inserted, so the generated graph is the subgraph *induced* by the selected nodes.

For each network we calculate the percentage of *star*, *chain*, *semi-clique* and *clique* motifs from the overall motif signature of the network. Thereafter, dividing the network in bins according to the percentage of the corresponding motif gives a

---

[5] http://snap.stanford.edu

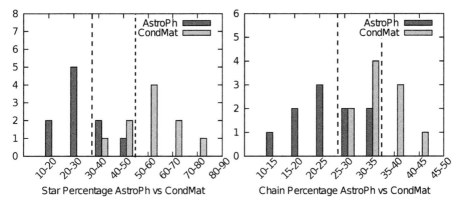

Fig. 8 Percentage of *star* (left) and *chain* (right) motifs for both condensed matters (green) and astro physics (red) subgraphs

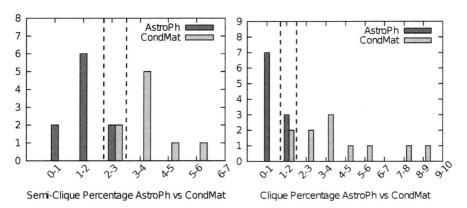

Fig. 9 Percentage of *semi-clique* (left) and *clique* (right) motifs for both condensed matters (green) and astro physics (red) subgraphs. In both cases there is a clear distinction between the two network types.

clear picture as displayed in Fig. 8 and Fig. 9, where the x-axis represents the bin size and the y-axis the number of networks in that bin.

In each of the four diagrams, we define a generous *grey zone*:

- Star: 30...50
- Chain: 25...35
- Semi-clique: 2...3
- Clique: 1...2

If an instance is above the specific grey zone in at least one of these four diagrams, it is regarded as an astrophysics network. On the other hand, if an instance is below the grey zone in at least one of the four diagrams, it is regarded as a condensed matter network. Since the grey zones are generously defined, this classification of instances outside the grey zone might be more or less safe. Despite this generosity, each of the 10 instances is indeed classified by that rule, for none of the instances is the classification contradictory, and each classification is correct.

While results are promising, further work is needed to fully leverage motif analysis for co-authorship networks. In this case study, we ware able to reasonably separate networks from two communities. This shows that community-specific habits for co-authorship are mirrored in the motif distribution.

3.5 Mailing Networks

Email networks are a special type of communication networks. The nodes are email addresses; two nodes are connected if, and only if, at least one email has been sent from one node to the other one. In particular, the edges are undirected. In this case study we consider two email networks, which are again available from the Stanford SNAP[6] Large Network Dataset Collection:

- The first network was generated using email data from a large European research institution, abbreviated here as *EU*. It comprises all incoming and outgoing emails over a period of 18 months. The network consists of 265,214 nodes and 420,045 edges.
- The second network is generated from a dataset of around half a million emails from the Enron email communication network. This data was originally made public by the Federal Energy Regulatory Commission. Here we abbreviate the network as *Enron*. It contains 36,692 nodes and 183,831 undirected edges. Here the big picture its not as clear as for the case study in Sect. 3.4. Reasonable grey zones do not allow a safe classification of all instances. Nonetheless, there is a strong tendency in each of the three pictures, which is promising for future investigations.

Again, we extract 10 connected subnetworks of 1,000 nodes from either network. In each of the 20 subnetworks, we count the number of the undirected motifs on four nodes, cf. Fig. 1. We use the same extraction principle as described in Section 3.4.

We then divide the networks in bins according to the percentage of the corresponding motifs, where the *x*-axis represents the bin size and the *y*-axis the number of networks in that bin. The corresponding histograms for the *star*, *chain* and *three-loop-out* motifs are displayed in Fig. 10

All three motifs together allow one to distinguish the *EU Enron* networks from one another, despite the fact that all network instances belong to the same network realm.That is possible because each single motif represents an additional distinction dimension.

---

[6] snap.stanford.edu

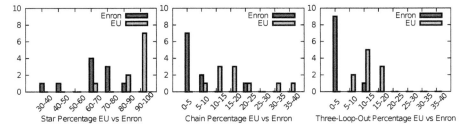

Fig. 10 Percentage of the *star*, *chain* and *three-loop-out* motifs (from left to right) for the *EU* (green) and *Enron* (red) networks. In all three cases there is a visible distinction between the two network types, although there is a region of overlap for all three motifs.

4 Conclusions and Outlook

The presented case studies demonstrate that motif signatures are a widely applicable method to classify subtypes of network types by a purely structural analysis.

In particular, this opens up new, promising perspectives for the analysis of networks from various realms. As it has become clear from results we obtained on networks from different realms, motif analysis is a general tool for network characterization and classification, and not in any way restricted to particular network classes.

We have demonstrated two uses of motif analysis for networks generated from natural language, and could successfully discriminate between coherent (real) and incoherent (n-gram-generated) language, as well as discriminate between verbs and other syntactic word classes. Further work on natural language networks should include more sophisticated language models such as the syntactic topic model (Boyd-Graber and Blei 2008), which explicitly models topicality, and studies involving other methods for inducing syntactic word classes, such as class-based n-gram models (Brown et al. 1992).

While the case study on network topologies showed a good discriminating power of network motifs, the last two case studies have shown that straightforward motif counting does not necessarily suffice. While showing promising indications, motif analysis should be complemented by additional features for discriminative tasks. Just as measures like the clustering coefficient or the transitivity measure unveil certain characteristics of networks, motif signatures provide yet another, more fine-grained fingerprint of network properties.

References

Aggarwal, C.C., Wang, H. (eds.): Managing and Mining Graph Data. Kluwer (2010)
Alon, U.: Network Motifs: Theory and Experimental Approaches. Nature Review Genetics 8, 450–461 (2007)

Biemann, C.: Unsupervised Part-of-Speech Tagging Employing Efficient Graph Clustering. In: Proceedings of the Student Research Workshop at COLING/ACL 2006, Sydney, Australia (2006)

Biemann, C., Bordag, S., Quasthoff, U.: Automatic Acquisition of Paradigmatic Relations using Iterated Co-occurrences. In: Proceedings of LREC 2004, Lisbon, Portugal (2004)

Biemann, C., Heyer, G., Quasthoff, U., Richter, M.: The Leipzig Corpora Collection – Monolingual corpora of standard size. In: Proceedings of Corpus Linguistic 2007, Birmingham, UK (2007)

Biemann, C., Roos, S., Weihe, K.: Quantifying Semantics Using Complex Network Analysis. In: Proceedings of COLING 2012, Mumbai, India (2012)

Bisk, Y., Hockenmaier, J.: An HDP Model for Inducing Combinatory Categorial Grammars. Transactions of the Association for Computational Linguistics 1, 75–88 (2013)

Boyd-Graber, J., Blei, D.M.: Syntactic Topic Models. In: Proceedings of Neural Information Processing Systems, Vancouver, British Columbia (2008)

Brinkmeier, M., Schäfer, G., Strufe, T.: Optimally DoS Resistant P2P Topologies for Live Multimedia Streaming. TPDS 20, 831–834 (2009)

Brown, P.F., Pietra, V.J.D., de Souza, P.V., Lai, J.C., Mercer, R.L.: Class-Based n-gram Models of Natural Language. Computational Linguistics 18(4), 467–479 (1992)

Clark, A.: Combining distributional and morphological information for part of speech induction. In: Proceedings of the Tenth Conference on European Chapter of the Association for Computational Linguistics, vol. 1, pp. 59–66 (2003)

Dorogovtsev, S.N., Mendes, J.F.F.: Language as an evolving word web. Proceedings of The Royal Society of London. Series B, Biological Sciences (2001)

Dunning, T.: Accurate methods for the statistics of surprise and coincidence. Computational Linguistics 19(1), 61–74 (1993)

Jin, N., Young, C., Wang, W.: Graph classification based on pattern co-occurrence. In: Proceedings of the 18th ACM Conference on Information and Knowledge Management, CIKM 2009, Hong Kong, China, pp. 573–582 (2009), doi:10.1145/1645953.1646027

Juszczyszyn, K., Kołaczek, G.: Motif-based attack detection in network communication graphs. In: De Decker, B., Lapon, J., Naessens, V., Uhl, A. (eds.) CMS 2011. LNCS, vol. 7025, pp. 206–213. Springer, Heidelberg (2011)

Koehn, P.: Statistical Machine Translation, 1st edn. Cambridge University Press, New York (2010)

Krumov, L., Andreeva, A., Strufe, T.: Resilient Peer-to-Peer Live-Streaming using Motifs. In: 11th IEEE World of Wireless, Mobile and Multimedia Networks (WoWMoM), pp. 1–8 (2010)

Krumov, L., Fretter, C., Müller-Hannemann, M., Weihe, K., Hütt, M.-T.: Motifs in co-authorship networks and their relation to the impact of scientific publications. European Physical Journal B 84(4), 535–540 (2011)

Krumov, L., Schweizer, I., Bradler, D., Strufe, T.: Leveraging Network Motifs for the Adaptation of Structured Peer-to-Peer-Networks. In: GLOBECOM, pp. 1–5 (2010)

MacQueen, J.: Some methods for classification and analysis of multivariate observations. In: Proceedings of the Fifth Berkeley Symposium on Mathematical Statistics and Probability, vol. I, pp. 281–297. Berkeley University of California Press (1967)

Manning, C.D., Schütze, H.: Foundations of Statistical Natural Language Processing. MIT Press, Cambridge (1999)

Miller, G.A., Charles, W.G.: Contextual Correlates of Semantic Similarity. Language and Cognitive Processes 6(1), 1–28 (1991)

Milo, R., Itzkovitz, S., Kashtan, N., Levitt, R., Shen-Orr, S., Ayzenshtat, I., Sheffer, M., Alon, U.: Superfamilies of evolved and designed networks. Science 303(5663), 1538–1542 (2004)

Milo, R., Shen-Orr, S., Itzkovitz, S., Kashtan, N., Chklovskii, D., Alon, U.: Network Motifs: Simple Building Blocks of Complex Networks. Science C 298(5594), 824–827 (2002), doi:10.1126/science.298.5594.824

Newman, M.E.J., Watts, D.J., Strogatz, S.H.: Random graph models of social networks. Proc. Natl. Acad. Sci. USA 99(suppl. 1), 2566–2572 (2002)

Quasthoff, U., Richter, M., Biemann, C.: Corpus Portal for Search in Monolingual Corpora. In: Proceedings of the Fifth International Conference on Language Resources and Evaluation, LREC, Genova, Italy, pp. 1799–1802 (2006)

Ramabhadran, B., Khudanpur, S., Arisoy, E. (eds.): Proceedings of the NAACL-HLT 2012 Workshop: Will We Ever Really Replace the N-gram Model? On the Future of Language Modeling for HLT, Montréal, Canada (2012)

Schmid, H.: Probabilistic Part-of-Speech Tagging Using Decision Trees. In: Proceedings of the International Conference on New Methods in Language Processing, Manchester, UK (1994)

Schreiber, F., Schwöbbermeyer, H.: Motifs in Biological Networks. In: Stumpf, M., Wiuf, C. (eds.) Statistical and Evolutionary Analysis of Biological Network Data, pp. 45–64. Imperial College Press/World Scientific (2010)

Shen-Orr, S.S., Milo, R., Mangan, S., Alon, U.: Network motifs in the transcriptional regulation network of Escherichia coli. Nature Genetics 31(1), 64–68 (2002)

Søgaard, A.: Unsupervised dependency parsing without training. Natural Language Engineering 18(Special Issue 02), 187–203 (2012), doi:10.1017/S1351324912000022

Strufe, T., Schäfer, G., Chang, A.: BCBS: An Efficient Load Balancing Strategy for Cooperative Overlay Live-Streaming. In: Proc. IEEE ICC (2006)

Wong, E., Baur, B., Quader, S., Huang, C.-H.: Biological Network Motif Detection: Principles and Practice. Briefings in Bioinformatics 13(2), 202–215 (2012)

Multidimensional Analysis of Linguistic Networks

Tanya Araújo and Sven Banisch

Abstract. Network-based approaches play an increasingly important role in the analysis of data even in systems in which a network representation is not immediately apparent. This is particularly true for linguistic networks, which use to be induced from a linguistic data set for which a network perspective is only one out of several options for representation. Here we introduce a multidimensional framework for network construction and analysis with special focus on linguistic networks. Such a framework is used to show that the higher is the abstraction level of network induction, the harder is the interpretation of the topological indicators used in network analysis. Several examples are provided allowing for the comparison of different linguistic networks as well as to networks in other fields of application of network theory. The computation and the intelligibility of some statistical indicators frequently used in linguistic networks are discussed. It suggests that the field of linguistic networks, by applying statistical tools inspired by network studies in other domains, may, in its current state, have only a limited contribution to the development of linguistic theory.

1 Introduction

Network analysis is an integral component in the study of complex systems. This is probably due to the generally accepted fact that complex systems are composed of elementary units and structures of mutual dependencies between those units which

---

Tanya Araújo
ISEG (Lisboa School of Economics and Management) of the University of Lisbon and Research Unit on Complexity in Economics (UECE), Portugal
e-mail: tanya@iseg.utl.pt

Sven Banisch
Max Planck Institute for Mathematics in the Sciences, Inselstrasse 22,
D-04103 Leipzig, Germany
e-mail: Sven.Banisch@mis.mpg.de

directly suggests a network representation. However, network-based analyses are nowadays quite common also in the analysis of systems where such a network representation is not always that intuitive. There may be many ways in which the elementary units and the links between them are conceived and the choices may depend strongly on the questions that a network analysis aims to address.

Here we discuss that case at the example of linguistic networks. Linguistic networks are characterized by a high level of abstraction compared to networks in other areas of research. While in power grid networks, the world wide web or even gene regulatory networks the nodes and links in between them are directly related to real processes taking place in the system – to electricity flow, web links or respectively biochemical reactions between DNA segments – this is not generally the case in linguistic networks which are often induced or synthesized from a linguistic data set for which a network perspective is only one out of several options for representation.

In this context, our main points are:

1. Although being different tasks, network design/induction and network analysis are strongly interdependent
2. Linguistic networks are special both in the design and in the analytical setting
3. There is a need for a framework for network construction and analysis with special focus on linguistic networks
4. A higher level of abstraction in network induction has an important bearing on network analysis, particularly on the choice of appropriate topological indicators and on the interpretation of their results
or
5. The higher is the abstraction level of network induction, the harder is the interpretation of the topological indicators used in network analysis

The paper is organized as follows: Section 2 presents our main arguments on the specificity of linguistic networks and on the consequent need for the identification of different abstraction levels when dealing with network design. These abstraction levels make up the first dimension of a framework for network construction and analysis. Section 3 is targeted at presenting the second dimension of such a framework, i.e., the statistical levels of network analysis. It also includes a detailed presentation of the statistical indicators most often used at each analytical level. In Section 4, we discuss on the intelligibility of these statistical indicators when applied to the analysis of linguistic networks. Section 5 provides examples in different fields of application together with a classification of the examples according with the introduced framework. Section 6 discusses the classification presented in Section 5, while Section 7 draws a conclusion on the paper as a whole.

2 Linguistic Networks Are Special

2.1 Three Types of Networks

Linguistic networks are characterized by a high level of abstraction compared to networks in other areas of research. They are usually induced or synthesized from a linguistic data set – typically a series of letters organized into words, sentences, may be paragraphs and so on – that does not obviously call for a network representation. This synthesis involves several design decisions and entities (i.e., words) linked in one design may not be linked in another. Moreover, there is no clear relation between the connections in such a network and the dynamic processes taking place in the system that created the data set. It is, for instance, in most cases not possible to state which kind of flow processes take place along the links of a linguistic network which makes the applicability of network measures that involve implicit assumptions about network flow problematic (see Borgatti (2005) and below). For the purposes of this paper, we differentiate roughly between three levels of abstraction by considering:

1. Abstraction Level-1: real networks of systems composed of elementary units which are explicitly linked or in between which real processes take place,
2. Abstraction Level-2: proximity networks of systems composed of elementary units and a well-defined measure of distance in between these units (shared features, correlation or similarity),
3. Abstraction Level-3: induced networks that are synthesized out of data bases (probably the outcome of a complex system) and in which the definition of elements and links is not explicit in the data.

Besides being of great importance at the very beginning of the network definition, the question whether a network or respectively a system falls into category one, two or three has, we believe, an important bearing on network analysis, particularly on the choice of appropriate topological indicators and on the interpretation of the results. As Borgatti (2005) has shown for centrality measures, the implicit assumptions about network flow that certain measures make, may challenge the interpretability of the measure. Namely, most of the well-known network structure indicators have been designed in order to describe the different aspects of transport phenomena where connectivity plays the important role and the mass conservation principle features the dynamics of network flows. Therefore, the situation is even worse if the dynamical processes between nodes are unspecified, as in linguistic networks.

The aim of describing networks at different levels of abstraction is not to present a rigorous classification of different systems or representations of systems. It is rather to show that the topological interpretation of linguistic networks is different from more concrete networks, because a rather abstract perspective has to be taken in deriving a network representation which leads to a certain degree of arbitrariness and a decrease in explanatory power.

For the first type of network (henceforward called type-1), it is relatively obvious how elements and connections in between them must be defined in order to obtain an operative description of the system. For instance, the system of air traffic is constituted of airports and flight connections in between them (Li and Cai 2004; Colizza et al. 2007; Opsahl et al. 2010; Konect 2014). A network which contains just that binary information (flight connection or not) can already be quite informative of the entire air traffic system as questions of efficiency, vulnerability etc. can be addressed at this level. Considering the amount of goods traveling from airport to airport or the number of people will quickly lead to a very detailed description of the whole system.

As another economic example, we mention networks of inter-bank dependencies (Huizinga and Nicodème 2004; Boss et al. 2004; McGuire and Tarashev 2006; Soramäki et al. 2007; Minoiu and Reyes 2011; Spelta and Araújo 2012). Payment systems allow banks to move money and securities between banks and other large financial institutions. Daily, a huge amount of capital flows arises from and depends upon the coordination of payments among banks. Such a close coordination engendered by payment systems creates a network of inter-bank dependencies, where banks are nodes and transactions are represented as links of the network. In this context, the main issue uses to be associated to the conditions that ensure network robustness since failures of a bank to make payments can trigger a cascade of liquidity losses.

Another example, already mentioned above, is the WWW (e.g., Albert et al. (1999), Broder et al. (2000), and Meusel et al. (2014). Web pages are connected to one another by hyperlinks and lot about the system can be understood on the basis of such structural information. However, the WWW is also an area in which the second type of network (type-2) can provide a complementary view on the system as a whole. Instead of hyperlinks as a direct reference from page to page, one could look at the similarity between the pages in terms of the information they provide. One way to measure the similarity between two pages is to compare frequencies with which words are used in the two pages, that is, vectorial semantics (Salton et al. 1975; Salton 1989; Landauer and Dutnais 1997) (for a related approach see also the chapter of Masucci et al. in this volume), or to compare the HTML structure of the documents (Mehler et al. 2007). Of course, the resulting similarity measure depends strongly on the features that are used to span the semantic space. Another way to construct a similarity network is to count the number of common references/hyperlinks or the geodesic distance in the type-1 network. That is to say, for networks of type two, there is already a certain amount of freedom in the definition of the network. It might so happen (even though this is not assumed the regular case) that two nodes close to one another in one network representation are distant in another depending essentially on the type of similarity measure used in the construction.

Some networks that go under the label of linguistic networks fall into this second category. For instance, there are a word networks based on the number of shared letters, phonemes or syllables (see, e.g., Soares et al. (2005), Mukherjee et al. (2009), Arbesman et al. (2010), and Yu et al. (2011)), and, at a higher level, there are interlanguage networks based on lexical similarity (see, e.g., Blanchard et al. (2011) and

Serva et al. (2011)). However, the number of ways to induce a linguistic network is huge and the resulting information represented by the network may differ greatly from one network to the other. This is because linguistics networks are often defined on the basis of linguistic data, such as corpora of texts or annotated dialog data so that many linguistic networks are of the third type, with still more freedom in defining a network, compared to the second network type.

Words, for instance, can be linked because they frequently share the same context (this is captured by co-occurrence networks), but the operationalisation of "sharing the same context" is far from unique (see below).

Two words could also be linked because it is likely that I think of the second when I hear the first (free association experiments; see, for instance, Nelson et al. (2004), Borge-Holthoefer and Arenas (2010), and Gravino et al. (2012)) or if I am asked to utter a sequence of words the two will probably appear together (verbal fluency experiments; see, for instance, Storm (1980), Lerner et al. (2009), Goñi et al. (2011), and Iyengar et al. (2012)). Free word association is probably the only example of a linguistic network of type-1. Although the links between any two words in any utterance sequence are not "hardwired", they are directly related to a real process taking place in the system (or the game) of free association. In this context, a sequence of associated words itself is naturally seen as dynamical process since it provides the unfolding of a flow of ideas. There, the network perspective comes naturally and the definition of nodes and links is relatively obvious: words are nodes and any link is defined as a connection between two spontaneously related words.

Still another way to come to word networks is to map the syntactic relations onto the network connections (syntax dependency networks, see Ferrer i Cancho et al. (2004) and Ferrer i Cancho et al. (2007)) or define a word network by considering explicitly the semantic relations (synonymy, hypernomy, etc.) as the connections in between words (Sowa 1991; Miller 1995). And so on. A further degree of freedom in defining a linguistic network concerns the choice of the elementary units. It is not explicit in the data what we should conceive as the nodes in a linguistic network and therefore models at the level of words, but also at the sentence or paragraph level are sometimes constructed.

The way how a network is defined in linguistics depends a lot on the question that is to be addressed by the network study. If one aims at understanding the structure of a language, syntax dependency network or co-occurence networks synthesized from large corpora of texts are natural choices. If the structure of the conceptual space is in the focus, free word association, verbal fluency experiments or explicit semantic relations are a more appropriate source for network induction. The structure of these different kinds of networks may be very different. The fact that so many different perspectives can be taken on the linguistic system may be seen has an indication of its increased complexity compared to other systems. There are several dimensions of analysis.

2.2 Network Induction

Network induction makes reference to the method by which networks are created on the basis of a certain data set or system. In research dealing with network models of type one, the method of "network induction" does not receive particular attention and many researchers may actually wonder about the notion of "inducing" a network. This is because in those systems the network representation is very natural and requires only a small level of abstraction from the real situation. It is relatively clear what should be considered as the elementary units of the system - the nodes - and also the relation between the nodes - the edges - are given either by real "hardwired" connections or by processes or flows between the elementary units. Design decisions, if any, concern the question whether the situation is best mapped onto a binary graph or a weighted and directed network.

This is already different for networks of the second type. While, in general, it is often clear what to conceive as the nodes in proximity networks, the nature of the connections (despite that they are similar with respect to a chosen set of features) it is not always clear. In functional brain networks, to make an example, different brain regions are linked if they are jointly activated in certain working tasks (see Sporns et al. (2005), Bassett and Gazzaniga (2011), and Menon (2011) and references therein). Even if it is natural to assume that there is exchange of information or signaling between the different regions during the working task, as a matter of fact, all that can be said on the basis of functional brain network is that there is a correlation of activity patterns across the different brain regions. The real processes (that probably exist) are obscured by the network representation and one would have to go to the micro level of neurons and synapses to obtain that information. That is going to the type-1 network.

Take the similarity graph between articles from the "bibliography on linguistic, cognitive and brain networks" compiled by Ramon Ferrer i Cancho (Ferrer i Cancho 2012) as another example for a type-2 network. This graph, one instance of which is shown in Fig. 1, is obtained on the basis of similar words used in the abstract of the articles. Nodes are labeled by the most frequent term in the respective abstract. While the picture illustrates nicely the structure of similarities between the articles - articles on linguistic networks are clustered into different modules depending on whether they deal with semantics, word networks, text systems or networks of languages as a whole and the articles dealing with functional brain networks form another rather independent cluster - it is not really clear whether linked articles indeed refer to one another and there is also no obvious functional relation or process that is represented by a connection. As we will discuss below, the lack of knowledge about the real relations between the elements has important consequences for the choices and interpretations of certain topological indicators.

The example of a similarity network shown in Fig. 1 makes also clear that various design decisions have to be taken in the creation of such a network. First, in this example, we have been interested in the thematic similarities of the publications as opposed to, for instance, stylistic similarities between different authors. Therefore, we disregarded functional words in the computation of the correlations and considered only words that do not appear in all the abstracts. If instead we were interested in the identification of authors and similarities in their writing style considering functional words only would probably give the better result (see, e.g., Peng and Hengartner (2002)). Second, once the correlation matrix is computed using the reduced ("cleaned") feature set, a network as shown in Fig. 1 is obtained only after thresholding the correlation matrix such that only strong correlations are preserved. This is also a rather decisive design decision to optimize the intelligibility of the system because, in effect, there is at least a small positive correlation between all pairs of abstracts. However, the complete graph resulting from that would not be very informative about the modularity structure in the network of articles. In fact it happens that when networks are induced from distance measures, the issue of deriving a sparse network from the complete one becomes the most important design decision. The less arbitrary choices (or the most endogenously based ones) usually define the threshold as the distance value used in the last step of the minimum spanning tree construction. In so doing we ensure the connectivity is preserved (the resulting network is necessarily connected).

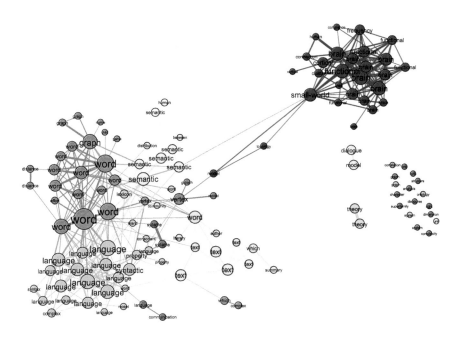

Fig. 1 A similarity network of publications from the "bibliography on linguistic, cognitive and brain networks" compiled by Ramon Ferrer i Cancho (Ferrer i Cancho 2012). The data has been retrieved on the third of December 2012, one week before the conference "Modeling Linguistic Networks" was held. It is constructed by computing the correlation between the abstracts of the articles on the basis of similar words they use. Node labels show the most frequent word in the respective abstract.

The induction process becomes more complex for type-3 networks, and some of the most popular linguistic networks in particular. The number of different options for network induction increases further and several design decisions must be taken from the very beginning of the network creation procedure. Sticking to word networks, one must first be clear about whether phonemic, syntactic, semantic or still other relations should be mapped onto the connections. Usually, this is determined by the research question to be addressed by the network study. Let us assume that the focus is on language structure. Then, we could map syntactic relations between words as they are realized in a set of sentences. This leads to the syntax dependency networks proposed by Ferrer i Cancho et al. (2004). But we could also decide to approach the question by considering shared context or word co-occurrences. Then an important design decision concerns the size of window that is considered as context. Words could be linked whenever they co-occur within at least one sentence, but one could also use a window of fixed size and consider co-occurrence within that window. Another way (considered more realistic in Solé et al. (2010)) is to consider the order in which words occur, most simply, we could say that there is an arrow from word A to word B if B follows A in some sentence (this gives rise to precedence or word-flow networks see, e.g., Grabska-Gradzinska et al. (2012) and below). Again, in all these cases, the question whether links or arrows are weighted or not requires a design decision. In other words, there is a large number of different ways to induce a word network from linguistic data and the data itself does not clearly suggest if one way is better than the other.

The need for a unifying framework for linguistic networks emerges from this diversity of network models. The fact that rather different networks can be obtained on the basis of the same data set seems to be quite unique in the network sciences and triggers, in fact, important epistemological questions concerning validity, comparability and interpretability of linguistic network studies. The comparison of different linguistic networks and a critical discussion of the relation of linguistic networks to networks in other fields are essential for the future development of the field of linguistic networks. First steps into that direction have already been taken (see, for instance, Choudhury et al. (2010), Zamora-López et al. (2011), and Gravino et al. (2012)).

3 Three Levels of Statistical Analysis

Closely related to network analysis and also grounded on a multilevel perspective is the differentiation between statistical levels of analysis (and the corresponding statistical tools). Here we identify three statistical levels inspired by statistical tools originally developed for the characterization of stochastic processes, and their application to the analysis of signals, i.e., signal processing on graphs.

3.1 A Brief Note on Signal Processing on Graphs

Signals evolving in time may be considered as signals in a forward-connected graph, the nodes being different points in time. The analysis of more general networks such as social and economic networks, linguistic networks and biological networks usually generates graphs with much more complex connections.[1]

Processing signals on graphs has been dealt with recently, in particular in the context of discovering efficient data representations for large high-dimensional systems and other dynamical systems (Miller et al. 2010; Shuman et al. 2013). Here we envision that, by analogy with signal processing on graphs, the statistical tools that have been developed for the characterization of stochastic process are mathematical devices that may be applied to the analysis of networks.

3.2 The Statistical Levels

When a phenomenon is measured with a set of statistical tools, what one registers is a sequence of values of some variable X

$$\cdots X_{-2} X_{-1} X_0 X_1 X_2 \cdots$$

which takes values in a space $\mathscr{X}$. The space $\mathscr{X}$ is called *state space* and the space of sequences $\mathscr{X}^{\mathbb{Z}}$ is referred to as *path space*. Statistical properties of the phenomenon may be described at three different levels, (Vilela Mendes et al. 2002):

1. by the expectation values of the observables;
2. by the probability measures on the state space $\mathscr{X}$;
3. by the probability measures on path space $\mathscr{X}^{\mathbb{Z}}$.

To obtain expectation values and probability measures we would require infinite samples and a law of large numbers. For any finite sample we obtain finite versions of the expectation values, of the probability on state space and of the probability on path space which are called the *mean partial sums*, the *empirical measures* (or empirical probability distribution functions - pdf's) and the measures on the *empirical process*.

The statistical levels represent successively finer levels of description of the statistical properties associated to the topological indicators used in network analysis. We will call these three types of description, respectively, level 1, level 2 and 3-statistical indicators.

---

[1] A first step in analyzing data in such networks is the construction of the appropriate signal transforms. For the forward-connected time graph the Fourier transform is a projection on the eigenvectors of the adjacency matrix. Therefore it is natural to construct transforms for general networks by projection on a basis constructed from the eigenvectors (or generalized eigenvectors) of some matrix relevant to that network.

1. Statistical Level-1 concerns the computation of the quantities related to averages values of one or more topological coefficients defined at the node level, as for instance: the average clustering coefficient, the network degree, the average path length, among others. At this level, the phenomena are described by the expectation values of the observables, i.e., one or more topological coefficients defined at the node level.
2. Statistical Level-2 concerns the computation of the quantities related to probability distribution functions of the above mentioned topological indicators computed at the node level. From this analysis it is possible to characterize power-law shapes in the distribution of the degree of the nodes or in any other indicator. Level-2 analysis allowed for the characterization of some important network regimes and mechanisms as, for instance, the scale-free regime and its associated preferential attachment mechanism.
3. Statistical Level-3 concerns the calculation of the probability measures on the path space, i.e., the cylinder measures on the empirical process. At this level, a phenomenon is described by the probability of a certain configuration of the network that represents the empirical process. Level-3 analysis allowed for the characterization of communities within the network structure.

Level-1 and level-2 analysis are the most common ones and their statistical indicators the most commonly quoted when a stochastic process is analyzed. However to the same expectation values for the observables or to the same pdf's, different processes may be associated. Therefore full understanding of the process requires the determination of the level-3 indicators.

It has been shown (Vilela Mendes et al. (2002), among many others) that the analysis and the reconstruction of a process involves two different but related steps.

- the first step is the identification of the *grammar* of the process, that is, the allowed transitions in the state space or the subspace in path space that corresponds to actual orbits of the system.
- the second step is the identification of the *measure*, which concerns the occurrence frequency of each orbit in typical samples.

Although largely independent from each other, this two features have a related effect on the constraints they impose on the statistical indicators.

In the social sciences and particularly in Economics and Finance, the application of such mathematical devices gave place to the description of a set of empirical findings which are usually called *stylized facts*.

3.3 Stylized Facts in Network Analysis

The notion of stylized facts was once introduced by Nicholas Kaldor (Kaldor 1956) and used thereafter as an encapsulation of regularities found in empirical researches of economic processes. In performing network analysis by means of applying statistical tools developed for the characterization of stochastic process, the concept of a stylized fact is adequate to identify the empirical evidence of recursive events and relations found in the description of real networks.

In Economics and Finance, some stylized facts use to be formalized as distributions describable by power laws. Others rely just on cross correlation values between stock returns. The main variable that is used to construct the statistical indicators is the differences of log-prices.

$$r(t,n) = \log p(t+n) - \log p(t) \tag{1}$$

sometimes called the n−days return. From the return series of different stocks, there are the following stylized facts:

1. cross correlations between stock returns,
2. non-linear dependencies in the trajectories through time (memory) of a stock return,
3. volatility memory, i.e., memory in the second moment of the volatility of the return of a stock.

Regardless the field of application, some stylized facts have been built on network properties or on the topological coefficients that characterize some network regimes. Most frequent examples are:

1. A power law signature of the distribution of the network degree (as in scale-free networks) are used to characterize heterogeneity, a notion that is used to identify the following different phenomena depending on the field of application, as for instance:

    - the Zipf law in linguistic networks,
    - heterogeneity in word free association in linguistic networks (Solé et al. (2010)),
    - systemic risk in financial networks ((Battiston et al. (2010) and Acemoglu et al. (2013)),
    - contagion phenomenon in epidemic networks (Newman et al. (2003)).

2. (Dis)Assortativeness of node degrees:

    - disassortative mixing in syntactic dependency networks (Ferrer i Cancho et al. (2004)),
    - assortative mixing in semantic networks (Ferrer i Cancho et al. (2004)),
    - capital flows in financial networks (Spelta and Araújo (2012)).

3. Phase transitions associated to characteristic values of topological coefficients:

    - phase transitions in early language acquisition in syntax networks (Solé et al. (2010)),
    - phase transition and systemic risk in financial networks (Acemoglu et al. (2013)).

4. Preferential attachment as the underlying mechanism of network growth:

   - preferential attachment in the evolution of scientific collaboration networks (Newman (2004)),
   - preferential attachment in financial networks (Podobnik et al. (2011)).

Here, some of these stylized facts are used to exemplify the three different levels of statistical analysis and more specifically, the different statistical characterizations that are performed in the analysis of linguistic networks. Several of these examples are also shown in Table 2 (Sect. 5).

3.4 Levels in the Statistical Analysis of Networks

We now look at a set of well-known network measures from the point of view of the statistical levels. Global network indicators typically map certain network properties onto a single value and correspond therefore to the first level (density, clustering coefficient, diameter, etc.). It is clear that a level-1 characterization is relatively rough because many different networks may give rise to the same measure. For instance, a random graph, a regular lattice as well as a scale-free network may well result in the same network density (average degree).

One possibility to differentiate between those cases is to compute a whole set of global level-1 indicators. An example of such a "multi-dimensional" characterization is the so-called small-world regime which is settled on the simultaneous observation, for the whole network, of a low value of the characteristic path length and a high clustering coefficient.

Another way to differentiate between networks that share the same level-1 characteristics is to include the second level of analysis. For instance, the differentiation between a random, a regular and a scale-free graph will be straightforward on the basis of the entire degree distribution. Other measures that are computed at the level of nodes (we may refer to $\mathscr{X}$ as node space) include degree, local clustering and various measures of node centrality.

The second aspect that this section aims to address is to point out that the computation of many of these indicators are in fact based on the third level of analysis. The network diameter (see below), for instance, defined as the longest path in the set of shortest paths between all node pairs in a network, requires the computation of all shortest paths between all node pairs which is related to the path space $\mathscr{X}^{\mathbb{Z}}$.

3.4.1 Node Degrees

Among the main variables that are used to characterize a network is the degree (k_i) of the network nodes (i). For each experimental sample (each network), two main statistical indicators are often computed: first, the average (over i) of k_i, and second a power law exponent (if it exists) which characterizes the shape of the distribution of $P(k)$. The average degree, or density, is computed as

$$k = \langle k_i \rangle = \frac{1}{n} \sum_1^n k_i \qquad (2)$$

with $\langle \ \rangle$ meaning the sample average and yielding the average degree (k) of the network.

Second, the degree distribution $P(k)$ of a network is defined as the fraction of nodes in the network with degree k. When the network has n nodes and n_k of them have degree k, then $P(k) = \frac{n_k}{n}$. In case the distribution $P(k)$ follows a power law

$$P(k) = k^{-\lambda} \qquad (3)$$

the shape of the distribution is often characterized by a constant λ (typically in the range $2 < \lambda < 3$) and the network is called scale-free.

While network density is probably the most typical level-1 indicator, the exponent λ characterizes the distribution of node degrees and is therefore (by Eq. (3)) related to the second level. Notice however that the computation of both k and λ involve the evaluation of all node degrees, that is $P(k)$. Notice also that many networks may give rise to a certain degree distribution and that the set of networks with the same average degree is even larger. As an additional degree characteristic one may therefore look at the assortativity of node degrees which assesses the degree correlation patterns between pairs of nodes. Despite the fact that assortativity characterizes the network by a single-value and should therefore be considered a level-1 indicator, it assesses characteristics of node pairs, that is, on the space $\mathscr{X}^2$.

3.4.2 Clustering Coefficient

The clustering coefficient is another typical example which is defined at the first and the second statistical level. Namely, in the global variant a single mean clustering value is used as a characterization of the network and this value contains no information about the contributions of individual nodes. On the other hand, the global clustering coefficient is obtained on the basis of a local clustering coefficient defined at the node level $\mathscr{X}$. Consequently, the computation of the global clustering coefficient, while being a first-level indicator, involves the computation of the distribution of the local clustering coefficient.

Notice, however, that the computation of clustering (local and therefore global) involves the computation of the relative frequency of triangles in the network. That is, it involves statistics at the level of triplets of nodes $\mathscr{X}^3$. In Table 1 we denote this as $\mathscr{X}^3 \to \mathscr{X} \to \mathbb{R}$ in order to make clear that the global clustering ($\mathbb{R}$) is based on local clustering ($\mathscr{X}$) which is computed at the as a statistic on node triplets ($\mathscr{X}^3$).

3.4.3 Average Path Length and Diameter

Two other rather typical global indicators (level-1 statistical measures) are the average path length and the network diameter. Both of them are based on network geodesics, that is on the computation of shortest paths between pairs of nodes. While the average path length informs about the average number of steps required to go from one node to another, the diameter is the longest of those. It is clear that both measures are based on the assumption that network trajectories follow shortest paths. It is also clear that both measures map from the third level ($\mathcal{X}^{\mathbb{Z}}$) to the first ($\mathbb{R}$ or respectively $\mathbb{N}$). There is also a node property associated to shortest paths, namely, eccentricity. See Table 1.

3.4.4 Centrality Measures

Centrality measures are a probably the most typical cases of per-node statistics (level-2). They are usually considered as a measure of importance of a node in the network. Various different measures have been proposed, such as degree, betweenness, closeness and eigenvector centrality. All of these measures assess a nodes position in the network with respect to a set of trajectories between pairs of nodes in the graph (see Borgatti (2005) and Borgatti and Everett (2006)). For instance, betweenness quantifies the number of shortest paths in the networks that traverse a given node. Eigenvector centrality, to make another example, is more related to random walks on the network and the respective stationary probability of traversal. Therefore, all of these measures (apart from degrees) define a mapping from the path space of a network to node properties ($\mathcal{X}^{\mathbb{Z}} \rightarrow \mathcal{X}$, see Table 1).

4 On the Intelligibility of Statistical Indicators in Linguistic Networks

4.1 Path-Based Measures

Global statistical indicators in network theory typically try to map certain network characteristics onto a single value in order to point out different network regimes with respect to the property at question. However, as shown above, the computation of most of these indicators involves the evaluation of statistics on the path space $\mathcal{X}^{\mathbb{Z}}$. For an overview, see Table 1 below. In this section, we follow the analysis of centrality measures and network flow due to Borgatti (2005) and suggest that, for the interpretation of such measures, it is important to be aware of the implicit assumptions that certain indicators make on network trajectories. Above all in the setting of linguistic networks.

Table 1 High-level network characteristics are mapped onto low-level network indicators

Measure	
density	$\mathscr{X} \to \mathbb{R}$
exp. degree dist.	$\mathscr{X} \dashrightarrow \mathbb{R}$
assortativity	$\mathscr{X}^2 \to \mathscr{X} \to \mathbb{R}$
avg. clustering	$\mathscr{X}^3 \to \mathscr{X} \to \mathbb{R}$
eccentricity	$\mathscr{X}^Z \to \mathscr{X}$
diameter	$\mathscr{X}^Z \to \mathscr{X} \to \mathbb{N}$
centralities	$\mathscr{X}^Z \to \mathscr{X}$
avg. path length	$\mathscr{X}^Z \to \mathscr{X} \to \mathbb{R}$

4.2 Links and Flows, Structure and Function

One of the main contributions of network science in the different areas is that it helps understanding the relation between the structure and the function of a system. For instance, it is now well-known that the connectivity patterns in functional brain networks of Schizophrenia or Alzheimer patients differ in important ways from the patterns in healthy people (e.g., Supekar et al. (2008), Bassett et al. (2009), He et al. (2012), and Zhao et al. (2012)). Moreover, and very importantly, the observed differences like the lack of small-worldness, or clustering enable plausible interpretations about the respective dysfunctions in the brain and we gain, in this way, more insight about the general functioning of that system. Likewise, in traffic networks, power grid networks or networks of inter-bank money transfer the network perspective allows to study the susceptibility to system failure in dependence of certain changes (such as removal of links or nodes) in the structure of the system. Various measures of vulnerability, robustness and stability have been proposed the consideration of which may have very important implications for the design of new, more stable infrastructures.

A better understanding of the relation between the structure of the various linguistic networks and the functioning of the linguistic system must also be at the heart of a unified and applicable theory of linguistic networks. However, as opposed to systems like inter-bank money transfer calling for a type-1 network representation with clear interpretations, in linguistic networks even the interpretation of the functions or processes related to or mapped onto the single connections is not always straightforward, and differs moreover from network type to network type. It is then even more difficult to think in terms of processes taking place in the network as a whole which are typically related to the functioning of the system.

The previous considerations have shown that the majority of statistical network indicators involve computations on the path space of the network and the applicability of these measures is challenged if the question of what flows through the network is undecidable. Borgatti (Borgatti 2005) has discussed these issues at the example of different centrality measures by relating the type of network flow implicit in the computation of the different measures to the flow in the real system to which those measures have been applied.

4.3 Types of Network Flow

In Borgatti (2005), Borgatti develops a typology of network flows based on two dimensions. The first one relates to different kinds of dynamical processes that flow along the links of a network. Borgatti (2005) proposed that, according to the trajectory, dynamical processes comprise the following types of flows:

1. Geodesics: shortest path to a target destination
2. Paths: no repetition of nodes or links
3. Trails: no repetition of links
4. Walks: no restriction

As a second dimension in this typology, Borgatti (2005) considers the mechanism of node-to-node transmission. The author differentiates:

1. parallel duplication
2. serial duplication, and
3. transfer.

The first one refers to a parallel copying mechanism as present, for instance, in news broad cast. Serial duplication, or copying, refers to the dyadic replication mode in which the information is passed in a serial manner from one node to only one other. As opposed to copying, where the sender does not "loose" the information passed to other nodes, transfer refers to processes in which some thing is transferred, that is, given away, from the sender to the receiver. For Borgatti, the purpose of considering these different kinds of flows is to match different measures of centrality to the different kinds of flows. It turns out that "the most commonly used centrality measures are not appropriate for most of the flows we are routinely interested in" (Borgatti (2005):55).

4.4 Flow in Linguistic Networks

As said, the specification of flow in linguistic networks is not generally straightforward. In networks of semantic relatedness, especially those obtained by word association experiments (Nelson et al. 2004), one could argument on the basis of cognitive processes by which concepts are linked even if a precise understanding of these processes is still lacking. In co-occurrence or precedence networks the situation becomes rather intricate, because it is, in fact, a flow of words that is used to construct a network representation. It is then true that "Paths on network (c) [reference to the co–occurence network] can be understood as the potential universe of sentences that can be constructed with the given lexicon" (Solé et al. (2010):21), because a text or any other verbal utterance is in fact a sequence of words and therefore naturally defines a trajectory on a word network. On the other hand, the set of trajectories on a co-occurrence network certainly contains a lot of "sentences" that are grammatically incorrect or do not make any sense.

Let us illustrate issues related to the interpretation of network flow with the drastic but simple example of precedence networks. We may conceive a text as a sequence of symbols $S = s_0 s_1 s_2 s_3 \ldots s_N$, where, depending on the problem of interest, the symbols s_i correspond to words, lemmata, part-of-speech (PoS) or even to a subset of items such as high-frequent, topic-related nouns. Here we consider words and denote, for convenience, the set of words as $s_i \in \{A, B, C, \ldots\}$. In a precedence network (or flow network) an arc from word A to B exists whenever a word pair $(s_i s_{i+1}) = (AB)$ is observed in the text S. In particular, we may put a weight on the arcs corresponding to the frequency with which the associated word pair occurs. In that case, the weighted adjacency matrix (say P) encodes the frequency of all word pairs that are observed in S.

Notice that the degree in such a network representation corresponds to the word frequency and the power law degree distribution is in essence due to the Zipf law. It is noteworthy, moreover, that such a clear interpretation of the network degree in terms of word use statistics is possible because the computation of degrees is based only on node characteristics ($\mathscr{X}$) and not on the network paths.

What kind of network flow (according to Borgatti (2005) in terms of geodesics, paths, trails and walks) can be associated to such a network? Clearly, sentences are not forced to follow shortest paths from a source to a target; a thinking in terms of sources and targets seems to be misplaced in that context. Also the repetition of nodes (words) and links (word pairs) is clearly possible and in fact rather likely. This would mean that sentences, seen as trajectories on word networks, are, in the setting of Borgatti (2005), best classified as walks.

Accordingly, we could normalize the frequency matrix P appropriately in order to contain the probabilities for all words to be followed by the other words, such that P defines a Markov process on $\{A, B, C, \ldots\}$. Clearly, such a "model" would be capable of generating all the sentences that have been in the original data S. However, it would generate much more. While it might occasionally generate sentences that make sense and are grammatically correct, this is clearly not the general case. The reason is that the precedence network representation (and co-occurrence more generally) is constructed without sensitivity to grammatical and semantic constraints that are at work in the construction of real sentences. All linguistic relations beyond those to the next word, are just not captured by a precedence network.

In fact, the grammatical constraints in the construction of real sentences which impose some (more complex) restrictions onto the trajectories are not accounted for by the typology proposed in Borgatti (2005). They cannot be accounted for because the information that is needed is just not available in a network representation and its incorporation requires further analysis of the original sequence S. If Borgatti (2005) shows that "centrality measures are not appropriate for most of the flows", the difficulty of interpreting linguistic network flow altogether challenges the appropriateness of centrality and other path-based measures and their interpretation in the case of linguistic networks.

5 Examples

Table 2 helps to illustrate with some examples the application of the framework for network construction and analysis. Although examples include different fields of application, the purpose of such a framework has a special focus on linguistic networks. The first column indicate the phenomenon at hand and the second column gives an example of the occurrence of the phenomenon. As earlier presented, both network induction and network analysis are performed at three different levels, these correspond to the third and fourth columns, respectively. We recall that while in network design leveling concerns three abstraction levels, in network analysis the three different levels are grounded on the statistical indicators used at each level. The last column in Table 2 provides the main bibliographic reference where the example was reported.

6 Discussion

We classify networks according their level of abstraction and network measures according to the statistical level on which they map. Obviously, the classification of different types of networks according to the level of abstraction they involve is not always clear-cut. Our aim here is not to present such a rigorous classification, but we rather aim at a heuristic perspective to show in what sense linguistic networks are particular. Clearly, in type-1 networks it is usually immediately clear which elements of the real system are encoded into a network description. For instance, speaking of "air traffic network" or "network of inter-bank money transfer" keeps very little space for interpretation: the first one maps air traffic onto a network the second money that is exchanged between banks. Of course, even in systems which call for a network representation (type-1) there is a relative amount of freedom in the design of such a representation. For instance, one could argue that there are many dimensions of (say) air traffic that could be considered, such as passengers, goods, or the number of flights, but it is always clear that a link indicates that something is transfered from one airport to the other, that something flows through the net.

Similarly, for type-2 the meaning of the links is very clear: they indicate the similarity of entities (nodes) with respect to certain features. However, the second type is different from the first especially because the links – encoding similarity – do not generally represent something that really happens between the entities. This is what similarity networks share with type-3 networks.

Indeed, going to the third abstraction level is not such a big step as the proposed classification might indicate. Words linked in a co-occurrence network, for instance, can be seen as words that appear in a similar context, and in that sense, co-occurrence is a kind of similarity. Co-occurrence networks are also nice examples to illustrate that the border between type-3 and type-1 is not always crystal

Table 2 Statistical and abstraction level for several examples of linguistic networks and networks in other fields of application of network theory. In some cases, the classification is not straightforward, see Discussion.

Phenomenon	Example	Sta	Abs	Ref
Small-World	USA power grid	1	1	Watts and Strogatz (1998)
Small-World	Film actors	1	1 or 3	Watts and Strogatz (1998)
Small-World	Air Traffic	1	1	Li and Cai (2004)
Diameter	WorldWideWeb links	1	2	Albert et al. (1999)
Degree, Clustering	Synctatic, Semantic and Co-occurrence	1	3	Solé et al. (2010)
Heterogeneity	Synctatic, Semantic and Co-occurrence	2	3	Solé et al. (2010)
Disassortative mixing	Syntatic nets	2	3	Ferrer i Cancho et al. (2004)
Heterogeneity	Syntatic nets	2	3	Ferrer i Cancho et al. (2004)
Heterogeneity	VWoolf co-occurrence nets	2	3	Ferrer i Cancho et al. (2004)
Degree	Vectorial Semantics	1	2	Salton et al. (1975)
TradeOff Centralities	Airport nets	1	1	Borgatti (2005)
Betweenness	Marriage in Renaissance Florentine	1	1	Borgatti (2005)
Clustering	Financial nets	1	2	Mantegna (1999)
Heterogeneity	Cross-border Debts	2	2	Spelta and Araújo (2012)
Density	Capital flow nets	1	1	Spelta and Araújo (2012)
Heterogeneity	Word free association	2	1	Borge-Holthoefer and Arenas (2010)
Systemic risk	Banking nets	2	1	Battiston et al. (2010)
Systemic risk	Board of Directors	2	3 (bi-partite)	Battiston et al. (2010)
Systemic risk	NYSE network	2	2	Battiston et al. (2010)
Communities	SFI scientists	3	1 or 3	Fortunato and Barthelemy (2007)

clear. On the one hand, we would consider word co-occurrence networks as type-3, corresponding to a high level of abstraction, due to the fact that a link between two words is not obviously related to a *process* in between the two words. As a matter of fact, there are several possibilities of defining a co-occurrence network depending on, for instance, the window size taken into account for the co-occurrence test, and words linked in one co-occurrence design may not be linked in another. On the other hand, however, there are co-occurrence networks in other domains – such as co-authorship – which we might see as networks of a lower abstraction level. In co-authorship networks, two scientists are linked if they are authors of the same article. Even if the specific forms of interchange between the authors is in most cases not visible, it is completely reasonable to consider that a link between the authors maps their conjoint activity and information exchange.

Concerning the statistical levels our main objective is to show that the three different characterizations of a phenomenon at question represent successively finer levels of description of the statistical properties of a network. The second aim is to make clear to which level the different measures relate, because this is important for whether there is a direct interpretation especially in the case of linguistic networks.

We admit that the example of precedence networks, for which we discuss these issues with some detail, is a very special construction and that the argumentation presented in Sect. 4 does not directly apply to all linguistic networks. We assume, however, that the larger class of co-occurrence networks suffers from similar problems concerning the linguistic interpretability of measures that characterize the path space $\mathscr{X}^Z$, because the relation between paths on networks and real word sequences (sentences) is unclear. This may be different for networks of semantic similarity.

In many areas, network representations have proven to be a useful explanatory device which can help to gain insight into the patterns of mutual interdependencies characteristic of complex systems. The flexibility of networks and their applicability to very different phenomena is one of the main reasons for their success and we believe that they are a useful metaphor even if the entities represented by it rely on indirect and observational evidence about the system at question. Accordingly, using networks as an abstraction of linguistic patterns, as a way to map and visualize dependencies between linguistic items may be appropriate and reasonable. In spite of that, however, the transfer of the theory developed for networks to the linguistic field requires a more careful consideration of the context in which this theory has originated and respectively a linguistic assessment of the underlying assumptions.

7 Concluding Remarks

Our contribution relies on highlighting the importance of recognizing network design and network analysis as interdependent tasks. In so doing we developed a framework for network construction and analysis with special focus on linguistic networks. According to such a framework, both network induction and network analysis are performed at three different levels. While in network design leveling concerns three abstraction levels, in network analysis the three different levels are grounded on the statistical indicators used at each level. Together with the introduction of the framework where network design and network analysis are identified as interdependent tasks, we argue that the level of abstraction at which a given network is induced has an important bearing on this network analysis, particularly on the choice of appropriate topological indicators and on the interpretation of their results. More precisely, we envision that the higher is the abstraction level of network induction, the harder is the interpretation of the topological indicators used in network analysis. We illustrate the framework using examples of linguistic networks as well as some other fields of application of network theory.

These considerations indicate that the field of linguistic networks, by applying well-known statistical tools inspired by network studies in other domains, may, in its current state, have only a limited contribution to the development of linguistic theory. A sophisticated analysis of what topological indicators represent as well as of what they miss is needed in order to advance into that direction. Most importantly, we do not yet have a clear understanding of the trajectories or dynamic processes on the different linguistic networks which makes the use of path-based measures (among them, centrality, average path length, etc.) problematic. On the other hand, the structural differences between linguistic networks of different types (e.g., Pustylnikov (2007)) are clearly indicative of their usefulness in a more applied context as tools for information retrieval and text classification. As soon as we can relate those structural differences to certain linguistic qualities, or show that they represent novel aspects that provide new knowledge of the language system, we may approach to a linguistic network theory. However, it may be that we must go beyond traditional network representation in order to achieve that.

In that regard, we envision the possibility to take the space of linguistic paths seriously in the definition of statistical indicators. Centralities, characteristic path length and other network indicators could be redefined at the level of paths (may be sentences) observed in the original data. Whether this provides new insight and in what sense the resulting measures deviate from their network variant are interesting questions for future research.

Acknowledgements. UECE (Research Unit on Complexity and Economics) is financially supported by FCT (Fundação para a Ciência e a Tecnologia), Portugal. This article is part of the Strategic Project: PEst-OE/EGE/UI0436/2014. SB acknowledges financial support of the German Federal Ministry of Education and Research (BMBF) through the project *Linguistic Networks* (http://project.linguistic-networks.net).

References

Acemoglu, D., Ozdaglar, A., Tahbaz-Salehi, A.: Systemic risk and stability in financial networks. Tech. rep. National Bureau of Economic Research (2013)

Albert, R., Jeong, H., Barabási, A.-L.: Internet: Diameter of the world-wide web. Nature 401(6749), 130–131 (1999)

Arbesman, S., Strogatz, S.H., Vitevitch, M.S.: Comparative Analysis of Networks of Phonologically Similar Words in English and Spanish. Entropy 12(3), 327–337 (2010), doi:10.3390/e12030327

Bassett, D.S., Bullmore, E.T., Meyer-Lindenberg, A., Apud, J.A., Weinberger, D.R., Coppola, R.: Cognitive fitness of cost-efficient brain functional networks. Proceedings of the National Academy of Sciences 106(28), 11747–11752 (2009)

Bassett, D.S., Gazzaniga, M.S.: Understanding complexity in the human brain. Trends in Cognitive Sciences 15(5), 200–209 (2011), doi:10.1016/j.tics.2011.03.006

Battiston, S., Glattfelder, J.B., Garlaschelli, D., Lillo, F., Caldarelli, G.: The structure of financial networks. In: Network Science, pp. 131–163. Springer (2010)

Blanchard, P., Petroni, F., Serva, M., Volchenkov, D.: Geometric representations of language taxonomies. Computer Speech & Language 25(3), 679–699 (2011), doi:10.1016/j.csl.2010.05.003

Borgatti, S.P.: Centrality and network flow. Social Networks 27(1), 55–71 (2005), doi:10.1016/j.socnet.2004.11.008

Borgatti, S.P., Everett, M.G.: A Graph-theoretic perspective on centrality. Social Networks 28(4), 466–484 (2006), doi:http://dx.doi.org/10.1016/j.socnet.2005.11.005

Borge-Holthoefer, J., Arenas, A.: Categorizing words through semantic memory navigation. The European Physical Journal B 74(2), 265–270 (2010) (English), doi:10.1140/epjb/e2010-00058-9

Boss, M., Elsinger, H., Summer, M., Thurner, S.: An empirical analysis of the network structure of the Austrian interbank market. Oesterreichesche Nationalbank Financial stability Report 7, 77–87 (2004)

Broder, A., Kumar, R., Maghoul, F., Raghavan, P., Rajagopalan, S., Stata, R., Tomkins, A., Wiener, J.: Graph structure in the web. Computer Networks 33(1), 309–320 (2000)

Choudhury, M., Ganguly, N., Maiti, A., Mukherjee, A., Brusch, L., Deutsch, A., Peruani, F.: Modeling discrete combinatorial systems as alphabetic bipartite networks: Theory and applications. Phys. Rev. E 81(3), 036103 (2010), doi:10.1103/PhysRevE.81.036103

Colizza, V., Pastor-Satorras, R., Vespignani, A.: Reaction-diffusion processes and metapopulation models in heterogeneous networks. Nature Physics 3(4), 276–282 (2007)

Ferrer i Cancho, R.: Bibliography on linguistic, cognitive and brain networks (2012), http://www.lsi.upc.edu/~rferrericancho/linguistic_and_cognitive_networks.html

Ferrer i Cancho, R., Mehler, A., Pustylnikov, O., Díaz-Guilera, A.: Correlations in the organization of large-scale syntactic dependency networks. In: TextGraphs-2: Graph-Based Algorithms for Natural Language Processing, pp. 65–72 (2007)

Ferrer i Cancho, R., Solé, R.V., Köhler, R.: Patterns in syntactic dependency networks. Phys. Rev. E 69(5), 051915 (2004), doi:10.1103/PhysRevE.69.051915

Fortunato, S., Barthelemy, M.: Resolution limit in community detection. Proceedings of the National Academy of Sciences 104(1), 36–41 (2007)

Goñi, J., Arrondo, G., Sepulcre, J., Martincorena, I., de Mendizábal, N.V., Corominas-Murtra, B., Bejarano, B., Ardanza-Trevijano, S., Peraita, H., Wall, D.P., Villoslada, P.: The semantic organization of the animal category: evidence from semantic verbal fluency and network theory. Cognitive Processing 12(2), 183–196 (2011) (English), doi:10.1007/s10339-010-0372-x

Grabska-Gradzinska, I., Kulig, A., Kwapien, J., Drozdz, S.: Complex network analysis of literary and scientific texts. International Journal of Modern Physics C 23(07), 1250051 (2012), doi:10.1142/S0129183112500519

Gravino, P., Servedio, V.D.P., Barrat, A., Loreto, V.: Complex Structures and Semantics in Free Word Association. Advances in Complex Systems 15(03n04), 1250054 (2012), doi:10.1142/S0219525912500543

He, H., Sui, J., Yu, Q., Turner, J.A., Ho, B.-C., Sponheim, S.R., Manoach, D.S., Clark, V.P., Calhoun, V.D.: Altered small-world brain networks in schizophrenia patients during working memory performance. PloS One 7(6), e38195 (2012)

Huizinga, H., Nicodème, G.: Are international deposits tax-driven. Journal of Public Economics 88(6), 1093–1118 (2004)

Iyengar, S.R., Veni Madhavan, C.E., Zweig, K.A., Natarajan, A.: Understanding Human Navigation Using Network Analysis. Topics in Cognitive Science 4(1), 121–134 (2012), doi:10.1111/j.1756-8765.2011.01178.x

Kaldor, N.: A model of economic growth. The Economic Journal 67(268), 591–624 (1956)

Konect: OpenFlights network dataset and US airports network dataset – KONECT (2014)

Landauer, T.K., Dutnais, S.T.: A solution to Plato's problem: The latent semantic analysis theory of acquisition, induction, and representation of knowledge. Psychological Review, 211–240 (1997)

Lerner, A., Ogrocki, P.K., Thomas, P.J.: Network Graph Analysis of Category Fluency Testing. Cognitive & Behavioral Neurology 22(1), 45–52 (2009)

Li, W., Cai, X.: Statistical analysis of airport network of China. Physical Review E 69(4), 046106 (2004)

Mantegna, R.N.: Hierarchical structure in financial markets. The European Physical Journal B-Condensed Matter and Complex Systems 11(1), 193–197 (1999)

McGuire, P., Tarashev, N.: Tracking international bank flows. BIS Quarterly Review, 27–40 (2006)

Mehler, A., Geibel, P., Pustylnikov, O.: Structural classifiers of text types: Towards a novel model of text representation. LDV Forum: Zeitschrift für Computerlinguistik und Sprachtechnologie; GLDV-Journal for Computational Linguistics and Language Technology 22(2), 51–66 (2007)

Menon, V.: Large-scale brain networks and psychopathology: a unifying triple network model. Trends in Cognitive Sciences 15(10), 483–506 (2011)

Meusel, R., Vigna, S., Lehmberg, O., Bizer, C.: Graph Structure in the Web – Revisited. Accepted paper at the 23rd International World Wide Web Conference (WWW 2014), Web Science Track, Seoul, Korea (April 2014)

Miller, B.A., Bliss, N.T., Wolfe, P.J.: Toward signal processing theory for graphs and non-Euclidean data. In: 2010 IEEE International Conference on Acoustics Speech and Signal Processing (ICASSP), pp. 5414–5417. IEEE (2010)

Miller, G.A.: WordNet: A Lexical Database for English. Communications of the ACM 38(11), 39–41 (1995)

Minoiu, C., Reyes, J.A.: Network Analysis of Global Banking: 1978-2009. International Monetary Fund (2011)

Mukherjee, A., Choudhury, M., Basu, A., Ganguly, N.: Self-organization of the Sound Inventories: Analysis and Synthesis of the Occurrence and Co-occurrence Networks of Consonants. Journal of Quantitative Linguistics 16(2), 157–184 (2009), doi:10.1080/09296170902734222

Nelson, D.L., McEvoy, C.L., Schreiber, T.A.: The University of South Florida free association, rhyme, and word fragment norms. Behav. Res. Methods. Instrum. Comput. 36(3), 402–407 (2004)

Newman, M.E.J., Barabási, A.-L., Watts, D.J.: The Structure and Dynamics of Networks. Princeton University Press, Princeton (2003)

Newman, M.E.J.: Detecting community structure in networks. The European Physical Journal B-Condensed Matter and Complex Systems 38(2), 321–330 (2004)

Opsahl, T., Agneessens, F., Skvoretz, J.: Node Centrality in Weighted Networks: Generalizing Degree and Shortest Paths. Social Networks 3(32), 245–251 (2010)

Peng, R.D., Hengartner, N.: Quantitative Analysis of Literary Styles. The American Statistician 56, 2002 (2002)

Podobnik, B., Valentinčič, A., Horvatić, D., Eugene Stanley, H.: Asymmetric Levy flight in financial ratios. Proceedings of the National Academy of Sciences 108(44), 17883–17888 (2011)

Pustylnikov, O.: Guessing Text Type by Structure. In: Proceedings of the 12th ESSLLI Student Session (2007)

Salton, G., Wong, A., Yang, C.S.: A vector space model for automatic indexing. Commun. ACM 18(11), 613–620 (1975), doi:10.1145/361219.361220

Salton, G.: Automatic text processing: the transformation, analysis, and retrieval of information by computer. Addison-Wesley Longman Publishing Co., Inc., Boston (1989)

Serva, M., Petroni, F., Volchenkov, D., Wichmann, S.: Malagasy Dialects and the Peopling of Madagascar. J. R. Soc. Interface 9(66), 54–67 (2011)

Shuman, D.I., Narang, S.K., Frossard, P., Ortega, A., Vandergheynst, P.: The emerging field of signal processing on graphs: Extending high-dimensional data analysis to networks and other irregular domains. IEEE Signal Processing Magazine 30(3), 83–98 (2013)

Soares, M.M., Corso, G., Lucena, L.S.: The network of syllables in Portuguese. Physica A: Statistical Mechanics and its Applications 355(2-4), 678–684 (2005), doi:10.1016/j.physa.2005.03.017

Solé, R.V., Corominas-Murtra, B., Valverde, S., Steels, L.: Language networks: Their structure, function, and evolution. Complexity 15, 20–26 (2010), doi:10.1002/cplx.20305

Soramäki, K., Bech, M.L., Arnold, J., Glass, R.J., Beyeler, W.E.: The topology of interbank payment flows. Physica A: Statistical Mechanics and its Applications 379(1), 317–333 (2007)

Sowa, J.F.: Principles of Semantic Networks. Morgan Kaufmann (1991)

Spelta, A., Araújo, T.: The topology of cross-border exposures:beyond the minimal spanning tree approach. Physica A: Statistical Mechanics and its Applications 391, 5572–5583 (2012)

Sporns, O., Tononi, G., Kötter, R.: The Human Connectome: A Structural Description of the Human Brain. PLoS Comput. Biol. 1(4), e42 (2005), doi:10.1371/journal.pcbi.0010042

Storm, C.: The Semantic Structure of Animal Terms: A Developmental Study. International Journal of Behavioral Development 3(4), 381–407 (1980), doi:10.1177/016502548000300403

Supekar, K., Menon, V., Rubin, D., Musen, M., Greicius, M.D.: Network analysis of intrinsic functional brain connectivity in Alzheimer's disease. PLoS Computational Biology 4(6), e1000100 (2008)

Vilela Mendes, R., Lima, R., Araújo, T.: A process-reconstruction analysis of market fluctuations. International Journal of Theoretical and Applied Finance 5(08), 797–821 (2002)

Watts, D.J., Strogatz, S.H.: Collective dynamics of small-world networks. Nature 393, 440–442 (1998)

Yu, S., Liu, H., Xu, C.: Statistical properties of Chinese phonemic networks. Physica A: Statistical Mechanics and its Applications 390(7), 1370–1380 (2011), doi:10.1016/j.physa.2010.12.019

Zamora-López, G., Russo, E., Gleiser, P.M., Zhou, C., Kurths, J.: Characterizing the complexity of brain and mind networks. Philosophical Transactions of the Royal Society A 369(1952), 3730–3747 (2011)

Zhao, X., Liu, Y., Wang, X., Liu, B., Xi, Q., Guo, Q., Jiang, H., Jiang, T., Wang, P.: Disrupted small-world brain networks in moderate Alzheimer's disease: a resting-state FMRI study. PloS One 7(3), e33540 (2012)

Semantic Space as a Metapopulation System: Modelling the Wikipedia Information Flow Network

A. Paolo Masucci, Alkiviadis Kalampokis,
Víctor M. Eguíluz, and Emilio Hernández-García

1 Introduction

The meaning of a word can be defined as an indefinite set of *interpretants*, which are other words that circumscribe the semantic content of the word they represent (Derrida 1982). In the same way each interpretant has a set of interpretants representing it and so on. Hence the indefinite chain of meaning assumes a *rhizomatic* shape that can be represented and analysed via the modern techniques of network theory (Dorogovtsev and Mendes 2013).

The semantic or conceptual space (SS hereafter) has already been investigated by different approaches. A common understanding within these approaches is that the SS is made up of words or concepts that are connected by certain relationships. Depending on the nature of these relationships different semantic webs have already been considered. In the psycholinguistics approach the SS is often extracted via *free word association* game and a network is constructed where two words are connected if they appear to be consecutive in a free word association experiment (Steyvers and Tenenbaum 2005; Borge-Holthoefer and Arenas 2010). Other semantic webs are generated through linguistics approaches (Montemurro and Zanette 2010). Among others, an interesting one is based on the dictionary, where the relationships between

---

A. Paolo Masucci
Centre for Advanced Spatial Analysis, University College of London,
90 Tottenham Court Road, W1T 4TJ, London, UK
e-mail: paolo_masucci@yahoo.it

Alkiviadis Kalampokis
Dipartimento di Scienze e Tecnologie, Università degli Studi di Napoli "Parthenope",
Centro Direzionale di Napoli - Isola C4, 80143 Napoli, Italy
e-mail: alkis@marine.aegean.gr

Víctor M. Eguíluz · Emilio Hernández-García
Instituto de Física Interdisciplinar y Sistemas Complejos IFISC (CSIC-UIB),
E-07122 Palma de Mallorca, Spain
e-mail: {victor,emilio}@ifisc.uib-csic.es

words are set to be of synonymy, antonymy, belonging to the same category or class, etc. (Sigman and Cecchi 2002; Samsonovic and Ascoli 2010). In all the mentioned cases a scale-free topology and small-world properties for the SS are found, suggesting an intrinsic self-organising nature of the SS (Sigman and Cecchi 2002). However it has been argued that networks derived by dictionaries and representing the so called *dictionary semantics*, characterised by scale-free distribution for the connectivity with exponents smaller than -2, reflect the properties of language use more than the properties of the SS (de Jesus Holanda et al. 2004; Violi 2001).

In contrast to the dictionary representation of the SS, it has been suggested that the meaning of a sign, where a sign can be a word, a concept, etc. can be recovered within an *encyclopedic model*, where every sign is specified by a set of other signs that interpret it (Eco 1986). "This notion of interpretants is fertile because it shows how semiotic processes, via continuous movements that refer a sign to other signs or sign chains, circumscribe the meanings in an asymptotic way. They never touch them, and make them accessible via other cultural units [...]. In this way an open system of connections between different signs is created that takes the shape of a rhizome (Deleuze and Guattari 1977)" (Eco 1986).

Hence, in its *encyclopedic semantics* acception, the SS can be interpreted as a metapopulation system where each page of an encyclopedia is a population of interpretants/words characterising some meanings. Then the structure of SS assumes a dynamical connotation, typical of population dynamics, where the different concepts are born and grow in time, exchanging and inheriting attributes from other concepts (it is interesting to notice how Deleuze and Guattari foresaw the very essence of the semantic machine not as a machine producing meaning, but as a machine producing its own structure (Deleuze and Guattari 1988)).

In this work, which summarizes results from Masucci et al. (2011b) and Masucci et al. (2011a), we attempt to extract the SS in its encyclopedic semantics acception. Following the semiotics rationale described above, we consider each entry of an encyclopedia as a population of interpretants and we measure the correlations between each pair of entries of that encyclopedia in terms of *directional semantic flows*. In particular we analyse a whole dump of Wikipedia. Wikipedia is not only the largest encyclopedia existing nowadays, but it is an *open encyclopedia* with its entries always growing in size and number, thus it represents well the idea of encyclopedic semantics expressed above. The resulting network is a directed network of semantic flows between the different concepts that are present in an encyclopedia and thus portrays a snapshot of the dynamics of meaning in that representation of the SS.

The concept of *information flow*, as it is used in this context, is introduced in Masucci et al. (2011b) to indicate the correlations between populations whose elements are defined by abstract attributes. Those populations can be social, biological or, as in this case, made of words. An encyclopedia could be considered as the recording surface of a collective phenomenon that acts to store and transmit knowledge. The term "information flow" is used in this context in order to stress that in those systems correlations are often caused by migration or inheritance of a part of a population to another one, in the sense that the content of each page finds its meaning in other pages that explain the underlying meaning of each concept contained in that page. Hence these correlations refer to a dynamical flow. Such a flow is measured by bits of information, via the Shannon entropy. In this sense the term "information flow" has to be considered technical and not a mere metaphor. It is important then not to confuse such terminology with the concept of information flow as the one introduced by Barwise (1997), which carries a complete different meaning. Thus the very term of information flow, which can be ethnic, genetic or, as in this case, semantic, gives an idea of such a dynamical process, where a movement of elements from a population to another one, corresponds to a flow of information in the attribute space where those elements are defined.

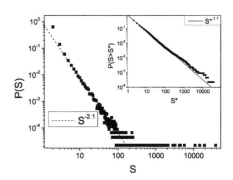

Fig. 1 Cluster size distribution of the semantic space. Cluster size distribution $P(S)$ of the semantic network at the percolation threshold. In the inset we show the cumulative distribution $P(S > S^*)$.

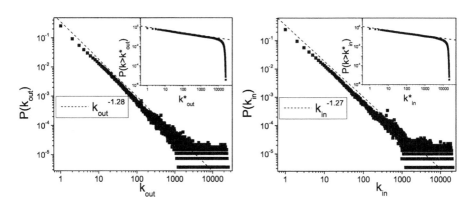

Fig. 2 Connectivity distribution of the semantic space. Out-degree distribution $P(k_{out})$ (left panel) and in-degree distribution $P(k_{in})$ (right panel) of the semantic network at the percolation threshold. In the insets the corresponding cumulative degree distributions $P(k > k^*)$ are displayed.

2 The Dataset

We consider the articles of a complete snapshot of English Wikipedia dated June 2008[1], consisting of $N \approx 2 \times 10^6$ entries. To process our dataset first of all we get rid of redirection pages. Then, in order to analyse the semantic content of the encyclopedia, we process the text, cleaning it of punctuation and of the so called *function words* like articles, pronouns, common adverbs, etc. (Hopper and Traugott 2003; Ferrer i Cancho and Solé 2001). In fact those words are very frequent in each page and often don't contribute to its semantic characterisation. After that we *lemmatize* the text, transforming the different words in their singular form or in their infinitive form if they are verbs[2]. The resulting set of lemmas defines the interpretants or *attribute* space where the different pages are defined and each Wikipedia page comes out to be defined by its lemmas frequency distribution and by its size.

In order to compute the directional semantic flow between the pages we use the method introduced in Masucci et al. (2011b). This method is very general and allows the extraction of a directed information flow network from a set of populations whose elements are defined by an n-dimensional symbolic attribute vector. It is based on the Jensen-Shannon divergence (Lin 1991) and within an information theory approach it is able to measure the amount of information flow within a set of populations of different sizes, defined in a symbolic attribute space. Moreover, using concepts derived from geographical segregation, the methodology in Masucci et al. (2011b) is able to infer the directionality of the information flow. More details are given in the Appendix.

The resulting network representing the SS, as we show below, displays scale invariant structures and small world properties, revealing a hierarchical SS, where the semantic clusters are strongly connected and communication between different areas of knowledge is fast.

3 Topology of the Semantic Space

To build the network, the directional semantic flow is measured between all the entry pairs. Then the entry pairs are ordered by the increasing values of their semantic distance, and a network of entries is defined considering two pages as linked when their semantic distance is smaller than a given threshold.

By increasing the value of the threshold we obtain a growing network where the first links to form are the strongest in a semantic sense. As the threshold is increased

---

[1] Wikipedia: Database download website. Available at http://en.wikipedia.org/wiki/Wikipedia:Database_download, accessed 2014 Sept.

[2] For the lemmatization process, we use a public dictionary containing lemmas, available from the Unitex website at http://infolingu.univ-mlv.fr/DonneesLinguistiques/Dictionnaires/telechargement.html (accessed 2014 June), the process consisting of retrieving the word in the dictionary and substituting it with the corresponding lemma.

Fig. 3 Connectivity distribution of the minimum spanning tree of the semantic space. Degree distribution $P(k)$ for the undirected minimum spanning tree for the whole network representing the semantic space. In the insets the cumulative degree distribution $P(k > k^*)$ is displayed.

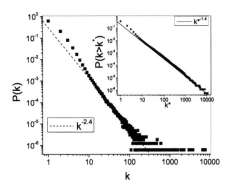

further, very well connected clusters form, each cluster representing different semantic areas. A significant threshold to analyse the network representing the SS is the *percolation threshold* (PT hereafter), when a giant cluster forms and a phase transition happens (Dorogovtsev and Mendes 2013; Stauffer and Aharony 1991).

The SS network reaches its PT at approximately 362000 pages, when the two main clusters merge to form a giant cluster of 57800 pages. At the PT the network has $L \approx 3 \times 10^8$ links with an average degree $\langle k \rangle \approx 1743$. The very large average degree means that the clusters are very densely connected. The network is composed by 44500 disconnected clusters showing scale invariant cluster size distribution, $P(S) \sim S^{-2.1}$, with a fat tail (Fig. 1).

The scale-free cluster size distribution is the first important property we find for the SS. It has been shown that in a random growing network at the PT the cluster size distribution decays faster than a power law, $P(S) \sim 1/(S^3 \ln^2 S)$ (Dorogovtsev and Mendes 2013; Kim et al. 2002). Then it can be argued that the scale-free behaviour we find for the cluster size distribution is not an effect of a random growing network, but a peculiar property of the SS.

As a matter of fact at this threshold almost each cluster represents a well defined semantic area. Hence the scale-free distribution implies a hierarchy between the semantic areas and gives us a picture of the structure of the SS.

The largest clusters, representing the greater body of the SS, are composed of large taxonomies, such as geographical places, biological species, etc... The largest cluster is made of 38500 pages and consists of geographical places of USA, such as villages, cities, rivers, etc... The second largest cluster is made of 19300 pages and consists of taxonomies of living species as animal, plants, insects, bacteria, etc...The third largest cluster is mainly made of Romanian geographical entries, the fourth by French cities and villages and so on. In each of these clusters the pages are very simple and have a structure very similar to each other. A typical example of these kinds of pages is the *Canarium Zeylanicum* page, which is in the second largest cluster: "Canarium Zeylanicum is a species of flowering plant in the frankincense family, Burseraceae, that is endemic to Sri Lanka.". The content word lemmas of this page are: "Canarium Zeylanicum specie flower plant frankincense family Burseraceae endemic Sri Lanka". This page easily connects with all pages

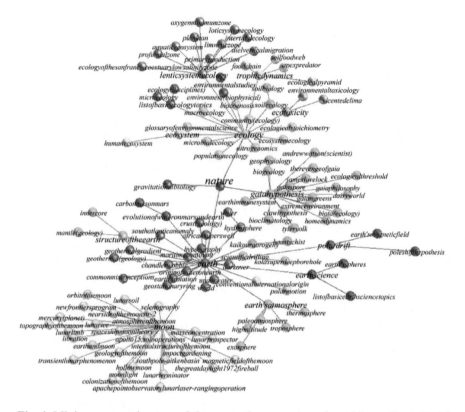

Fig. 4 Minimum spanning tree of the semantic space. A portion of the undirected minimum spanning tree of the network representing the semantic space in the neighbourhood of the entry *nature* until its third neighbour. The nodes represent different Wikipedia entries, while the edges represent a semantic flow between the different entries. The colour partition is based on the nodes modularity classes. Figure realised with the open source software Gephi (Bastian et al. 2009).

containing "specie flower plant endemic Sri Lanka", hence forming a taxonomy with other pages as for example the *Mastixia Nimali* page: "Mastixia Nimali is a species of plant in the Cornaceae family. It is endemic to Sri Lanka.". It is interesting to notice how the passage from a taxonomic page to another resides in the mutation of a few rare words.

Then clusters are found at all scales of magnitude, consisting of any semantic area one can possibly think of and generally the greater is the complexity of the page, the smaller is the cluster which it belongs to. There are clusters of football players, small clusters of different kind of bicycles, ethnicity clusters, language family clusters, singers, technology, religions, etc...

The out-degree and in-degree distribution of the network at the PT are scale-free with a very slow decay, characterised by exponents: $\gamma_{out} \approx -1.28$ and $\gamma_{in} \approx -1.27$ (see Fig. 2). The distributions are scale invariant until very large scales where a sharp cut-off appears, revealing that the SS is characterised by structures at all the scales. The giant component of the network has a directed diameter $d = 20$ that is of the order of the logarithm of the cluster size. Moreover its average clustering coefficient is $\langle C \rangle = 0.87$, that is larger than the clustering coefficient of a random network of the same size, $\langle C \rangle_{RAND} = 0.17$, revealing local *small-world* properties of the SS (Watts and Strogatz 1998).

If the description at the cluster level represents the main body of the SS, the *minimum spanning tree* (hereafter MST) represents its backbone. The MST of a weighted network is an acyclic graph that has all the vertices of the network and that minimises the sum of the distances between the pages (Macdonald et al. 2005). It represents the skeleton of the network and in a sense it represents how semantic information best flows throughout the SS.

We compute the undirected MST of the complete network of Wikipedia via the Prim's algorithm (Prim 1957). The degree distribution of the MST is scale free with exponent -2.4 and a fat tail (see Fig. 3). Again the scale-free behaviour of the degree distribution tells us about the hierarchical structure of the MST of the SS. If we glimpse at Fig. 4, where a small portion of the MST centred on the Wikipedia entry *nature* is shown, we can have a rough idea of how this hierarchy organises itself. A very general concept, such as "nature", hasn't got a lot of connections, but it is an important bridge for the semantic flow between less complex concepts. Those less complex concepts, such as "earth", "ecology", etc. are in general more connected and eventually form taxonomies, which are hubs in the MST.

From what has been said, we can draw a general picture of the SS as a space whose body is mainly composed of simple concepts that are densely clustered in taxonomies or classifications. Then, at higher levels, more complex concepts form, creating smaller semantic clusters. This hierarchy goes further, in a scale-free fashion, until the more general and elaborated concepts emerge and those create an architecture of semantic flow channels that spans through the whole SS.

The values of the exponents of the degree distributions are too large to be explained by standard growing network models based on preferential attachment (Albert and Barabási 2002). For the character of the system and its statistical properties, the emerging topology of the SS is more likely to be represented by a new class of models of stochastic *content-based networks* of the type presented in Balcan et al. (2007), Mungan et al. (2005), and Bergmann et al. (2003), with the difference that in the case of Wikipedia the correlations generated by the Zipfean distribution of content words (Ferrer i Cancho and Solé 2001) play an important role on the topology of the system as it is explained below. This observation relates the topology of the SS to a wider range of biological phenomenology (Bergmann et al. 2003).

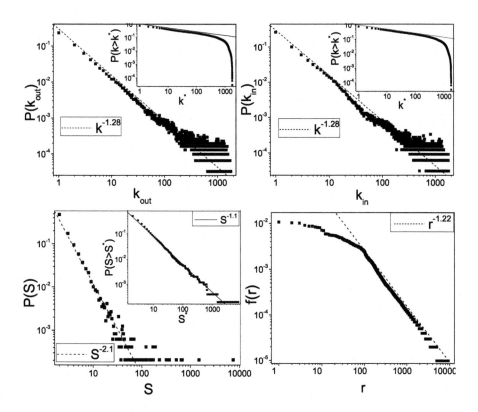

Fig. 5 The stochastic model representing the semantic space. Numerical results of the stochastic model representing the semantic space. This is a simulation of a *toy-model* for an encyclopedia of 5×10^4 pages with size l log-normally distributed, with first moment $\bar{l} = 20$ and second moment $l_\sigma = 0.5$. The parameters of the model are $\alpha = 0.001$, $p = 0.7$ and $M = 5$. In the top panels the out-degree distribution $P(k_{out})$ (left panel) and the in-degree distribution $P(k_{in})$ (right panel) of the semantic network at the percolation threshold are shown. The corresponding cumulative distributions $P(k > k^*)$ are displayed in the insets. In the bottom left panel we show the cluster size distribution $P(S)$ at the percolation threshold, the relative cumulative distribution $P(S > S^*)$ is displayed in the inset. In the right bottom panel we show the frequency-rank distribution $f(r)$ for the words in the model.

> ### *The Heaps'-like law*
> Written human language displays a fascinating puzzle of empirical regularities. Among them, the Heaps' law states that the vocabulary V of a written text is a function of its size L, $V(L) \propto L^\beta$, with $0 < \beta < 1$. The Heaps' law is strictly related to the Zipf's law for the word frequency distribution. The Heaps' law has been analytically and algorithmically derived from the Zipf's law (Leijenhorst and Van der Weide 2005; Serrano et al. 2009). In Zanette and Montemurro (2005) and Gerlach and Altmann (2013) there is a derivation of the Zipf's law from the Heaps' law, even if it is not straightforward.
>
> In network theory a written text can be represented by a network whose vertices are the words and two vertices are linked if they are adjacent in the text (Dorogovtsev and Mendes 2001; Masucci and Rodgers 2006). A convenient way to model a growing text in network theory is to assume that at each time step t a new word and a fraction of old words $2\alpha \times t$ is introduced in text, possibly preserving the *Eulerianity* of the system (Masucci and Rodgers 2007). Hence in this representation the discrete time t represents the size V of the vocabulary. At each time step, $2\alpha \times t + 1$ words are introduced in the text and the size L of the text is a quadratic function of the vocabulary size, $L(t) = \alpha t^2 + t$, so that at large sizes $L \sim V^2$.

4 Modelling the Semantic Space

The complexity of the system we are considering is large, since it relates the phenomenology of page writing on different topics to the topology of the macroscopic system of the SS.

Here we present a descriptive model that is able to catch the properties of the SS at three different levels of complexity. In particular it is able to reproduce the scale-free cluster size distribution, the exponents for the out and in-degree distribution and the Zipf's law for the word frequency distribution (Zipf 1949). It is a growing stochastic model of content-based network generated by a copy and mutation algorithm. This is intended on one hand to create a multiplicative process *a la* Simon (Simon 1955), to reproduce the scale-free cluster distribution and on the other hand to create, via mutation, very well connected taxonomic clusters to reproduce the very low exponent of the degree distribution. In the overall process a Heaps'-like law (Heaps 1978) for the text growth is imposed to produce correlations between pages and to allow a phase transition and this finally generates the Zipf's law for the word frequency distribution (see the coloured box for more details about the Heaps'-like law).

The Model

We start the model with a few pages of random words. Then at each time-step:

1- we generate a new page as explained at point a.

2- We create M new pages copying M old pages picked up randomly from the old pages and mutate them as explained at point b.

a- When we generate a new page we first extract its size l from a log-normal distribution, with first moment $\bar{l}$ and second moment l_σ. Then we fill the page with some new words from a potentially infinite vocabulary and some old words picked at random from the already written pages. To establish the balance between new and old words we assume a variation of the Heaps' law, frequently used in network theory (Dorogovtsev and Mendes 2013), that states that the growth of the length L of a written text is a quadratic function of the size of the vocabulary t, $L(t) = \alpha t^2 + t$. Then we assign to the page a random number m, so that $1 < m < l$, representing the size of the invariant part of the page.

b-When we mutate a page we keep unchanged the first $m-1$ words of the page, that is the page invariant part. For the last $l-m+1$ words of the page, each word is changed with probability p and it is kept unchanged with probability $1-p$, where p is a random number between 0 and 1. When we change a word we substitute it with an old or a new word considering the balance between vocabulary and text size as in point a.

The important parameters of the model are M and α. M regulates the exponent of the cluster size distribution that goes, increasing the value of M, from -3 to -2. Moreover, increasing M, more connected clusters form and this increases the exponent of the degree distribution from -1.5 to -1. The coefficient α regulates the amount of correlations between the different semantic clusters and this gives the possibility to tune the point of percolation of the system.

The generation of new pages in the model is intended to mimic the appearance of new pages in the encyclopedia, which are formed partially by a new vocabulary and partially by words which are inherited by already existing pages. Besides, the generation of pages via the mutation mechanism allows creating the different growing taxonomies and in this way mimicking the phenomenology observed in the real encyclopedia, where different pages belonging to the same taxonomy differ only by a few rare words.

In Fig. 5, we show that with an opportune choice of the parameters[3] this model can generate a synthetic semantic space with the desired properties.

---

[3] For the page length distribution we used a log-normal distribution, which is the actual Wikipedia article length distribution, but we rescaled the parameters to the size of our toy model. The parameters M, α and p have been tuned manually to obtain the required exponents. In particular the M parameter is strictly related to the cluster size distribution exponent, while α and p are related to the percolation threshold of the system. The exact parameter values are given in the caption of Fig. 5.

5 Discussion

Nowadays understanding the topology of the SS and the dynamics of meaning is a fundamental issue in many fields of knowledge and technology (Menczer 2002). We can think about its value for understanding the dynamics of language and its evolution (Skyrms 2010; Lieberman et al. 2007; Fitch 2007; Petersen et al. 2012; Amancio, Oliveira Jr, and Fontoura Costa 2012), or its relevance in psycholinguistics and psychology (Ratner et al. 1999). Also, apart for being one of the main research topics in semiotics, linguistics and philosophy (Eco 1986), it has recently been a hot topic in artificial intelligence and robotics (Baronchelli et al. 2010). Moreover, there is an active effort in the information systems community to develop semantic-based web research tools (Bizer et al. 2009; Amancio, Oliveira Jr, and Costa 2012; Navigli 2009).

The results of this research shed light upon the topology of the SS, represented as an encyclopedic semantics. The empirical research is unique in two fundamental aspects: the first one is the content analysis of a whole snapshot of Wikipedia, the second one is that this analysis is directional.

The empirical analysis reveals interesting properties of the SS. On one hand, we can observe that the SS cluster size distribution is scale-free. This observation relates the SS to a wide range of scale-free phenomena (Albert and Barabási 2002). On the other hand, we find a peculiar behaviour of the degree distribution. The latter observation relates the SS to a recently observed class of molecular biology phenomena, such as protein networks and more in general of genomic interaction networks (Balcan et al. 2007; Bergmann et al. 2003) and draws a bridge between phenomena whose underlying mechanism is a language, words for the SS and the DNA alphabet for the genomic networks.

Moreover, the fact that the out and in-degree distribution of the SS are very similar has not to be considered in any way a trivial result, which would in some sense make the directional analysis a useless tool. Instead, it helps to contrast the topology of the SS with other knowledge system networks such as citation networks, or axiomatic systems such as the Principia Mathematica by B. Russell, or the Ethics by Spinoza. These present a different out and in-degree distributions (namely, exponential in-degree and scale-free out-degree distributions) which seem to be characteristic of formal systems of knowledge (Masucci 2011). On the other hand, the fact that the SS displays similar scale-free out and in-degree distributions does not mean that the actual out and in- degrees for individual entries are the same. In fact, we find consistent deviations from a linear relationship when correlating k_{in} and k_{out} at each node. These evidences help to depict the highly rhizomic nature of the SS.

Models of content-based network are powerful tools and recently they have attracted the attention of the scientific community especially for their relation to the so called *hidden variable graphs* (Ramasco and Mungan 2008; Sinatra et al. 2010). The presented stochastic model is descriptive and wants to individuate a few simple mechanisms that are able to reproduce some interesting statistical behaviours of the SS. In particular, it is able to capture the statistical properties of Wikipedia at three levels of complexity. At microscopic level it can reproduce the Zip's law for the word frequency distribution, at mesoscopic level it generates the scale-free distribution for the semantic cluster size and finally at macroscopic level it can reproduce the exponents of the out and in-degree distribution. Interestingly enough, we find again that the description of the SS at the model level resembles mechanisms of DNA evolution, where the process involved is of copy and mutation. Such a genetic-semantic parallel does not lack of depth. The fact that the SS structural properties can be reproduced by a model based on copy and mutation mechanisms highlights the organic nature of the SS, where structures form through an articulation of relationships, which are able to generate a hierarchical and efficient knowledge system.

It is straightforward to relate the presented analysis with previous works on Wikipedia based on the analysis of the network of hyperlinks connecting the Wikipedia different entries (Muchnik et al. 2007). In particular one of the questions arising from those works is if the Wikipedia hyperlink network has any relation to the underlying semantic network. Some attempts to answer this question are exposed in Menczer (2002), where a positive correlation is found between hyperlinked pages and their semantic content. In the light of this research we can say that the topology of the SS is drastically different from the ones obtained by the hyperlinks analysis. In the latter case the exponents of the degree distribution are smaller than -2 and linear preferential attachment is recovered (Muchnik et al. 2007; Capocci et al. 2006), revealing a dynamics based on popularity. However this should not be a surprise since the network of hyperlinks is superimposed to the SS of the encyclopedia, so that it does not reflect the topology of the SS, but the structures locally imposed by the writers of the different entries.

We also notice that the topological properties of the SS are different from the ones obtained for dictionary semantics (Sigman and Cecchi 2002; Jesus Holanda et al. 2004). The topological properties of dictionary networks, characterised by scale-free distribution with exponents smaller than -2, seem to be based again on a popularity mechanism and to reflect properties of language use more than the properties of the SS (Violi 2001). In contrast we find that the architecture of SS is scale invariant, hierarchical, it has small-world properties, but it is not associated to a *rich get richer* mechanism for the degree distribution.

In fact the SS structure is keen to be interpreted as an emerging property of a content-based network, where the Zipf's distribution of the content words is a key feature for the resulting topology.

Other efficient approaches to text semantics are based on vector space models (Salton 1989), which are capable to retrieve efficiently the similarity between words, phrases and documents, and in this way to generate a symmetrical distance matrix from a list a text entries. Even if these approaches are very interesting, they don't allow to our knowledge to generate a dynamical system of attribute flows, which allows the creation of a directional network. A direct comparison of vector space model with the Jensen-Shannon Divergence (JSD) is beyond the scope of this research. Anyway a detailed description of the JSD is given in the Appendix.

Acknowledgements. APM was partially funded by the EPSRC SCALE project (EP/G057737/1). VME and EHG acknowledge financial support from MINECO (Spain) and FEDER through projects MODASS (FIS2011-24785) and INTENSE@COSYP (FIS2012-30634).

Appendix: Information Flow between Populations Defined in a Symbolic Attribute Space

In literature there are different ways to compare probability distribution functions (Rényi 1961). A convenient one for the kind of systems we want to study is the Jensen-Shannon divergence (JSD hereafter) (Lin 1991). As we better explain below, we choose it because it is framed in information theory, it takes into account the different sizes of the populations and the probability distributions don't have to be *absolutely continuous* in each other domain (Grosse et al. 2002).

Given two probability distributions $P = \{p_1, p_2, ...\}$ and $Q = \{q_1, q_2, ...\}$ of a discrete random variable, the JSD between P and Q is defined as:

$$JSD(P\|Q) \equiv H(\pi_1 P + \pi_2 Q) - \pi_1 H(P) - \pi_2 H(Q) \qquad (1)$$

where π_i are weights, that is $\pi_1 + \pi_2 = 1$ and $H(P) = -\sum_i p_i \ln p_i$ is the Shannon entropy measured in *nats* (Shannon 2001).

JSD was introduced in Lin (1991) and its properties are well reviewed in Grosse et al. (2002). For our purposes the most important feature of the JSD is that the two distributions we want to compare have not to be absolutely continuous in each other domain, as it happens for instance in the case of the *Kullback-Leibler* divergence (Lin 1991). In fact we want to compare distributions of attributes that are not necessarily shared by all the populations of the system. Moreover the JSD embeds a weighting system for the different distributions and it was demonstrated in Grosse et al. (2002) that the optimal choice for the weights is the statistical weight of the samples. This feature is necessary in order to compare populations that are different in size. Hence if the number of the elements of the population defined by the distribution P is n_1 and the number of elements of the population defined by the distribution Q is n_2, we define $\pi_i \equiv n_i/(n_1 + n_2)$.

It has been demonstrated that the square root of JSD defines a *metric* in the case of populations of the same size, $\pi_1 = \pi_2 = 1/2$, while for different population sizes the triangular inequality has not been demonstrated yet (Briët and Harremoës 2009). Moreover we have that $0 \leq JSD(P\|Q) \leq -\pi_1 \ln \pi_1 - \pi_2 \ln \pi_2 \leq \ln 2$. $JSD(P\|Q) = 0 \Leftrightarrow P = Q$ and $JSD(P\|Q) = -\pi_1 \ln \pi_1 - \pi_2 \ln \pi_2$ if and only if P and Q have disjoint domains.

JSD measures the information flow between two distributions in terms of their shared elements and non-shared elements. To understand the meaning of the JSD we can refer to the example of the two probability distributions P and Q defined in a certain attribute space showed in Fig. 6. P is defined on an attribute domain D_P, while Q is defined on a certain attribute domain D_Q. Let us call $X = D_P \cup D_Q$ and suppose that $J = D_P \cap D_Q \neq \emptyset$ is the joint attribute domain of the two distributions, while $D = X - J$ is the disjoint attribute domain of the distributions. Then Eq. (1) can be split in the two different domains: $JSD(P\|Q) = JSD(P\|Q)_J + JSD(P\|Q)_D$, where $JSD(P\|Q)_D = H(\pi_1 P + \pi_2 Q)_D - \pi_1 H(P)_D - \pi_2 H(Q)_D = -\pi_1 \ln \pi_1 \sum_D p_i - \pi_2 \ln \pi_2 \sum_D q_i$. Then the contribution given to the JSD by the disjoint domains is a statistical measure quantifying the non-shared attribute distribution sizes.

For the part of the joint domain we have that $JSD(P\|Q)_J = -\sum_J (\pi_1 p_i + \pi_2 q_i) \ln(\pi_1 p_i + \pi_2 q_i) + \pi_1 \sum_J p_i \ln p_i + \pi_2 \sum_J q_i \ln q_i$. $JSD(P\|Q)_J$ is the entropy of the weighted sum of the two distributions minus the weighted sum of the entropy of the distributions, measured in the shared part of the attributes domain. From an informational point of view we can say that if the sum of the distributions is broader than the single distributions, it results in a large value of the divergence. Otherwise if the weighted sum of the distribution has a larger informative value, hence a smaller entropy than the one of the single distributions, then we obtain a small divergence from the shared part of the attribute domain.

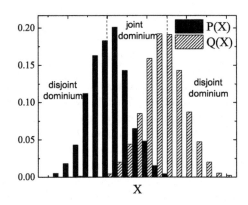

Fig. 6 An example of two probability distributions $P(X)$ and $Q(X)$ defined in an attribute domain X where a fraction J of the domain is shared by the two distributions

The only issue we get through applying the JSD to a system composed by many populations is that its maximum value depends on the population size. That means that we can find cases where the JSD of two uncorrelated distributions is smaller than the one of two correlated ones. To avoid this problem we introduce a new index D defined as the JSD normalised to its maximum value:

$$D(P\|Q) \equiv \frac{JSD(P\|Q)}{-\pi_1 \ln \pi_1 - \pi_2 \ln \pi_2}. \qquad (2)$$

$D(P\|Q)$ has the same properties of $JSD(P\|Q)$ with the difference that $0 \leq D(P\|Q) \leq 1$, where $D(P\|Q) = 0 \Leftrightarrow P = Q$ and $D(P\|Q) = 1 \Leftrightarrow J = \emptyset$.

Directionality

$D(P\|Q)$ as $JSD(P\|Q)$ is a symmetric quantity in its arguments, that is $D(P\|Q) = D(Q\|P)$. Hence it doesn't give information about the directionality of the interaction. In order to infer directionality for the information flow we borrow a rationale from sociology, in particular from the idea of *geographical segregation*.

Geographical segregation is a concept that is widely used in many areas of science, such as sociology (Duncan and Duncan 1955; Schelling 1969), economy (Hutchens 2004), geography (Crooks 2010), physics and biology (Balloux and Lugon-Moulin 2002). It refers to the inequality between population attribute distributions inside of a metapopulation. In particular a population inside of a metapopulation is said to be segregated in respect to some attributes if those attributes are found with a consistent probability in that population and are not found with a significative probability in the other populations of the system.

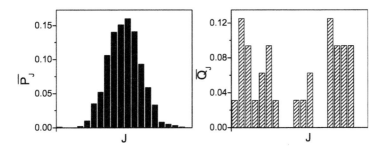

Fig. 7 Following the notation of the example of Fig. 6, here we show the renormalized distribution of the shared elements $\overline{P}_J$ and $\overline{Q}_J$, in the joint domain J. In this particular example it is evident that $\overline{P}_J$ is a peaked distribution, while $\overline{Q}_J$ is uniformly distributed. Hence the distribution $\overline{P}_J$ is more segregated than $\overline{Q}_J$.

There are many indexes in literature to measure geographical segregation (Mora and Ruiz-Castillo 2003). A popular one is the *Theil's segregation index* (Theil and Finizza 1971; Fuchs 1975) and it is based on information theory. The Theil index is the difference between the total entropy of the system in respect to some attributes and the weighted sum of the entropy of the different populations and it is defined as $T \equiv H_T - \sum_i w_i H_i$, where H_T is the total Shannon entropy of the system, H_i is the entropy of population i and w_i is its statistical weight. If T is close to 0 it means that those attributes are not segregated in the system, but they are distributed more or less uniformly through it. If T is consistently larger than 0, it means that those attributes are segregated in one or more populations of the system.

The Shannon entropy is a well defined measure to estimate the amount of inequalities represented by a probability distribution. It is large when the attribute frequency distribution is uniform and it increases with population size. In our case a large entropy for an attribute ensemble represents the fact that different attributes are equally mixed and it is a hint of small segregation in the attribute space. Otherwise a small value of Shannon entropy is associated to a large inequality between attribute frequencies and to a small number of different attributes and it is an evidence of segregation for the population in the attribute space, where exchanges with other populations are a few. Then, in general terms, if an information flow is detected between two populations we can argue that the origin of the information is in the most segregated distribution, where the Shannon entropy is smaller (see Fig. 7).

Hence, given two distributions P and Q between which an information flow is detected, to infer the directionality of the flow we first consider the inequality of the two distributions in the joint domain. To do that we consider the distributions $\overline{P}_J$ and $\overline{Q}_J$, that are the distributions of the elements of P and Q that belong to the joint domain J, with their frequencies renormalized to unity in J (see Fig. 7).

The number of attributes shared by two distributions is the same for both distributions, hence the entropy H measured over the joint domain J depends only on the relative frequencies of the attributes. In particular more peaked distributions have smaller entropy than broader distributions. Then we have to take into account the fact that the population sizes are different. In particular it is important to understand which is the ratio of the shared elements within the whole population. To do that we define the index $\mu_P \equiv \frac{\sum_J p_i}{\sum_X p_i}$, $\mu_Q \equiv \frac{\sum_J q_i}{\sum_X q_j}$ and we have $0 < \mu_{P,Q} \leq 1$. If for a certain distribution μ is close to one, it means that the shared attributes are the dominant part of that sample. Then an estimator for the information flow directionality between P and Q can be defined as

$$I(P \to Q) \equiv -\text{sign}\left[\frac{H(\overline{P}_J)}{\mu_P} - \frac{H(\overline{Q}_J)}{\mu_Q}\right]. \tag{3}$$

If $I(P \to Q) = +1$ the information carried by the attributes in the joint domain of P is larger than the information carried by the attributes in the joint domain of Q. Then we can infer an information flow from the attribute distribution P to Q. Otherwise, if $I(P \to Q) = -1$, we can infer an information flow from the attribute distribution Q to P.

References

Albert, R., Barabási, A.-L.: Statistical mechanics of complex networks. Reviews of Modern Physics 74(1), 47 (2002)

Amancio, D.R., Oliveira Jr., O.N., da F. Costa, L.: Unveiling the relationship between complex networks metrics and word senses. EPL (Europhysics Letters) 98(1), 18002 (2012)

Amancio, D.R., Oliveira Jr., O.N., da Fontoura Costa, L.: Identification of literary movements using complex networks to represent texts. New Journal of Physics 14(4), 043029 (2012)

Balcan, D., Kabakçıoğlu, A., Mungan, M., Erzan, A.: The information coded in the yeast response elements accounts for most of the topological properties of its transcriptional regulation network. PLoS One 2(6), e501 (2007)

Balloux, F., Lugon-Moulin, N.: The estimation of population differentiation with microsatellite markers. Molecular Ecology 11(2), 155–165 (2002)

Baronchelli, A., Gong, T., Puglisi, A., Loreto, V.: Modeling the emergence of universality in color naming patterns. Proceedings of the National Academy of Sciences 107(6), 2403–2407 (2010)

Barwise, J.: Information flow: the logic of distributed systems. Cambridge University Press (1997)

Bastian, M., Heymann, S., Jacomy, M., et al.: Gephi: an open source software for exploring and manipulating networks. In: ICWSM, vol. 8, pp. 361–362 (2009)

Bergmann, S., Ihmels, J., Barkai, N.: Similarities and differences in genome-wide expression data of six organisms. PLoS Biology 2(1), e9 (2003)

Bizer, C., Lehmann, J., Kobilarov, G., Auer, S., Becker, C., Cyganiak, R., Hellmann, S.: DBpedia-A crystallization point for the Web of Data. Web Semantics: Science, Services and Agents on the World Wide Web 7(3), 154–165 (2009)

Borge-Holthoefer, J., Arenas, A.: Categorizing words through semantic memory navigation. The European Physical Journal B-Condensed Matter and Complex Systems 74(2), 265–270 (2010)

Briët, J., Harremoës, P.: Properties of classical and quantum Jensen-Shannon divergence. Physical Review A 79(5), 052311 (2009)

Capocci, A., Servedio, V.D.P., Colaiori, F., Buriol, L.S., Donato, D., Leonardi, S., Caldarelli, G.: Preferential attachment in the growth of social networks: The internet encyclopedia Wikipedia. Physical Review E 74(3), 036116 (2006)

Crooks, A.T.: Constructing and implementing an agent-based model of residential segregation through vector GIS. International Journal of Geographical Information Science 24(5), 661–675 (2010)

Deleuze, G., Guattari, F.: Rhizom, vol. 67. Merve (1977)

Deleuze, G., Guattari, F.: A thousand plateaus: Capitalism and schizophrenia. Bloomsbury Publishing (1988)

Derrida, J.: Margins of philosophy. University of Chicago Press (1982)

Dorogovtsev, S.N., Mendes, J.F.F.: Evolution of networks: From biological nets to the Internet and WWW. Oxford University Press (2013)

Dorogovtsev, S.N., Mendes, J.F.F.: Language as an evolving word web. Proceedings of the Royal Society of London. Series B: Biological Sciences 268(1485), 2603–2606 (2001)

Duncan, O.D., Duncan, B.: A methodological analysis of segregation indexes. American Sociological Review, 210–217 (1955)

Eco, U.: Semiotics and the Philosophy of Language, vol. 398. Indiana University Press (1986)

Ferrer i Cancho, R., Solé, R.V.: Two Regimes in the Frequency of Words and the Origins of Complex Lexicons: Zipf's Law Revisited. Journal of Quantitative Linguistics 8(3), 165–173 (2001)

Fitch, W.T.: Linguistics: an invisible hand. Nature 449(7163), 665–667 (2007)

Fuchs, V.R.: A note on sex segregation in professional occupations. Explorations in Economic Research 2(1), 105–111 (1975)

Gerlach, M., Altmann, E.G.: Stochastic model for vocabulary growth in natural languages. Physical Review X 3(2), 021006 (2013)

Grosse, I., Bernaola-Galván, P., Carpena, P., Román-Roldán, R., Oliver, J., Eugene Stanley, H.: Analysis of symbolic sequences using the Jensen-Shannon divergence. Physical Review E 65(4), 041905 (2002)

Heaps, H.S.: Information retrieval: Computational and theoretical aspects. Academic Press, Inc. (1978)

Hopper, P.J., Traugott, E.C.: Grammaticalization. Cambridge University Press (2003)

Hutchens, R.: One Measure of Segregation*. International Economic Review 45(2), 555–578 (2004)

de Jesus Holanda, A., Pisa, I.T., Kinouchi, O., Martinez, A.S., Ruiz, E.E.S.: Thesaurus as a complex network. Physica A: Statistical Mechanics and its Applications 344(3), 530–536 (2004)

Kim, J., Krapivsky, P.L., Kahng, B., Redner, S.: Infinite-order percolation and giant fluctuations in a protein interaction network. Physical Review E 66(5), 055101 (2002) van Leijenhorst, D.C., Van der Weide, T.P.: A formal derivation of Heaps' Law. Information Sciences 170(2), 263–272 (2005)

Lieberman, E., Michel, J.-B., Jackson, J., Tang, T., Nowak, M.A.: Quantifying the evolutionary dynamics of language. Nature 449(7163), 713–716 (2007)

Lin, J.: Divergence measures based on the Shannon entropy. IEEE Transactions on Information Theory 37(1), 145–151 (1991)

Macdonald, P.J., Almaas, E., Barabási, A.-L.: Minimum spanning trees of weighted scale-free networks. EPL (Europhysics Letters) 72(2), 308 (2005)

Masucci, A.P., Kalampokis, A., Eguíluz, V.M., Hernández-García, E.: Wikipedia information flow analysis reveals the scale-free architecture of the semantic space. PloS One 6(2), e17333 (2011a)

Masucci, A.P.: Formal versus self-organised knowledge systems: A network approach. Physica A: Statistical Mechanics and its Applications 390(23), 4652–4659 (2011)

Masucci, A.P., Kalampokis, A., Eguíluz, V.M., Hernández-García, E.: Extracting directed information flow networks: an application to genetics and semantics. Physical Review E 83(2), 026103 (2011b)

Masucci, A.P., Rodgers, G.J.: Network properties of written human language. Physical Review E 74(2), 026102 (2006)

Masucci, A.P., Rodgers, G.J.: Multi-directed Eulerian growing networks. Physica A: Statistical Mechanics and its Applications 386(1), 557–563 (2007)

Menczer, F.: Growing and navigating the small world web by local content. Proceedings of the National Academy of Sciences 99(22), 14014–14019 (2002)

Montemurro, M.A., Zanette, D.H.: Towards the quantification of the semantic information encoded in written language. Advances in Complex Systems 13(02), 135–153 (2010)

Mora, R., Ruiz-Castillo, J.: Additively decomposable segregation indexes. The case of gender segregation by occupations and human capital levels in Spain. The Journal of Economic Inequality 1(2), 147–179 (2003)

Muchnik, L., Itzhack, R., Solomon, S., Louzoun, Y.: Self-emergence of knowledge trees: Extraction of the Wikipedia hierarchies. Physical Review E 76(1), 016106 (2007)

Mungan, M., Kabakloğlu, A., Balcan, D., Erzan, A.: Analytical solution of a stochastic content-based network model. Journal of Physics A: Mathematical and General 38(44), 9599 (2005)

Navigli, R.: Word sense disambiguation: A survey. ACM Computing Surveys (CSUR) 41(2), 10 (2009)

Petersen, A.M., Tenenbaum, J.N., Havlin, S., Eugene Stanley, H., Perc, M.: Languages cool as they expand: Allometric scaling and the decreasing need for new words. Scientific Reports 2 (2012)

Prim, R.C.: Shortest connection networks and some generalizations. Bell System Technical Journal 36(6), 1389–1401 (1957)

Ramasco, J.J., Mungan, M.: Inversion method for content-based networks. Physical Review E 77(3), 036122 (2008)

Ratner, N.B., Gleason, J.B., Narasimhan, B.: An introduction to psycholinguistics: what do language users know. In: Gleason, J.B., Ratner, N.B. (eds.) Psycholinguistics. Harcourt Brace College, Philadelphia (1999)

Rényi, A.: On measures of entropy and information. In: Fourth Berkeley Symposium on Mathematical Statistics and Probability, pp. 547–561 (1961)

Salton, G.: Automatic Text Processing: The Transformation, Analysis, and Retrieval of Information by Computer. Addison-Wesley (1989)

Samsonovic, A.V., Ascoli, G.A.: Principal semantic components of language and the measurement of meaning. PloS One 5(6), e10921 (2010)

Schelling, T.C.: Models of segregation. The American Economic Review, 488–493 (1969)

Serrano, M.A., Flammini, A., Menczer, F.: Modeling statistical properties of written text. PloS One 4(4), e5372 (2009)

Shannon, C.E.: A mathematical theory of communication. ACM SIGMOBILE Mobile Computing and Communications Review 5(1), 3–55 (2001)

Sigman, M., Cecchi, G.A.: Global organization of the Wordnet lexicon. Proceedings of the National Academy of Sciences 99(3), 1742–1747 (2002)

Simon, H.A.: On a class of skew distribution functions. Biometrika, 425–440 (1955)

Sinatra, R., Condorelli, D., Latora, V.: Networks of motifs from sequences of symbols. Physical Review Letters 105(17), 178702 (2010)

Skyrms, B.: Signals: Evolution, learning, and information. Oxford University Press (2010)

Stauffer, D., Aharony, A.: Introduction to percolation theory. Taylor and Francis (1991)

Steyvers, M., Tenenbaum, J.B.: The Large-Scale Structure of Semantic Networks: Statistical Analyses and a Model of Semantic Growth. Cognitive Science 29(1), 41–78 (2005)

Theil, H., Finizza, A.J.: A note on the measurement of racial integration of schools by means of informational concepts (1971)

Violi, P.: Meaning and experience. Indiana University Press (2001)

Watts, D.J., Strogatz, S.H.: Collective dynamics of 'smallworld' networks. Nature 393(6684), 440–442 (1998)

Zanette, D., Montemurro, M.: Dynamics of text generation with realistic Zipf's distribution. Journal of Quantitative Linguistics 12(1), 29–40 (2005)

Zipf, G.K.: Human behavior and the principle of least effort. Addison-Wesley Press (1949)

Are Word-Adjacency Networks Networks?

Katharina Anna Zweig

Abstract. This article discusses the question of whether word-adjacency relationships are well-represented by a complex network. The main hypothesis of this work is that network representations are best suited to analyze *indirect effects*. For an indirect effect to occur in a network, a *network process* needs to exist that uses the network to exert an indirect effect, e.g., the spreading of a virus in a social network after a small group of persons were infected. Given any sequence of words, it can be represented by a so-called *word-adjacency network* by representing each word by a node and by connecting two nodes if the corresponding words are directly adjacent or at least close to each other in this sequence. It can be easily seen that the result of a speech production process gives rise to a *word-adjacency network* but it is unlikely that speech production uses an underlying word-adjacency network—at least not in any easily describable way. Thus, the results of clustering algorithms, centrality index values, and the results of other distance-based measures that quantify indirect effects cannot be interpreted with respect to speech production.

1 Introduction

Is it a reasonable question to ask whether something called a *network* should actually be represented by a complex network? In general, a *complex network* represents a *relationship* between entities in the form of a *relation* between *nodes*. Mathematically, any kind of relationship can be represented by a complex network; certainly, words that are directly adjacent to each other in some given text are in a clearly defined relationship that can easily be represented as a network (cf. Mehler 2008); however, representing a data set as a complex network is normally a preparation step to apply network analytic methods to it. And the main statement is that for

---

Katharina Anna Zweig
Graph Theory and Complex Network Analysis, Department of Computer Science,
TU Kaiserslautern, Gottlieb-Daimler-Str. 48
e-mail: zweig@cs.uni-kl.de

some complex networks, the resulting values cannot be meaningfully interpreted. The question discussed in this article is whether the *representation* of the *word-adjacency* relationship belongs to this latter kind of complex networks. It can, of course, be argued that word-adjacency networks have already been used successfully in network analytic projects, e.g., by Milo et al. (2004). This is true as long as word-adjacency networks have been used to find so-called *universal structures*, a question of interest mainly for statistical physicists that try to identify *universal laws*, which are thought to be valid in all kinds of complex systems. However, if the complex network is used to understand the specific context of speech planning, the word-adjacency network is likely to be misleading. To understand the difference between the **context-free search for universal structures** and a **contextual interpretation of complex networks**, it is instructive to summarize the recent history of social and complex network analysis.

1.1 Perspectives of Network Analysis

The field of network analysis is an evolving field of study, situated at the border of many different disciplines, foremost sociology and physics. Sociology started to develop methods to analyze social networks in the 1950s; in the late 1990s, physicists started to apply methods from statistical physics to large networks from any kind of complex systems. Both fields have very different aims and perspectives when they analyze networks; these are shortly sketched in the following.

1.1.1 The Shortest History of Social Network Analysis

Blau defined sociology as the field concerned with understanding how the differentiation between people (e.g., by gender, race, status) leads to social structures (Blau 1974). In the beginning, sociology was then concerned with direct statistics of people in a population, e.g., their age, gender, and income, and the analysis of possible correlations between these parameters. In the 1950s, Moreno and others started to look at interactions between people as another parameter to explain social positions (Freeman 2004). This approach evolved into the field of *social network analysis* and was already well-developed in the middle of the 1990s when Wasserman and Faust published their classic textbook on "Social Network Analysis—Methods and Applications". Borgatti et al. (2009) state that social network analysis is mainly concerned with two types of research questions: the first regards the connection between network structure and the social environment they create, and the second focuses on the question of how individuals perceive their social network. A typical social network analysis question might, for example, ask how the communication network of a company and the position of an individual in this network influence his or her income. Another question would be how different communication structures influence the company's overall performance. Social network analysis is thus interested in the context of a social network and tries to correlate external parameters with the knowledge of the network's structure.

1.1.2 The Shortest History of Complex Network Analysis

In the late 1990s, another discipline started to work with network analytic tools, the field of (statistical) physics. Watts and Strogatz published their paper on small-worlds in 1998, closely followed by a paper by Barabási and Albert on the "Emergence of scaling in random network" in 1999. In contrast to social network analysis, the newly christened *complex network analysis* based on statistical physics is not limited to social relations: it tries to understand structures common to all complex systems, be it between proteins, film actors, websites, or animals. The interest of physicists in networks is a consequence of their expertise in the analysis of macroscopic behavior based on local interactions of atoms and molecules in matter like magnets and gases. The pressure and temperature of a gas depends on the distribution of the individual properties of its atoms, and the magnetic field of a piece of matter depends on both, the local interaction between the individual's spin and the external magnetic field. Spin interactions are analyzed based on ideal models in which either all atoms are placed on regular grids (called *lattices*, resulting in a so-called *Ising model*) or one in which all pairs of atoms have a random chance of interacting (*spin glasses*). This historical background is why the second sentence of Watts and Strogatz' paper claims that: "Ordinarily, the connection topology is assumed to be either completely regular or completely random." (Watts and Strogatz 1998). In their ground-breaking paper, they then show that real-world networks show a totally different structure: while they are locally densely connected as expected in a completely regular lattice, the average distance of their nodes is as small as if they were connected in a totally random fashion. Watts and Strogatz already observed this special structure in three very different types of complex networks (a neural network, a co-actor network, and the power grid of the USA.) and in the following it was shown for virtually every complex network. This so-called *small-world* structure thus proved to be a *universal structure*. By definition, a *universal structure* and the *universal laws* which produce it, can be found in every context, they constitute a **context-free finding**, which is valid for a very broad range of different situations.

In summary, the statistical physic's approach to network analysis is to find universal structures that can be found in many different complex networks from very different complex systems. This approach can and should deal with large networks and large sets of networks; the actual method to detect a universal structure is not depending on the context. The classic social network analysis project focuses on a small to medium network to better understand how the individual is embedded in the network; these projects need a deep understanding of the context in which the network emerges. I have observed that network analysis goes awry, when a network is analyzed with a physicist's approach in mind but when the results are later interpreted in a contextual manner like in sociology. It can be shown that word-adjacency networks were successfully used in the context-free quest for universal structures—but that they do not lend themselves to a contextual interpretation as I will argue below.

In the following, I will first give the necessary graph theoretic definitions in Sect. 2, before I generalize a concept by Borgatti regarding the relationship between network flows and centrality indices to a more general relationship between network flows and any kind of walk-based method in Sect. 3. The special case of word-adjacency networks is discussed in Sect. 4. The article closes with a summary in Sect. 5.

2 Definitions

A graph G is a combination of a set of vertices V and a set of edges $E \subseteq V \times V$, i.e., E is a *relation*. If necessary for disambiguitation, $E(G)$ and $V(G)$ denote the edge and vertex set of graph G. $|V(G)|$ is defined as the *order* of G, $|E(G)|$ as its *size*. The graph can be *directed*, i.e., (v,w) might be in E while the converse is not necessarily true. If the graph is undirected, i.e., there is no direction in the relation, edges are written as sets: $\{v,w\}$

The *degree* $deg(v)$ of a node is defined as the number of edges it is contained in: $deg(v) = |\{(v,v_j) \in E\}|$; equivalently, let $N(v)$ denote the set of neighbors w of v, i.e., (v,w) or $(w,v) \in E$, then $deg(v) = |N(v)|$.

A *walk* is a sequence of nodes $v_1, v_2, \ldots, v_k$ such that for any two subsequent nodes v_i, v_{i+1}, $1 \leq i < k$ there is an edge (v_i, v_{i+1}) in $E(G)$; it can thus also be defined as a sequence of edges $e_1 = (v_1, v_2), \ldots, e_{k-1} = (v_{k-1}, v_k)$. The length of a walk in a graph is defined as the number of edges in it; the walk with minimal length between two nodes v, w is called a *shortest path* between v and w, and the length of it is called the *distance* $dist(v,w)$ of v and w. A *trail* is a walk that does not use any edge twice; a *path* is a walk that does not use any node twice.

A *triad* is any subset of three nodes $\{A, B, C\}$. There are in total 4 different triads in an undirected network: an empty and a fully connected one, one with one edge, and one containing two edges. If only connected triangles are regarded in the directed case, there are 13 different triangles (Milo et al. 2004).

2.1 Definition of Word-Adjacency Networks

Word-adjacency networks belong to the large class of *word co-occurrence* networks. Given a set of words W and a list of k corpora $C = \{c_1, c_2, \ldots, c_k\}$, the *undirected co-occurence network* is defined as $G = (W, E(W,C))$ where $\{w_i, w_j\} \in E(W,C)$ if w_i and w_j co-occur in at least one corpus.

Word-adjacency networks make the special requirement that words do not only co-occur in any large text but that they are very close to each other. Let in general k be the allowed number of words between two words such that they will be considered to be *adjacent*. In the strictest sense of *adjacency*, k is set to 0, i.e., w_i and w_j need to be directly following one another in at least one of the texts such that $(w_i, w_j) \in E(W,C)$. Given a large body of text T as a sequence of words, the following procedure will then build the word-adjacency network as a special case of a *word co-occurrence network*: Given T, let $C(T,k)$ be all subsequences of $k+2$ words in T and let $W(T)$ be all single words in T. Then, the word-adjacency network is given by the word co-occurrence network of $W(T)$ and $C(T,k)$.

There are multiple variants of the initial concept of word-adjacency networks, for examply, by thresholding the number of times two words need to co-occur before a link is built or by stemming before building $W(T)$ and $C(T,k)$.

3 Walk-Based Methods and Network Flows

When searching for a universal structure, any kind of method from social network analyis can be applied to any given network—the main question is only if a common structure can be found in all of these different networks. The *interpretation* of why this structure emerges and what its function is in the system of interest is not of prime concern. However, in the social sciences, the interpretation of any finding is most important and depends strongly on the context of the network. In this case, not every method can be meaningfully applied to all networks; the interpretation of any method depends strongly on the research question and the specific network representation—a situation that Dorn, Lindenblatt, and Zweig have called the *trilemma of social network analysis* (Dorn et al. 2012).

How can the result of a network analytic method depend on the research question? The choice and application of a centrality index to a network may serve as an example for this connection. There are many ways to define the *centrality* of a node in a network: the simplest is to rank the nodes by the number of edges they have (*degree centrality*) or by the inverse of the total distance to all other nodes (*closeness centrality*):

$$C_C(v) = \frac{1}{\sum_{w \in V} dist(v,w)}. \tag{1}$$

Another very popular centrality index is the *betweenness centrality* which is defined as:

$$C_{betw}(v) = \sum_{s \neq v, t \neq v} \frac{\sigma_{st}(v)}{\sigma_{st}}, \tag{2}$$

where $\sigma_{st}(v)$ is the number of shortest paths between s and t, containing v and σ_{st} is the number of all shortest paths between them. Note, that despite the very common description of the betweenness centrality saying that it was "counting the number of all shortest paths containing v" it has a complicated normalization per pair of s and t. The description is, however, correct for the *stress centrality*:

$$C_S(v) = \sum_{s \neq v, t \neq v} \sigma_{st}(v). \tag{3}$$

Next to these four, there are at least three dozens different ways to compute the centrality of a node in a network (Koschützki, Lehmann, Peeters, et al. 2005; Koschützki, Lehmann, Tenfelde-Podehl, et al. 2005). In his paper "Centrality and Network Flow", Borgatti brings order to chaos by stating that a centrality index gives the *expected* importance of a node with respect to a certain type of *network flow* or *network process*:

> A key claim made in this paper is that centrality measures can be regarded as generating expected values for certain kinds of node outcomes (such as speed and frequency of reception) given implicit models of how traffic flows, and that this provides a new and useful way of thinking about centrality. (Borgatti 2005)

For example, if the network process of interest works only between direct neighbors, then the degree centrality points to the nodes that have the most influence. If the network process of interest flows only along shortest paths, if it furthermore flows in a serial manner, and between all pairs of nodes, then the node with the highest closeness centrality is the first one which is finished. However, if the centrality index is not matched to the research question, Borgatti warns: " It is shown that the off-the-shelf formulas for centrality measures are fully applicable only for the specific flow processes they are designed for, and that when they are applied to other flow processes they get the 'wrong' answer."

This insight can be easily generalized to all network analytic methods that are designed to measure an *indirect effect*. In this article, an effect is defined as *indirect* if node v changes its state and if node w not directly connected to it also changes its status and there is a sequence of events along a walk in the network that causes this status change of w. Thus, by definition, indirect effects rely on a network process by which these indirect effects are mediated. In the following, a *network process* is defined as:

1. It transports an object or data over entities that are interacting with each other in a certain time; it thus has a *dynamic dimension*.
2. The transport may be *stochastic* or *deterministic*;
3. The process uses certain walks in the network, i.e., it is important to model this *set of used walks*;
4. As already discussed by Borgatti (2005), the process may use only one walk at a time or be able to use multiple walks at the same time, i.e., we differentiate serial vs. parallel working network processes;
5. Following Borgatti (2005), the object or data transported by the process can behave differently while being transported: it may be at only one place at any given time; it may be splittable like money, or it may be copied during the transfer like a virus or information that is spread.

A method is said to be a *walk-based method* if it relies on measuring the length of walks or the frequency with which walks are used. It is obvious that all centrality indices presented above belong to this kind of measure. However, many other methods are explicitly or implicitly based on walks. The clustering coefficient $cc(v)$ of a node v in an undirected graph is, e.g., normally defined as:

$$cc(v) = \frac{2 \cdot e(v)}{deg(v) \cdot (deg(v) - 1)}, \qquad (4)$$

where $e(v)$ is the number of edges between v's neighbors. However, let $P_2(v, w)$ denote the number of walks of length two between v and w, then it can also be rewritten as:

$$cc(v) = \frac{\sum_{w \in N(v)} P_2(v, w)}{deg(v) \cdot (deg(v) - 1)}. \qquad (5)$$

Similarly, most clustering procedures rely on walks of some sort which are devised to identify groups of nodes which are tightly connected—in essence, all the more interesting network analytic methods are walk-based methods in the sense of the definition given above.

3.1 Models of Walks

So far, there are not very many different models of walks used by a network process. Mainly there are the following three models:

1. The set of all walks, trails, or paths of a given length k or up to length k.
2. The set of all *shortest* paths of a given length k or up to a given length k;
3. The set of all random walks of length k or up to length k; a walk is a *random walk* if, at any given node v, the next node in the walk is chosen uniformly at random among all neighbors of v.

It is clear that no network process will strictly adhere to any of these models. Thus it is necessary to find out whether any chosen model of walks is close enough to approximate the walks really used by the network process of interest. For example, data packets are usually routed on shortest paths in the internet unless they find that the queue of the next router is already congested; in the latter case, they will take a small detour. Under very small load, i.e., a very small number of packets to route through the system, the betweenness centrality is a good predictor of the sensitive points in the system, as shown by Holme (2003). However, Holme also showed that by increasing the load just a little bit, the process will already deviate so markedly from using shortest paths that the betweenness centrality is not anymore able to detect the most important nodes in the network.

Applying a walk-based method to some complex network assumes that a network process exists which is using the network whose preferred walks are well approximated by the model of walks used in the method. If no such network process is known that uses a given network to exert indirect effects—it might be more harmful to represent a relation as a network than it does good. The question is now whether word-adjacency networks are the basis for any kind of network process whose walks through the network follow one of these classical models of walks.

4 Word-Adjacency Networks in the Literature

The first to bring word adjacency networks to the attention of the network analysis community was Ferrer i Cancho's and Solé's paper on "The small world of human language" (Cancho and Solé 2001), which will be discussed in more detail below. Another milestone in increasing the word-adjacency network's popularity is the work by Milo et al. (2004), in which they used different word-adjacency networks to analyze their so-called *triad significance profile*. For this analysis, four directed word-adjacency networks ($k = 0$) were built from texts in four different languages and the occurrence of the 13 different types of connected, directed triads were counted. Then, their occurrence was compared to the expected number of the respective triad in a suitably randomized network model by computing the so-called *z-score*. The *z-score* quantifies the significance of a finding with respect to an expected normal distribution; a triad can then either occur as often as expected or be significantly over- or underrepresented. The research showed that all word-adjacency networks gave rise to the same kind of profile, i.e., the same kind of triads were over- or underrepresented in all four cases. This is an interesting result as the languages were not only Indo-European but included Japanese as well. Moreover, Milo et al. showed that the triad significance profile was also significantly different from that of networks originating from other complex systems. This kind of research question and result is typical for the physics-based approach discussed above: Given a large set of networks from very different complex systems, is there any kind of pattern that is common to all of them or that helps to classify data into groups? Again, this approach is context-free and without aiming at a contextual interpretation of the finding. It simply shows that there is a universal structure within word-adjacency networks, without interpreting it.

Cancho and Solé (2001) start similarly by stating that they found that word-adjacency networks are so-called *small-worlds*, i.e., those that have a small average distance and a reasonably large clustering coefficient in comparison to random graphs of the same size and order (Watts and Strogatz 1998). However, according to the above, the average distance of a graph is a walk-based method. It is thus instructive to state explicity to what kind of network process the average distance implicitly refers to:

1. The network process uses only shortest paths;
2. All nodes of the network want to exert an indirect effect onto all other nodes of the network and with the same frequency.
3. The network process is serial, i.e., if there is a path of nodes $a - b - c - d$, then a talks to them independently but via the shortest path.

The question is whether there is any network process on word-adjacency networks that fulfills these criteria. Ferrer i Cancho and Solé interpret the finding as follows:

> In spite of the huge number of words that can be stored by a human, any word in the lexicon can be reached with fewer than three intermediate words, on average. If a word is reached during communication, jumping to another word requires very few steps. Speed during speech production is important and can be more easily achieved if intervening words are close to each other in the underlying structure used for the construction of sentences. (Cancho and Solé 2001).

It is thus indicated that the speech planning process uses the word-adjacency network to navigate. While it is clear that a word-adjacency network is the result of a speech production process, does that mean that the word-adjacency network is used for speech planning or speech production? And if it is: is it the basis for a speech planning process that uses shortest paths in a serial manner—at least approximately? Only in this case does it make sense to interpret the average distance with respect to the speech production. The following argument make this rather unlikely: An average distance smaller than 3 means that there is a combinatorial explosion of reachable words in distance 1, 2, and 3, starting from almost every word. It is an old finding that a high number of possible choices increases the reaction time rather than to decreasing it (Hick's law (Hick 1952)). In that sense, a "large" network would be better for speech planning and a small-world network is unlikely to be the basis for speech production.

If speech planning does not use the word-adjacency network, is there any other network process that uses the word-adjacency network in a meaningful sense? Of course, reading any kind of text uses some walks of this network by definition—sentences are sequences of adjacent words, after all. But the question is whether this set of walks can be modeled easily by any of the above models. Certainly, they are not shortest paths in the word-adjacency network. Meaningful sentences with k words also make only a tiny fraction of all the possible walks of k words in the word-adjacency network, so that they are also not likely produced by a simple random walk over the word-adjacency network—not even approximately so. If there is thus no simple set of walks that approximates how humans build sentences based on a word-adjacency network, it might then be better not to combine the sentences into a complex network but to use *walk-based statistics* instead. A walk-based statistics takes a set of observed walks, i.e., sequences of nodes used by some process, and analyzes it directly without combining them into a network. For example, it was claimed that there are so-called *anomalous airports*, i.e., regional, small airports that are much more central as stop-overs than expected. This was analyzed by using the betweenness centrality on a network that connects two airports if there is at least one scheduled flight between them (Guimerá et al. 2005). Travellers use this network to fly from A to B, where sometimes they have to take a stop-over in one or more airports. However, for this process of interest it does not make sense to use the original betweenness centrality (Dorn et al. 2012). As explained above, the betweenness centrality assumes that there is a process using shortest paths between all pairs of nodes in the network. Not suprisingly, however, travellers have distinct preferences regarding their destinations and their origin is mainly determined by where they live. For example, 40% of all possible origin-destination airports in the USA are not once requested in a three-month period, as Dorn et al. state. Dorn et al. thus proposed to use only the requested pairs of airports based on purchased tickets, and to simply measure the betweenness centrality on the corresponding paths between airports. Similarly, the *stress centrality* can be easily adapted to a set of walks as defined by the sentences of a given text without producing a complex network first. In that sense, the "new" stress centrality could just count how often a word is contained in the set of walks defined by the sentences. However, one can quickly see that such a walk-based statistics is identical to a word frequency statistic and does not necessarily need any network representation.

Note that the case is different with other word-based networks like the word association network: here, two words are connected if people associate them with each other. Many processes seem to be based on word-association, e.g., creativity, which is classically measured by using association tests (Gough 1976). It is less clear how to describe the set of *association chains* ('walks') mathematically, which makes it difficult to apply out-of-the-box network measures like the above named centrality indices; it might also not be meaningful to find out what the 'most central' word in a word-association network is since most walks (association chains) will have a small 'horizon' (length) in the sense of Friedkin (1983). However, since a process is known that uses the network in a transitive way, walk-based methods can be applied if the set of walks is adapted accordingly.

5 Summary

Without any question, the statistical analysis of co-occurrence and co-location of words in texts has produced interesting and productive insights in the last decades. Word-adjacency networks are also interesting in the search for universal laws and universal structures as demonstrated by Milo et al. (2004). The question raised in this article is whether we can learn more about speech or thought related processes by representing text in the form of word-adjacency networks and by applying walk-based network analytic methods to it. While it is possible to represent the word-adjacency relationship as a complex network, there does not seem to be any such process that uses most of the possible walks in this network in a transitive way. Walk-based methods from network analysis should thus not be applied and interpreted with respect to speech or thought generation unless a process is identified which at least approximately uses the walks as defined in the respective method. Finally, if walk-based network analytic methods cannot be applied to a word-adjacency network, a network representation of a sequence of words is not necessary and might otherwise seduce one to apply methods to it that are not appropriate for its analysis. Therefore, word-adjacency networks should not be represented as networks until a convincing network process using it in a meaningful and easily describable way is defined.

References

Barabási, A.-L., Albert, R.: Emergence of Scaling in Random Networks. Science 286(5439), 509–512 (1999)

Blau, P.M.: Presidential Address: Parameters of Social Structure. American Sociological Review 39(5), 615–635 (1974)

Borgatti, S.P.: Centrality and Network Flow. Social Networks 27, 55–71 (2005)

Borgatti, S.P., Mehra, A., Brass, D.J., Labianca, G.: Network Analysis in the Social Sciences. Science 323, 892–895 (2009)

Cancho, R.F.I., Solé, R.V.: The Small World of Human Language. Proceedings of the Royal Society of London B 268, 2261–2265 (2001)

Dorn, I., Lindenblatt, A., Zweig, K.A.: The Trilemma of Social Network Analysis. In: Proceedings of the 2012 IEEE/ACM International Conference on Advances in Social Network Analysis and Mining, Istanbul (2012)

Freeman, L.C.: The Development of Social Network Analysis - A Study in the Sociology of Science. Empirical Press, Vancouver (2004)

Friedkin, N.E.: Horizons of Observability and Limits of Informal Control in Organizations. Social Forces 62, 54–77 (1983)

Gough, H.G.: Studying creativity by means of word association tests. Journal of Applied Psychology 61(3), 348–353 (1976)

Guimerá, R., Mossa, S., Turtschi, A., Amaral, L.A.N.: The Worldwide Air Transportation Network: Anomalous Centrality, Community Structure, and Cities' Global Roles. Proceedings of the National Academy of the Sciences 102, 7794–7799 (2005)

Hick, W.E.: On the Rate of Gain of Information. Quarterly Journal of Experimental Psychology 4(1), 11–26 (1952)

Holme, P.: Congestion and Centrality in Traffic Flow on Complex Networks. Advances in Complex Systems 6, 163 (2003)

Koschützki, D., Lehmann, K.A., Peeters, L., Richter, S., Tenfelde-Podehl, D., Zlotowski, O.: Centrality Indices. In: Brandes, U., Erlebach, T. (eds.) Network Analysis. LNCS, vol. 3418, pp. 16–61. Springer, Heidelberg (2005)

Koschützki, D., Lehmann, K.A., Tenfelde-Podehl, D., Zlotowski, O.: Advanced Centrality Concepts. In: Brandes, U., Erlebach, T. (eds.) Network Analysis. LNCS, vol. 3418, pp. 83–111. Springer, Heidelberg (2005)

Mehler, A.: Large Text Networks as an Object of Corpus Linguistic Studies. In: Lüdeling, A., Kytö, M. (eds.) Corpus Linguistics. An International Handbook of the Science of Language and Society, pp. 328–382. De Gruyter, Berlin (2008)

Milo, R., Itzkovitz, S., Kashtan, N., Levitt, R., Shen-Orr, S., Ayzenshtat, I., Sheffer, M., Alon, U.: Superfamilies of Evolved and Designed Networks. Science 303, 1538–1542 (2004)

Wasserman, S., Faust, K.: Social Network Analysis – Methods and Applications, revised, reprinted. Cambridge University Press, Cambridge (1999)

Watts, D.J., Strogatz, S.H.: Collective dynamics of 'smallworld' networks. Nature 393, 440–442 (1998)

Part III
Syntax

Syntactic Complex Networks and Their Applications

Radek Čech, Ján Mačutek, and Haitao Liu

Abstract. We present a review of the development and the state of the art of syntactic complex network analysis. Some characteristics of such networks and problems connected with their construction are mentioned. Relations between global network indicators and specific language properties are discussed. Applications of syntactic networks (language acquisition, language typology) are described.

1 Introduction

Syntax is considered to be a key component of human language. Its properties, origin, cognitive status etc. have been discussed intensively for decades by researchers from different branches of science and it has caused tough controversies among them. Despite a huge number of arguments it has been difficult, or still impossible, to find a generally acceptable criterion or method which can help to solve fundamental problems considering syntax, especially its origin.

A theory of complex networks emerged at the turn of the millennium (Barabási and Albert 1999; Barabási 2002) and its rapid and successful development appeared to be a useful tool for an explanation of system properties in many branches of science. It is not surprising that a use of this theory for an explanation of some fundamentals of syntax was tempting. And indeed, just after the beginning of the

---

Radek Čech
Department of Czech Language, Faculty of Arts,
University of Ostrava, Reální 5, Ostrava, CZ-70103, Czech Republic
e-mail: cechradek@gmail.com

Ján Mačutek
Department of Applied Mathematics and Statistics,
Comenius University, Mlynská dolina, Bratislava, SK-84248, Slovakia
e-mail: jmacutek@yahoo.com

Haitao Liu
Department of Linguistics, Zhejiang University, Hangzhou, CN-310058, China
e-mail: lhtzju@gmail.com

to study syntax properties by methods of complex network analysis Ferrer i Cancho et al. (2005) brought promising explanation considering syntax origin. Specifically, they introduced complex network based model of language which takes into account (1) relationships between words and objects, (2) relationships among words related to the same object, and (3) Zipf's law; a property of the model (namely, connectedness) represents a precondition for syntax evolving, according to the authors (for more details see Sect. 3). Afterwards, Solé (2005) presented the approach of Ferrer i Cancho et al. in slightly changed form in popular science article in *Nature*, one of the most prestigious scientific journals. Especially the article by Solé represents some kind of "great expectations" (cf. its title: "Syntax for free?") which could bring the use of complex network analysis in language analysis.

After almost a decade, it seems reasonable to critically review the development of syntactic complex network analysis and to try to answer the following questions: What are the results of the application of complex network theory to syntax analysis? Has the application met the expectations? What kind of explanation has complex network analysis of syntax brought? Which new problems have emerged? What is the actual scope of syntax network analysis now? What are the perspectives? In this paper, we attempt to track main aspects of the development of syntactic network analysis and to summarize the results of this scientific endeavor. Our article follows a review presented by Mehler (2007). It can be considered as a complement to a more general overview on network analysis (Baronchelli et al. 2013).

The review is organized as follows: first, main characteristics of syntactic networks are introduced in Sect. 2; then, an early development of syntactic network analysis is presented (Sect. 3) and an impact of syntax on network properties is discussed (Sect. 4); further, important problems related to data preprocessing (e.g., coordination and lemmatization) are discussed in Sect. 5; next, Sect. 6 is dedicated to applications of syntactic networks in language typology and language acquisition; and the article is finalized by Conclusions (Sect. 7).

2 Basic Characteristics of Syntactic Networks

A network is a set of nodes and links. Nodes represent some entities while links represent relationships among nodes. As for syntactic network, nodes usually denote either particular wordforms (e.g., *sing, sings, sang, sung*) or lemmas (in this case all word forms are represented by the canonical form, e.g. *sing, sings, sang, sung* are represented by the single lemma *sing*) (cf. Čech and Mačutek 2009).

Links denote the so-called dependencies, i.e., syntactic relationships between pairs of words. For instance, there are four syntactic relationships in the sentence

(1) *Peter gave Mary the pen,*

specifically, between pairs of words *Peter – gave, gave – Mary, gave – pen,* and *the – pen*. The notion of dependency expresses the fact that

> one wordform must depend on another for its linear position and its grammatical form,

cf. Mel'čuk (2003, p. 188). This approach to syntax is called the dependency grammar formalism Mel'čuk (2003) and Hudson (2007). It is the only syntactic formalism which has been so far exploited for syntactic network analysis. Other ones (e.g., phrase structure or construction grammar) cannot be excluded, in principle; however, due to the lack of linguistic reasoning or interpretation they have not been used, to our knowledge.

A sentence structure reflecting syntactic dependencies between pairs of words can be described by a tree graph (all nodes in the tree graph must be connected and no cycles are allowed in this type of graph, which is in accordance with the syntactic dependency formalism), see Fig. 1.

Fig. 1 The structure of sentence (1). Links between words represent the syntactic dependency relationships, arrows express the direction of the dependency. However, there is no general agreement among linguists regarding the direction; thus, one can find dependency formalisms using opposite direction (from modifier to head).

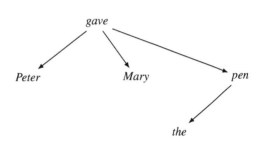

A syntactic dependency complex network is constructed by accumulating sentence structures and, thus, the network is an emergent property of these structures. Specifically, the network contains all words which occur in a text corpus and words are linked if the words appear syntactically linked at least once in a sentence of the corpus (Ferrer i Cancho et al. 2004; Ferrer i Cancho 2005a). Figure 2 shows an example of a small syntactic network containing 50 lemmas.

3 Early Development of Syntactic Complex Network Analysis

The early development of syntactic complex network analysis can be characterized as an endeavor to show that (1) syntactic properties of human language follow some universal patterns that are not observable by an analysis of particular sentences (Ferrer i Cancho et al. 2004; Ferrer i Cancho 2005a; Solé et al. 2010) and (2) that some language properties which are modelled well by complex network represent a necessary precondition for development of full syntax, cf. Ferrer i Cancho et al. (2005) and Solé (2005). As for the former, the patterns emerge only if the language is analyzed from a global point of view as a complex system containing huge number of language units and interrelations among them. Of course, it is nothing new to see the language as a complex system – F. de Saussure (de Saussure 1979) was probably

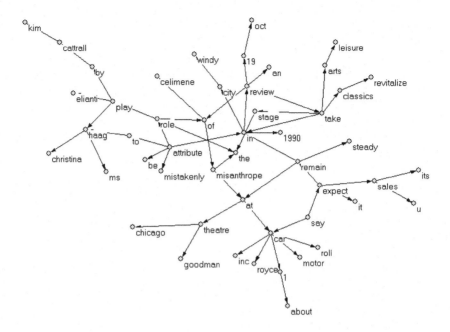

Fig. 2 The network containing the first 50 lemmas from the Penn TreeBank (Surdeanu et al. 2008).

the first one to stress this aspect. What is new, indeed, is the global point of view which is represented by the complex network theory (cf. Newman 2011). In other words, applying network models to language reveals that language networks share the same statistical characteristics, for instance, small world structure, degree distribution, betweenness centrality etc. Moreover, these statistical characteristics appear in networks based on different language units (phonemes, syllables, words) and different kind of relationships among the units (co-occurrence, collocation, syntactic dependency, semantic relationships). Therefore, these statistical characteristics are considered to be universal patterns which could be candidates for new linguistics universals (Ferrer i Cancho et al. 2004; Ferrer i Cancho 2005a; Solé et al. 2010; Choudhury and Mukherjee 2009). Focusing on syntax, this line of thinking brings an important implication. If syntactic networks display the same global characteristics as other linguistic networks, a uniqueness or specificity of syntax is cast into doubt (cf. Nowak et al. 2000, Hauser et al. 2002, and Fitch and Hauser 2004). Consequently, syntax should be ruled by the same (or similar) general principles as other language properties. Further, these principles should be rooted in a general cognitive faculty of human mind (Ferrer i Cancho 2005a, p. 68):

> the structure of global syntactic dependency networks mirrors the structure of brain. It is obvious that the brain is made of millions of neurons connected through synapses but the similarities go beyond mere physical resemblance. The activation of different brain areas shows the small-world phenomenon and a power degree distribution (Eguíluz et al.

2005; Grinstein and Linsker 2005). (...) While no one has ever found a rewriting rule in the brain of a human, the web organization of the brain at many levels, with linguistic networks on top, cannot be denied.

Naturally, this kind of findings opens questions on the origin of syntax. Complex network analyses of many different systems (World Wide Web, social networks, biochemical networks, ecological networks, neural networks etc.) reveal that a complex structure of these systems is a result of self-organization which is based on relatively simple principles, e.g., continuous growing of the system and preferential attachment (Barabási and Albert 1999). Specifically, the outcome of self-organized phenomena is a scale-free power-law distribution of degrees (i.e., numbers of links connected to each node). Analogously, the origin of syntax can be considered to be an outcome of the same principles.

As we noted in Sect. 1, Ferrer i Cancho et al. (2005) used the complex network as a model of a certain combinatorial property of words, namely, their connectedness, which is considered to be a necessary precondition for full syntax. According to Ferrer i Cancho et al. (2005), the connectedness arises naturally from the Zipf's law, independently of details of the linguistics setting. Even though it is stated that the model does not correspond to the complexity of human language, only a "small step" from the model to full syntax and full human language is supposed, cf. Ferrer i Cancho et al. (2005, p. 562)

> For various reasons, our grammar is not a grammar in the strict sense, but rather a protogrammar, from which full human language can *easily*[1] evolve (...) although our model is obviously much simpler than present-day languages, it provides a basis for the astronomically large number of sentences that human speakers can produce and process.

This aspect of the approach of Ferrer i Cancho et al. (2005) is strongly emphasized by Solé who links the property of the model to emerging of *syntactic rules*, cf. Solé (2005)

> ...sometimes illogical and quirky appearance of syntactic rules might be nothing but a by-product of scale-free network architecture. (...) ... Zipf's law could have been a precondition for syntax and symbolic communication.

According to us, a difference between 1) a conception of the complex network model as a necessary *precondition* for syntax, which is the most important outcome of the study (Ferrer i Cancho et al. 2005), and 2) a direct relationship between complex network properties and full syntax claimed by Solé (2005), is fundamental. In other words, the impact of complex network properties to syntax evolving radically differs, if one (1) consider these properties of complex network to be necessary, but not sufficient condition for full syntax, or (2) consider them to be both necessary and sufficient condition.

Regardless of the scope of the interpretation, the reference to the Zipf's law represents one of the most important aspects of the attempt to explain the origin of syntax by network analysis. The Zipf's law states that the relationship between the

---
[1] Emphasized by the authors of this review.

frequency of a word in a text and its rank is approximately linear when plotted on the double logarithmic scale, which means that the word frequency distribution is a power law. In Ferrer i Cancho et al. (2004) it is shown that the relationship between word frequency and word degree in all observed syntactic networks is approximately linear and, more interestingly, both distributions, those of word frequencies as well as of word degrees, have approximately the same exponent. Based on these observations, Ferrer i Cancho concludes that word degree distribution could be a consequence of word frequency and asks a fundamental question (Ferrer i Cancho 2005a, p. 66):

> If word degree is a consequence of the Zipf's law for word frequencies, a pressing question is: what is the origin of that law?

Since Zipf (1949) it has been known that power law distribution of word frequencies could be explained by general communication principles, such as the principle of least effort (Zipf 1949) or communication requirements (Köhler 1986; Köhler 2005). Roughly speaking, these principles are based on the idea that a general communication strategy is to minimize the cost of word usage, on the side of speaker, and the cost of word perception, on the side of hearer. These competitive strategies lead to an equilibrium that has the main impact on the form of the language system in general. The Zipf's law represents one kind of the equilibrium, cf. Ferrer i Cancho (2005b) and Ferrer i Cancho and Solé (2003).

To sum up, the early development of syntactic network analysis reveals a relationship between frequency of word and its syntactical properties expressed by the degree distribution. However, the mere relationship between these two phenomena does not explain the emerging of syntactic rules, i.e., if there is a linear correlation between the word frequency and word degree in a syntactic complex network, another question appears: what is the role of syntax in syntactic networks, if the word degree can be seen as a consequence of word frequency? Attempts to answer this question are presented in the next section.

4 Role of Syntax in Syntactic Dependency Complex Networks

Traditionally, syntax is considered to be, roughly speaking, a set of rules which govern the behavior of words in a sentence. We emphasize that the rules should be understood as probabilistic, not deterministic. The aim of syntax as a science is to describe a sentence structure, the character of rules and, sometimes, to explain why both, the structure and rules, are as they are. Regardless of different approaches to syntax, there is a general agreement among linguists about the hierarchical sentence structure – it means that "behind" the linearity of a sentence there are grammatical relationships between words. This fact is clearly illustrated on the example of highly inflected languages which have a flexible word order, cf. six grammatically well-formed Czech sentences expressing the English sentence *Peter beats Paul*:

(2)   *Petr bije Pavla* [Petr$_{\text{noun-subject-nominative}}$ – bije$_{\text{verb}}$ – Pavla$_{\text{noun-object-accusative}}$]

(3) *Pavla bije Petr* [Pavla$_{\text{noun-object-accusative}}$ – bije$_{\text{verb}}$ – Petr$_{\text{noun-subject-nominative}}$]

(4) *Petr Pavla bije* [Petr$_{\text{noun-subject-nominative}}$ – Pavla$_{\text{noun-object-accusative}}$ – bije$_{\text{verb}}$]

(5) *Pavla Petr bije* [Pavla$_{\text{noun-object-accusative}}$ – Petr$_{\text{noun-subject-nominative}}$ – bije$_{\text{verb}}$]

(6) *Bije Petr Pavla* [Bije$_{\text{verb}}$ – Petr$_{\text{noun-subject-nominative}}$ – Pavla$_{\text{noun-object-accusative}}$]

(7) *Bije Pavla Petr* [Bije$_{\text{verb}}$ – Pavla$_{\text{noun-object-accusative}}$ – Petr$_{\text{noun-subject-nominative}}$]

In all these instances, the object of the sentence *Pavla (Paul)* is determined by its accusative form which is a result of the syntactic rule; the object of the sentence is not determined by the word order. Similarly, syntactic rules also determine the dependency of *Pavla (Paul)* on the verb. Consequently, according to the dependency grammar formalism, the structure of sentences (2)–(7), i.e. the relationships between each pair of words, is expressed by the graph in Fig. 3, in accordance with dependency grammar formalism.

Fig. 3 The hierarchical structure of sentences (2)–(7).

but not by graphs in Fig. 4 (of course, there are more possibilities of non-correct graphs).

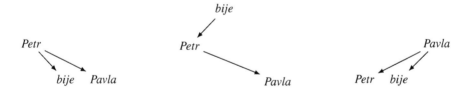

Fig. 4 Examples of non-correct hierarchic representations (i.e. representations violating syntactic rules) of the structure of sentence *Petr bije Pavla*.

Remember, a syntactic dependency complex network is constructed by accumulating sentence structures (cf. Sect. 2) and it was assumed (Solé 2005) that

> sometimes illogical and quirky appearance of syntactic rules might be nothing but a by-product of scale-free network architecture.

If Solé's statements were true, one should find radical differences between a network based on "proper" syntactic rules, on the one hand, and a network based on relationships between words which are not ruled by syntax (e.g, if relationships are generated randomly), on the other.

The idea to investigate the role of syntax in syntactic dependency networks was introduced by Liu and Hu (2008). According to them, the fact that language networks built on different principles are small-world and scale-free just like other real networks opens a problem with respect to a relation between these global network indicators and specific language properties. They focused on syntax and asked the following questions:

> if all language networks have properties such as small-world and scale-free: Could they be viewed as a general feature of a language network? What role does syntax play in such a syntactic (language) network? If dependencies are built by randomly linking words in the same sentence, would the network still follow the properties similar to the syntactic one? Can the local (micro) syntactic analysis in a sentence be reflected in the global (macro) properties of a language network?

As an attempt to answer these questions, Liu and Hu (2008) com-pared properties of three networks: 1) a syntactic network generated from a treebank based on the dependency grammar formalism; 2) a network based on randomly generated relationships between each pair of words within a sentence (sentences were taken from the treebank): from each sentence one word was randomly selected as a root (in accordance with the dependency grammar formalism) and for each of the remaining words a word of the same sentence was randomly selected as a governor; 3) a network based on randomly generated relationships between each pair of words from a sentence which, however, respect the principle of continuity (also called projectivity) of a tree representing a sentence structure; this condition is added because discontinuous (or non-projective) syntactic trees are rather exceptional in the natural language (Ferrer i Cancho 2006), so, this network should express more similar properties to a syntactic network than the totaly random network.

A comparison of the networks was focused on five global network characteristics, namely, the average path length, which is defined as the shortest distance between a pair of vertices; the diameter, which is defined as the longest shortest path in a network; the average degree, i.e. the average number of links of a node; the clustering coefficient, i.e. the probability that two nodes which are neighbors of a given node are neighbors themselves; and the degree distribution. The analysis revealed that all networks displayed very similar global network characteristics and all of them were scale-free and small-world. Consequently, the fact that a network is small-world or scale-free cannot alone explain the role of syntax in the network. Obviously, this result seriously undermines the Solé's statement which considers the emergence of syntactic rules to be a by-product of the scale-free network architecture (Solé 2005, similarly also Solé et al. 2010). The properties of small-worldness and scale-freeness in random syntactic networks can be explained by frequency characteristics of words (the Zipf's law), not by syntax. A mere presence of these statistical indicators in a syntactic network is not a consequence of syntactic properties of language and, therefore, it cannot be considered to be a candidate for a *syntactic* language universal. Results of Liu and Hu (2008) reveal that network properties can be considered, at most, to be just the necessary *precondition* for the emergence of syntax (in the sense of a set of rules). Consequently, it is not appropriate to suppose that full syntax can "easily evolve" as soon as the precondition is satisfied (Ferrer i Cancho et al. 2005; Solé 2005).

Interestingly, a similar attempt to build a random network (i.e., without predefined syntactic rules) was presented by Corominas-Murtra et al.[2] (2009, 2010). The creation of a random network was based on two principles: 1) the frequency of words followed the Zipf's law, and 2) the length of a sentence corresponded to real language data. The relationships between words were generated randomly. The resulting random network had similar typical global network properties, such as the clustering coefficient, the degree distribution or the average path lengths, like real language syntactic networks. This fact was interpreted as a proof of emerging syntax properties in a language evolution "for free" and led the authors to more general arguments on the language evolution: specifically, its non-adaptive nature and innateness of syntax.

However, there are several problems in the analysis. First, to interpret the emergence of network properties as a result of innateness of syntax is not appropriate for the following reasons: 1) there is not a clear and empirically proved connection between the emergence of network properties and innateness; 2) both small number of children and only one language (English) is analyzed; the results from so tiny a sample cannot be interpreted in such a general way, for obvious reasons.

Further, the authors do not take into account a potential impact of the length of the samples on network properties; in other words, the emergence of observed network characteristics could be a side-effect of the fact that as time goes on, children speak more and the length of their transcripts increases. Next, the model is based on the assumption that the exponent of the Zipf's law is equal one and that it remains constant as a child evolves. However, this assumption is not correct (cf. Baixeries et al. 2013); the change of the exponent can lead to different network characteristics. Yet another problem is the absence of statistical testing.

Finally, in the light of the study Liu and Hu (2008), conclusions presented in Corominas-Murtra et al. (2009) and Corominas-Murtra et al. (2010) seem to be not acceptable because the global network properties emerge even if syntactic rules are "deleted" by randomness involved in the process of the network creation. In other words, the emergence of these global network properties is a result of frequency characteristics of language, and does not have to do anything with syntax.

On the other hand, Liu and Hu (2008) obtained results which show that there are some differences between syntactic and random networks (e.g., a syntactic network has a lower average degree and clustering coefficient). Even though these differences are too small for classifying syntactic and random networks as different types of complex networks, their existence indicates some influence of syntax on these indicators. This means that a use of complex networks for a syntactic analysis has to be focused on more fine-grained network properties in order to be useful. Moreover, it should not be forgotten that a syntactic network analysis (as well as any language network analysis) should be based linguistically which means that it is necessary to

---

[2] Both papers, i.e., Corominas-Murtra et al. (2009) and Corominas-Murtra et al. (2010), contain the same data and introduce the same procedure; actually, in Corominas-Murtra et al. (2009) one can find a more thorough discussion and a more general explanation.

explain observed network properties with regard to certain language characteristics or (in the better case) a linguistic theory. In other words, to say that, for instance, "syntax is small-world" without linguistic grounds can be misleading, as the analysis by Liu and Hu (2008) uncovered. The same conclusion was presented in the analysis by Liu et al. (2010) focused on an impact of different annotation schemes used for capturing syntactic coordination on global network properties.

In accordance with the approach from Liu and Hu (2008) and Liu (2008), which emphasizes the need for a linguistic explanation of network properties, the analysis of the role of syntax in syntactic complex networks in Čech et al. (2011) focused on verb characteristics. Contradicting previous studies that revealed the linear relationship between frequency and degree in a syntactic network in general, it was hypothesized that verbs should play a central role (expressed by the degree of node) in a syntactic network not due to their frequency but because of their syntactic property called valency (Allerton 2005; Liu 2008).

The paper Čech et al. (2011) starts with the well-known idea on the relationship between the shape of a network (its topological properties) and its functionality (Caldarelli 2007) and, further, it was deduced how a function of a verb in a sentence should influence network characteristics. Specifically, verb valency determines, besides other things, the number of words obligatory dependent on the verb in a sentence and it plays a decisive role in the sentence structure. Moreover, a verb is present at least once in each sentence due to its syntactic function to be the predicate of the sentence – this fact guarantees a relatively high frequency of verbs in any language sample. However, it should be emphasized that verbs are not the most frequent part of speech (nouns have the highest frequency, at least in languages used for the analysis). Based on these properties of verbs (i.e., valency and a relative high frequency caused by its function in a sentence), it was predicted (Čech et al. 2011, p. 3616) that

> verbs should play an important role in the network expressing syntactic relationships in the language. In other words, it is predicted that verbs will occur among the most important elements of the network.

The importance of an element was determined by its degree.

Six languages (Catalan, Czech, Dutch, Hungarian, Italian, and Portuguese) were used for testing the hypothesis. The results reveal that proportions of verbs (with regard to other parts of speech) in histogram bins of the ranked distributions of degrees tend to decrease while the proportions of verbs (again with regard to other parts of speech) in histogram bins of the rank-frequency distribution of lemmas are more or less constant and clearly tend to attain lower values than verb proportions in the case of degrees. Differences between rank-degree and rank-frequency distributions are statistically significant. Thus, the results do not falsify the hypothesis and allow to state that the topology of a syntactic dependency network is significantly affected by syntax of the language, at least in the case of verb valency.

To sum up, studies focused on the analysis of the role of syntax show that it is not acceptable to interpret and explain syntax properties of human language by global network properties, such as the small-worldness and scale-freeness. Further, a network should be used as a tool for a linguistically well-grounded research in order to avoid some mistakes which can be caused, for instance, by a non-proper analogy with another kind of network analyses. However, one has to bear in mind that all studies focused on this topic represent only first steps in an unexplored area.

5 Preprocessing of Data for a Syntactic Complex Network Analysis – Pitfalls to be Avoided

Any complex network represents a relatively simple model of an observed system. In many cases, it is not difficult to determine both units, which are represented by nodes of the network, and relationships, which are represented by links connecting the nodes: e.g., World Wide Web, sexual relationships among people, a co-occurrence network of word forms. However, the analysis of syntactic networks is not the case. Even though the dependency grammar formalism brings general principles for sentence parsing, the variability of particular parsing systems is rather large. Further, the majority of syntactic networks analyses uses syntactic treebanks as the source of language data; the treebanks are language corpora containing a syntactic annotation which is usually processed automatically. One should keep in mind that there is not a unique annotation scheme for automatic parsing and that different annotation schemes lead to different results (Boyd and Meurers 2008). To illustrate that these differences are not negligible for network analysis, we present various approaches to coordination (Liu et al. 2010) and the problem of lemmatization.

Coordination (Crysmann 2006, p. 183) is

> a combination of like or similar syntactic units into some larger group of the same category or status, typically involving the use of a coordinating conjunction, such as *and* or *or*, to name just two. The units grouped together by means of a coordinating conjunction are usually referred to as conjuncts (or conjoints).

This phenomenon is a difficult point especially for dependency syntax, in which binary asymmetrical relations are basic elements (Lobin 1993; Osborne 2003; Temperley 2005). The problem is that all members of a coordination group fill one syntactic "slot", in fact. For instance, there is one accusative object (*Mary*) in the sentence (8)

(8) *I see Mary,*

while in the sentence (9)

(9) *I see Peter, Paul, and Mary,*

the accusative object – again one(!) object as well as in sentence (9), from the syntactic point of view – is represented by three members (*Peter, Paul, Mary*). As the development of syntactic studies reveals, it is impossible to find "the best" annotation scheme of this phenomenon which would be broadly accepted among linguists.

Following Liu et al. (2010), we can parse sentence (9) in three different ways, as is presented in Figure 5. It should be emphasized that each parsing is linguistically well grounded (Tesnière 1959; Schubert 1987; Mel'čuk 1988; Liu and Huang 2006; Hudson 2007).

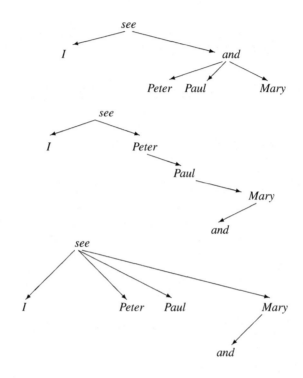

Fig. 5 Three different representations of the structure of (9).

We remind again that a syntactic dependency network is constructed by accumulating sentence structures, such as are in Fig. 5. At the first sight it is obvious that syntactic networks based on different annotation schemes will have different structures. Specifically, the first approach (Fig. 5 on the top) will "favor" (with regard to the node degree) the conjunction, the approach from Fig. 5 in the middle will lead to a more uniform distribution of links in the network, while graph in the bottom will "favor" verbs." Naturally, the choice of an annotation scheme can strongly influence tests of hypotheses, for instance the analysis of verbs presented in Sect. 4.

Not only differences among approaches to syntactic relationships can significantly influence a syntactic network analysis. In addition, the annotation of linguistic units, which are represented by nodes in a syntactic network, should always be clearly presented. There are relatively few problems if one uses word forms. However, often word forms are not suitable units for the analysis, for many reasons. For instance, if one aims at working with word as a semantic unit, the use of lemmas is a more reasonable approach. At the first sight, the use of lemmas is unproblematic; a lemma represents the canonical form for all word forms, so, for instance, the lemma *sing* represents the word forms *sing, sings, sang, sung*. However, the situation becomes more problematic when polysemy enters into the play; cf. different meanings of the word *school* in sentence (10), (11), and (12):

(10) *I visited her school*

(11) *This linguistic school has influenced the history of science in the U.S.A.*

(12) *My experience is drawn from the school of life*

Can all these occurrences of the word *school* be represented by the unique lemma? Or by two lemmas, one denoting a building and the other denoting an abstract notion? Or even by three different lemmas because the meaning of *school* seems not to be the same in sentences (11) and (12)? There are good linguistic reasons to follow all of these three approaches. This fact is reflected by different methods of lemmatization: some corpora annotate differences caused by polysemy (but the question is how fine-grained the annotation is), while others not. As in the case of coordination, it is obvious that different lemmatizations lead to different network characteristics.

Coordination and lemmatization are obviously not the only phenomena which can be (and actually they are) annotated differently in particular corpora – in fact, they were used just for an illustration of a more general problem. We emphasize the impact of the annotation scheme because this factor seems to be, in our opinion, neglected in language complex network analyses. To avoid shortcomings of this kind, it is necessary to know details of language data used for the analysis, especially if one's goal is to use complex networks for comparative studies (e.g., language typology). The cases of coordination and lemmatization show that the problem of preprocessing data for the network analysis is not trivial and that it needs to be taken seriously. In the ideal case, in any analysis of this kind a technical report should be enclosed which would provide an opportunity to critically analyze presented results.

6 Applications of Syntactic Complex Networks to Language Typology and Acquisition

Syntactic networks are not applied only to syntax studies. The complex network theory (Newman 2011) offers many ways how to analyze global network properties (not only small-world and scale-freeness) which can be used to compare individual systems modeled by the network. There are currently two main directions in the applications of syntactic network analysis: language typology and language acquisition.

6.1 Language Typology

Despite the problems related to the role of syntax (we mean full syntax properties of present languages) in syntactic complex networks presented in Sect. 4, an application of global network characteristics to language typology seems to be fruitful. It is no surprise, if one realizes that some global network characteristics are connected to word frequency and the Zipf's law (see Sect. 4); many indexes based on word frequency were used for language typological studies (Popescu, Altmann, et al. 2009; Popescu, Mačutek, and Altmann 2009; Popescu et al. 2010; Popescu et al. 2011). However, syntactic networks express more than frequency characteristics and, consequently, their analysis could enhance typological characteristics of languages.

Studies focused on a typological interpretation of syntactic network properties take global characteristics of syntactic networks and exploit a cluster analysis to evaluate (dis)similarities among observed languages (Liu and Li 2010; Liu and Xu 2012). Satisfactory results of these studies justify this approach, even though some linguistic reasons (e.g., the impact of word frequencies vs. syntax properties) causing the differences remain unclear. From a linguistic point of view, a more interesting approach is represented by studies which add an explanation of a linguistic meaning of particular global network characteristics. Specifically, Čech and Mačutek (2009) analyze in detail a potential impact of grammar on differences between lemma and word form syntactic networks (the average degree and clustering coefficient are scrutinized) and try to determine which properties should be influenced by a typological character of language, on the one hand, and which by language usage (e.g., genre differences), on the other. Further, Liu and Xu (2011) compare lemma and word form syntactic networks of 15 languages and show how differences among global network characteristics reflect morphological variation degrees and a morphological complexity. Finally, Abramov and Mehler (2011) explain and discuss the linguistic meaning of particular global network characteristics thoroughly whenever it is possible and, consequently, offer a deeper insight to the issue (11 languages were used for the analysis).

It is important to note that there are great similarities between language typology (i.e., language classification) and text classification; one conducts text classification when the texts in question are from different genres of the same language and language classification or typological identification of languages when the texts in question are from different languages. This fact can have a significant impact on the results of a complex network based language typology. Methodologically, there is a fundamental problem related to a usage-based language typology analysis in general, in the previous studies (Liu and Li 2010; Liu and Xu 2011; Liu and Xu 2012; Abramov and Mehler 2011). Specifically, the syntactic dependency networks in these studies are based on language data which are not necessarily consistent in semantic content and genre. The basic assumption of language classification based on syntactic dependency networks is that the topological similarities and differences of these networks (manifested by their complex network parameters) reflect the similarities and differences of the corresponding languages. However, heterogeneity in the semantic content and genre of the language data selected, which is independent of the similarities and differences of the languages, may also contribute to the topological similarities and differences of the corresponding syntactic dependency networks and thus may affect the results of language classification. Therefore, a more desirable type of language data for complex network based language classification are parallel texts (i.e., a collection of texts with the same semantic content but in different languages, e.g., a novel plus its translations in different languages), which are consistent in both semantic content and genre (c.f. Kelih 2009). However, a requirement to analyze parallel texts is constrained by technical reasons, up to now; even though it is possible to annotate language data automatically, tools for the annotation are usually not easily available and a manual annotation of syntax for a network analysis is almost impossible because a huge amount of data is needed. In addition, an existence of parallel treebanks (e.g., the Prague Czech-English Dependency Treebank 2.0; SMULTRON - Stockholm MULtilingual Treebank; GRUG Parallel Treebank) is rather exceptional and their range (expressed by the number of languages) is too small.

6.2 Language Acquisition

The complex network theory enables not only to model and interpret real systems from a global point of view but also to analyze a global dynamic behavior of the systems. Consequently, the use of network analysis for modeling language acquisition should be no surprise, because language acquisition is nothing else than a dynamic process. What is more surprising is a relatively rare application of the network analysis in this branch of science; one could expect that a successful application of this methodology to modeling other dynamic systems should trigger its usage in this research area as well. Moreover, existing results are promising (Ninio 2006; Ke and Yao 2008; Corominas-Murtra et al. 2009; Corominas-Murtra et al. 2010; Hills et al. 2009). According to us, both an unfamiliarity of scientists focused on the language acquisition with the network theory and a relative technical difficulty (especially in the case of syntactic networks) are the main reasons of this state.

Ninio (2006) was the first who tried to use the network theory for modeling language acquisition focusing on syntax, to our knowledge. Despite an interest of syntax, her approach cannot be interpreted as an analysis of syntactic networks in the sense as is presented here (cf. Sect. 2). Specifically, she uses bipartite networks (this kind of network contains different classes of nodes, here one class represents speakers while the other words) to model language behavior of both mothers and children and then observes distributions of words, verbs etc. Discovered similarities of distributions (all are power-law) lead her to a conclusion as follows (Ninio 2006, p. 141):

> Application of Complexity Theory to language development sheds new light on the stance of the learner vis-à-vis the linguistic environment. It sees language as a network of speakers and the speech items they produce which children join when they, too, start to produce similar items. Developmental data shows that children act just like

Google when it searches the Web: they pick popular items, but only if their content is relevant for them. The results support a view of children as free agents exercising Preferential Attachment when they develop their minds and acquire knowledge in a social environment.

Corominas-Murtra et al. (2009, 2010) focused on dynamics of a large-scale organization of the use of syntax. Particularly, they used language data from child's corpora, parsed them and, finally, modeled a development of syntactic networks based on this data. As a result, they observed dynamics of the syntax behavior of children between theirs 22 and 28 months. Their analysis reveals both two different regimes of children's syntactic behavior and a sharp transition between these regimes. Specifically, a tree-like organization of a syntactic network before the transition (around 24 month) is suddenly replaced by a much larger, heterogeneous network which has global properties similar to adults networks. Further, the transition is accompanied by a strong reorganization of the network; in the pre-transition stage degenerated lexical items, such as *it*, are words with the highest degree while after the transition functional items, such as *a* or *the*, replaced them. Even though author's conclusions regarding the innateness are not convincing, according to us, and despite some methodological problems of the analysis (cf. Sect. 4), the study shows that syntactic network analysis can bring interesting findings which contribute to a better understanding of the process of language acquisition, at least heuristically.

7 Conclusion

Syntax is one of the main components of the human language system. However, due to the lack of means, the emergence of syntax was difficult to study in the past. Nowadays, complex networks provide a feasible tool. Therefore, in recent years, the construction of syntactic networks and investigations of their properties have become important and interesting fields in language research. Promising results on syntactic global network characteristics were achieved in some branches of linguistics, especially in language typology and language acquisition. On the other hand, the development revealed also several pitfalls (some of which are, admittedly, already at least partially solved):

1. There are some exaggerated interpretations of (almost ubiquitous) small-worldliness and scale-freeness of language networks (cf. Solé 2005, Solé et al. 2010, Corominas-Murtra et al. 2009, and Corominas-Murtra et al. 2010).

    The problem is more of a historical than a scientific character, as studies of random language complex network clearly show (Liu and Hu 2008) that the global properties of language complex networks are a consequence of word frequencies rather than syntax. Therefore, the complex network properties are a necessary, not a sufficient condition for syntax. However, given that the paper Solé (2005) was published in one of the most prestigious scientific journals (*Nature*), it remains to be influential.

2. Automatic parsing is problematic in case of coordination – there are several linguistically substantiated possibilities which result in dramatically different representations, cf. Sect. 5.
3. Similarly, semantics also has an impact on syntactic complex networks (cf. the example of polysemy in Sect. 5).

Consequently, we allow ourselves to summarize some challenges which a researcher using syntactic complex network faces:

1. Particular characteristics of syntactic networks need more in-depth linguistic interpretations.
2. Properties of full syntax of present-day language and their impact on network characteristics should be analyzed in detail.
3. Either a universal syntactic dependency-based parsing formalism, which would be a basis for a more detailed study on syntactic networks, must be searched for, or, at least, one should take into account several possibilities of parsing and then compare results.
4. The relationship between syntactic and cognitive networks should be investigated.

These open problems are symptoms of the development, rather than indications that complex networks should not be used to study the human language syntax. We are convinced that the analysis of global syntactic dependency networks is a helpful tool in language research. It can contribute – under the condition that the results obtained be linguistically interpretable and interpreted – to a deeper understanding of basic and important issues of the human language and cognition, such as, for example, the emergence of syntax, syntactic relations, and connections between syntax and cognition.

Acknowledgements. The authors are thankful to Ramon Ferrer-i-Cancho and to the anonymous referees. Their comments significantly improved the article.

Radek Čech, Ján Mačutek and Haitao Liu were supported by the Czech Science Foundation (grant no. P406/11/0268 Historical semantics), by the VEGA grant agency (grant no. 2/0036/12) and by the National Social Science Foundation of China (grant no. 11&ZD188), respectively.

References

Abramov, O., Mehler, A.: Automatic language classification by means of syntactic dependency networks. Journal of Quantitative Linguistics 18, 291–336 (2011)

Allerton, D.J.: Valency grammar. In: Brown, K. (ed.) The Encyclopedia of Language and Linguistics, pp. 4878–4886. Elsevier (2005)

Baixeries, J., Elvevåg, B., Ferrer i Cancho, R.: The evolution of the exponent of Zipf's law in language ontogeny. PLoS ONE 8(3), e53227 (2013)

Barabási, L.-A.: Linked: The New Science of Networks. Perseus, Cambridge (2002)

Barabási, L.-A., Albert, R.: Emergence of scaling in random networks. Science 286, 509–512 (1999)

Baronchelli, A., Ferrer i Cancho, R., Pastor-Satorras, R., Chater, N., Christiansen, M.H.: Networks in cognitive science. Trends in Cognitive Sciences 17, 348–360 (2013)

Boyd, A., Meurers, D.: Revisiting the impact of different annotation schemes in PCFG parsing: A grammatical dependency evaluation. In: Proceedings of the Workshop on Parsing German, pp. 24–32 (2008)

Caldarelli, G.: Scale-Free Networks: Complex Webs in Nature and Technology. Oxford University Press, Oxford (2007)

Čech, R., Mačutek, J.: Word form and lemma syntactic dependency networks: a comparative study. Glottometrics 19, 85–98 (2009)

Čech, R., Mačutek, J., Žabokrtský, Z.: The role of syntax in complex networks: local and global importance of verbs in a syntactic dependency network. Physica A: Statistical Mechanics and its Applications 390, 3614–3623 (2011)

Choudhury, M., Mukherjee, A.: The structure and dynamics of linguistic networks. In: Ganguly, N., Deutsch, A., Mukherjee, A. (eds.) Dynamics on and of Complex Networks: Applications to Biology, Computer Science and the Social Sciences, pp. 145–166. Birkhäuser, Boston (2009)

Corominas-Murtra, B., Valverde, S., Solé, R.V.: The ontogeny of scale-free syntax networks: Phase transition in early language acquisition. Advances in Complex Systems 12, 371–392 (2009)

Corominas-Murtra, B., Valverde, S., Solé, R.V.: Emergence of scale-free syntax networks. In: Nolfi, S., Mirolli, M. (eds.) Evolution of Communication and Language in Embodied Agents, pp. 83–101. Springer, Heidelberg (2010)

Crysmann, B.: Coordination. In: Brown, K. (ed.) The Encyclopedia of Language and Linguistics, pp. 183–196. Elsevier, Oxford (2006)

de Saussure, F.: Cours de linguistique générale. Payot, Paris (1979)

Eguíluz, V.M., Chialvo, D.R., Cecchi, G.A., Baliki, M., Apkarian, A.V.: Scale-free brain functional networks. Physical Review Letters 94, 018102 (2005)

Ferrer i Cancho, R.: The structure of syntactic dependency networks: insights from recents advances in network theory. In: Altmann, G., Levickij, V., Perebyinis, V. (eds.) Problems of Quantitative Linguistics, pp. 60–75. Ruta, Chernivtsi (2005a)

Ferrer i Cancho, R.: Zipf's law from a communicative phase transition. European Physical Journal B 47, 449–457 (2005b)

Ferrer i Cancho, R.: Why do syntacitc links not cross? Europhysics Letters 76, 1228–1235 (2006)

Ferrer i Cancho, R., Riordan, O., Bollobás, B.: The consequences of Zipf's law for syntax and symbolic reference. Proceedings of the Royal Society of London Series B 272, 561–565 (2005)

Ferrer i Cancho, R., Solé, R.V.: Least effort principle and the origins of scaling in human language. Proceedings of the National Academy of Sciences of the USA 100, 788–791 (2003)

Ferrer i Cancho, R., Solé, R.V., Köhler, R.: Patterns in syntactic dependency networks. Physical Review E 69 (2004)

Fitch, W.T., Hauser, M.D.: Computational constraints on syntactic processing in a nonhuman primate. Science 303, 377–380 (2004)

Grinstein, G., Linsker, R.: Synchronous neural activity in scale-free network models versus random network models. Proceedings of the National Academy of Sciences of the USA 102, 9948–9953 (2005)

Hauser, M.D., Chomsky, N., Fitch, W.T.: The faculty of language: what is it, who has it and how did it evolve? Science 298, 1569–1579 (2002)

Hills, T.T., Maouene, M., Maouene, J., Sheya, A., Smith, L.: Longitudinal analysis of early semantic networks: preferential attachment or preferential acquisition? Psychological Science 20, 729–739 (2009)

Hudson, R.A.: Language Networks: The New Word Grammar. Oxford University Press, Oxford (2007)

Ke, J., Yao, Y.: Analyzing language development from a network approach. Journal of Quantitative Linguistics 15, 70–99 (2008)

Kelih, E.: Slawisches Parellel-Textkorpus: Projektvorstellung von "Kak zakaljalas' stal' (KZS)". In: Kelih, E., Levickij, V.V., Altmann, G. (eds.) Methods in Text Analysis, pp. 106–124. Ruta, Chernivtsi (2009)

Köhler, R.: Zur linguistischen Synergetik. Struktur und Dynamik der Lexik. Quantitative Linguistics, vol. 31. Brockmeyer, Bochum (1986)

Köhler, R.: Synergetic linguistics. In: Altmann, G., Köhler, R., Piotrowski, R.G. (eds.) Quantitative Linguistics. An International Handbook, pp. 760–775. de Gruyter, Berlin (2005)

Liu, H.: The complexity of Chinese syntactic dependency networks. Physica A: Statistical Mechanics and its Applications 387, 3048–3058 (2008)

Liu, H., Hu, F.: What role does syntax play in a language network? EPL 83 (2008)

Liu, H., Huang, W.: A Chinese Dependency Syntax for Treebanking. In: Proceedings of the 20th Pacific Asia Conference on Language, Information and Computation, pp. 126–133. Tsinghua University Press, Beijing (2006)

Liu, H., Li, W.: Language clusters based on linguistic complex networks. Chinese Science Bulletin 55, 3458–3465 (2010)

Liu, H., Xu, C.: Can syntactic network indicate morphological complexity of a language? EPL 93 (2011)

Liu, H., Xu, C.: Quantitative typology analysis of Romance languages. Poznań Studies in Contemporary Linguistics 48, 597–625 (2012)

Liu, H., Zhao, Y., Huang, W.: How do local syntactic structures influence global properties in language networks? Glottometrics 20, 38–58 (2010)

Lobin, H.: Koordinationssyntax als prozedurales Phänomen. Narr, Tübingen (1993)

Mehler, A.: Large text networks as an object of corpus linguistic studies. In: Lüdeling, A., Kytö, M. (eds.) Corpus Linguistics. An International Handbook, pp. 328–382. de Gruyter (2007)

Mel'čuk, I.A.: Dependency syntax: Theory and Practice. State University of New York Press, Albany (1988)

Mel'čuk, I.A.: Levels of Dependency in Linguistic Description: Concepts and Problems. In: Ágel, V., Eichinger, L.M., Eroms, H.W., Hellwig, P., Herringer, H.J., Lobin, H. (eds.) Dependency and Valency. An International Handbook of Contemporary Research, vol. 1, pp. 188–229. de Gruyter, Berlin (2003)

Newman, M.E.J.: Networks: An Introduction. Oxford University Press, Oxford (2011)

Ninio, A.: Language and the Learning Curve: A New Theory of Syntactic Development. Oxford University Press, Oxford (2006)

Nowak, M.A., Plotkin, J.B., Jansen, V.A.A.: The evolution of syntactic communication. Nature 404, 495 (2000)

Osborne, T.: The Third Dimension: A Dependency Grammar Theory of Coordination for English and German. PhD thesis. Pennsylvania State University (2003)

Popescu, I.-I., Altmann, G., Grzybek, P., Jayaram, B.D., Köhler, R., Krupa, V., Mačutek, J., Pustet, R., Uhlířová, L., Vidya, M.N.: Word Frequency Studies. de Gruyter, Berlin (2009)

Popescu, I.-I., Čech, R., Altmann, G.: The Lambda-Structure of Texts. RAM-Verlag, Lüdenscheid (2011)

Popescu, I.-I., Mačutek, J., Altmann, G.: Aspects of Word Frequencies. RAM-Verlag, Lüdenscheid (2009)

Popescu, I.-I., Mačutek, J., Kelih, E., Čech, R., Best, K.-H., Altmann, G.: Vectors and Codes of Text. RAM-Verlag, Lüdenscheid (2010)

Schubert, K.: Metataxis: Contrastive Dependency Syntax for Machine Translation. Foris, Dordrecht (1987)

Solé, R.V.: Syntax for free? Nature 434, 289 (2005)

Solé, R.V., Corominas-Murtra, B., Valverde, S., Steels, L.: Language networks: Their structure, function and, evolution. Complexity 15, 20–26 (2010)

Surdeanu, M., Johansson, R., Meyers, A., Màrquez, L., Nivre, J.: The CoNLL-2008 shared task on joint parsing of syntactic and semantic dependencies. In: CoNLL 2008: Proceedings of the 12th Conference on Computational Natural Language Learning, pp. 159–177 (2008)

Temperley, D.: The Dependency Structure of Coordinate Phrases: A Corpus Approach. Journal of Psycholinguistic Research 34, 577–601 (2005)

Tesnière, L.: Eléments de la syntax structurelle. Klincksieck, Paris (1959)

Zipf, G.K.: Human Behavior and the Principle of Least Effort. An Introduction to Human Ecology. Addison Wesley, Cambridge (1949)

Function Nodes in Chinese Syntactic Networks

Xinying Chen and Haitao Liu

Abstract. Based on two syntactic dependency networks derived from two Chinese treebanks of different registers, a statistical study is conducted regarding word frequency and distributions. We chose three grammatical (function) words as our research objects and analyzed their network features, including degree, out-degree, in-degree, closeness, in-closeness, out-closeness and betweenness. Then we removed these three word nodes from the networks so as to see what consequences may follow in the number of vertices, average degree, average path length, diameter, the number of isolated vertices, domain and density. The results showed that all three function words are central nodes of the Chinese syntactic networks but have different status, since their influence to the overall structure is quite different. The research provides not only a new way for the study on Chinese function words but also a method for examining the influence of node characteristics to a complex network.

1 Introduction

Network approaches are attracting increasing interest in contemporary linguistic research, which is mainly due to two reasons: Firstly, language is physiognomically a network and modeling of language should follow this guiding principle, and secondly, computational tools that have proven to be successful in sociology and computer science may prove to be valid in the research of language networks.

The key interest of the network approach to linguistic research is that it provides a new way to analyze language systems. A central assumption of modern linguistics is that language is a system (De Saussure 1916; Kretzschmar 2009). This widely

---

Xinying Chen
School of International Study, Xi'an Jiaotong University, Xi'an, China
e-mail: chenxinying@mail.xjtu.edu.cn

Haitao Liu
Department of Linguistics, Zhejiang University, Hangzhou, CN-310058, China
e-mail: lhtzju@gmail.com

accepted point of view, however, has remained purely theoretic due to the absence of an operational methodology, until corpora and modern network analysis tools appeared. As a system, language is expected to include rules that cannot be predicted directly on the basis of the linguistic units (e.g., words, phrases, characters, or syllables). So looking at some specific words (or the relationship between them) may not be an efficient way for discovering the global features of a language system. Modeling language as a network provides an operational way for observing the macroscopic features of a language system and the relationships between the units and the whole system. For example, the network approach can be used for determining the function or status of some units, such as words, in the language system as a whole.

Secondly, the network approach agrees with long-standing linguistic theories and finds supports in empirical data. De Saussure (1916) emphasized the importance of the relationship between language units which can be understood as links in networks. Then Lamb and Newell's stratificational grammar (Lamb and Newell 1962) put forward a syntactic system of which the central idea is that *a language is a network of relationships*. Hudson (2007) even named his book on word grammar *The Language Network*. The idea is not new and it is widely accepted. In addition to the assumption that languages are networks, linguistic research based on authentic language data is become more and more popular since great efforts, monetary and personal, go into all kinds of data-driven natural language processing, ranging from machine translation to text classification. Many tools have been developed to facilitate collecting language data, building corpora and parsers, annotating and detecting errors. More and more authentic language data are available for all kinds of linguistic research. All these data can be used as sources of inducing language networks, given the language network approach a solid data foundation.

So far, much research has been carried out, mainly concerned with the structure of syntactic dependency networks (Ferrer i Cancho 2005; Liu 2008; Chen and Liu 2011; Čech et al. 2011), the patterns in syntactic dependency networks (Ferrer i Cancho et al. 2004; Chen et al. 2011), language development or language evolution (Ke and Yao 2008; Mukherjee et al. 2013; Mehler et al. 2011), language clustering and linguistic categorization (Liu 2010; Liu and Cong 2013; Gong et al. 2012; Abramov and Mehler 2011), manual and machine translation (Amancio et al. 2008; Amancio et al. 2011), word sense disambiguation (Christiano and Raphael 2013), communication and interaction (Banisch et al. 2010; Mehler et al. 2010), the structure of semantic networks (Borge-Holthoefer and Arenas 2010; Liu 2009), phonetics (Arbesman et al. 2010; Yu et al. 2011), morphology (Čech and Mačutek 2009; Liu and Xu 2011), parts of speech (Ferrer i Cancho et al. 2007), Knowledge Networks (Allee 2007), cognitive networks (Mehler et al. 2012).

Works on Chinese language include networks at different levels: networks taking as nodes the Chinese characters (Li and Zhou 2007; Peng et al. 2008), words and phrases (Li et al. 2005), phoneme and syllables (Yu et al. 2011; Peng et al. 2008), syntactic structure (Liu 2008; Liu 2010; Chen and Liu 2011; Chen et al. 2011), semantic structure (Liu 2009), etc.

In general, the language network research, including those devoted to Chinese language, is developing rapidly in recent years. But there is surely vast area to be explored, with abundant issues awaiting investigation. Up till now, it seems that most of the language networks studies put a heavy emphasis on common features of various networks, such as 'small world' (Watts and Strogatz 1998) and 'scale-free' (Barabási and Bonabeau 2003) features, treating alike different levels of language and different concerns on which the networks are built. At the same time, many language networks were built without proper guide of a specific linguistic theory, such as the co-occurrence networks of words, characters, or phrases (Li and Zhou 2007; Peng et al. 2008; Li et al. 2005; Liu and Sun 2007), lacking a strong connection to existing linguistic theories and research. But as more and more linguists get involved in the study of language networks, this situation is gradually changing.

In this paper, we present a function words study which is based on Chinese dependency syntactic networks, aiming to find different properties of language networks or different behavior models of specific nodes in language networks rather than focusing on the common features.

2 The Chinese Dependency Networks for This Study

The basic idea underlying dependency networks is very simple: instead of viewing the trees as linearly aligned on the sentences of the corpus, we fuse together each occurrence of the same word to a unique node, thus creating a unique and (commonly) connected network of words, in which the tokens are the vertices and dependency relations are the edges or arcs. This connected network is then ready to undergo common network analysis with tools like UCINET (Borgatti et al. 2002), PAJEK (Nooy et al. 2005), NETDRAW (Borgatti 2002), or CYTOSCAPE (Shannon et al. 2003).

In reality, extracting a network from a dependency treebank is slightly more complicated, as we have to use some heuristics to fuse together only the words that belong to the same lexeme (same category, near meaning). We refer to Liu (Liu 2008) for a description of multiple ways of network creation from dependency treebanks.

For the present work, we used a Chinese treebank of 37,024 tokens, which is composed of 2 sections of different styles:

- "新闻联播" xin-wen-lian-bo 'news feeds', hereinafter referred to as XWLB, is a transcription of a famous Chinese TV news program. The style of text is quite formal. The section contains 17,061 words.
- "实话实说" shi-hua-shi-shuo 'straight talk' (name of a famous Chinese talk show), hereinafter referred to as SHSS, is more colloquial, containing spontaneous speech appearing in interviews of people of various social backgrounds. The section contains 19,963 words. The text of this section was based on the transcription provided by the program group of SHSS. The transcription consists in plain text that includes the marks of speaking persons. For our study, we

used only the text that contains the conversation sentences without the marks of speaking persons.

Both sections have been annotated manually as described in Liu (2006). Table 1 shows the file format of this Chinese dependency treebank, which is similar to the CoNLL dependency format, although a little more redundant (double information on the governor's POS) to allow for easy exploitation of the data in a spreadsheet and converting to language networks. The data can be represented as a dependency graph as shown in Fig. 1.

Table 1 Annotation of a sample sentence in the Treebank: *zhe-shi-yi-ge-bo-luo* 'this is a pineapple'

Sentence	Dependent			Governor			Dependency
Order	Order	Character	POS	Order	Character	POS	type
S1	1	这 *zhe*	pronoun	2	是 *shi*	verb	subject
S1	2	是 *shi*	verb	6	.	punctuation	main governor
S1	3	一 *yi*	numeral	4	个 *ge*	classifier	compl. of classifier
S1	4	个 *ge*	classifier	5	菠萝 *boluo*	noun	attributer
S1	5	菠萝 *boluo*	noun	2	是 *shi*	verb	object
S1	6	.	punctuation				

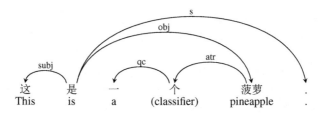

Fig. 1 The graph of the dependency analysis of *zhe-shi-yi-ge-bo-luo* 'this is a pineapple'

The POS tagging and dependency annotation is done on the transcribed texts. As the treebank are built with texts of very different styles, the research findings, despite the limited size of the treebank, may be seen as generally applicable to the language as a whole, not register specific. Another benefit of the double nature of the data is that we can do comparative work based on these 2 sections.

With words as nodes, dependencies as arcs, and the frequency of the dependencies as the value of arcs, we can build a network. For example, the dependency tree in Fig. 1 can be converted to a network as shown in Fig. 2 (excluding punctuation).

Following the same principle, our Chinese treebank can be converted into a network as shown in Fig. 3, a visualization that gives an impression of the global structure of the treebank.

Function Nodes in Chinese Syntactic Networks 191

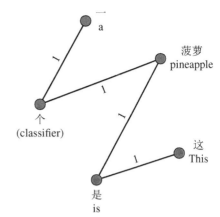

Fig. 2 Network of *zhe-shi-yi-ge-bo-luo* 'this is a pineapple'

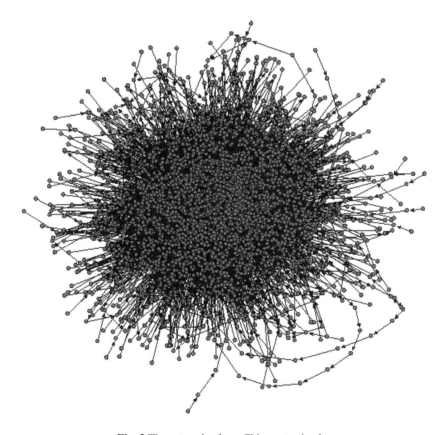

Fig. 3 The network of our Chinese treebank

The resulting network has the following properties: it is fully connected and there are no isolated vertices, it is a *small word* and has a *scale-free* structure. As we mentioned before, only limited language characteristics can be directly derived from this big picture. What we need to do is to look into the structure, concentrating on some specific words in this big network, an approach that has brought about some interesting findings (Čech et al. 2011). The first step is to decide on the words we wanted to look into: function words.

3 Chinese Function Words

Chinese is an isolating language, relying primarily on function words and word order rather than rich morphological information to encode functional relations between elements (Levy and Manning 2003). This means that there is (with the exception of plural) no distinction between word forms and lemmas. In the following, therefore, we simply use the term 'word'. Function words are generally words that express grammatical relationships among other words with a sentence or specify the attitude or mood of the speaker (Klammer et al. 2000), having little lexical meaning. In Chinese, function words include prepositions, conjunctions, and auxiliary and modal particles (Yu 1998).

As in any language, function words distinguish themselves not only by their syntactic properties, but also simply by their frequency. The words we are interested in are among the most common Chinese words: 的 *de* 'ablative cause suffix or possessive particle similar to the English genitive marker 's', 在[1] *zai* '(to be located) in or at', 了 *le* 'perfective aspect marker or modal particle intensifying the preceding clause'.

We compared the frequent function words in XWLB, SHSS, and the *Modern Chinese Frequency Dictionary* and found that there are three function words that appear in all these three resources, namely *de*, *zai* and *le*. The frequency information of these three function words is shown in Table 2[2] and these three function words are the objects we want to observe in the language networks.

The differences in word distribution between the two kinds of texts are mostly due to the lexical poverty of spontaneous speech (SHSS), resulting in higher frequencies (of the smaller number of types) in SHSS. Moreover, the notably higher relative frequency of *le* in SHSS can be explained by the fact that the usage of *le* as an intensifier is typical for spontaneous oral language. Inversely, *zai* can be omitted before locatives in oral Chinese.

---

[1] In Chinese, *zai* may be a verb, adverb or preposition. Here we only refer to the preposition.
[2] Considering the size of XWLB and SHSS, we only paid attention to the function words whose frequency is in the top 30 of all words that have shown in these transcriptions.

Table 2 The frequency information of the three function words *de*, *le* and *zai*. *R*: rank, F_1: frequency, *W*: function word, F_2: frequency in 10,000, MCFD: Modern Chinese Frequency Dictionary*.

XWLB			SHSS			MCFD		
R	F_1	*W*	*R*	F_1	*W*	*R*	F_2	*W*
1	930	*de*	1	1,051	*de*	1	69,080	*de*
3	223	*zai*	6	429	*le*	2	26,342	*le*
4	202	*le*	21	124	*zai*	6	13,438	*zai*

* In Chinese, *le* and *zai* also can be content words even though these phenomena are not common. The Modern Chinese Frequency Dictionary doesn't distinguish these difference but we believe the deviation of the data won't change the fact that these two function words are among the most common Chinese words.

4 Chinese Function Words in the Language Networks

4.1 Network Properties of Chinese Function Words

With the XWLB and SHSS syntactic networks, we studied several important network parameters: *degree, out-degree, in-degree, closeness, in-closeness, out-closeness, and betweenness*.

The degree of a vertex (a word) refers to the number of its neighbors. This variable specifies the number of different word types which are connected with a specific word. The directions of the arcs make the distiction between in-degree and out-degree. The in-degree of a word node is the number of arcs it receives while the out-degree is the number of arcs it sends. Reformulated linguistically, the in-degree reflects the number of governors of a word, and the out-degree reflects the number of a word's dependents.

The term 'closeness' is short for *closeness centrality*. It is measured by the inverse of the sum of the lengths of the paths to every other vertex. The larger the sum, the smaller the closeness. The same holds for in-closeness and out-closeness, save that in-closeness only counts the paths that point to the vertex in question, while the out-closeness only counts the paths that point away from it. In linguistics, these indices may describe the constructive complexity of language units that include a specific word, the vertex, because, as shown in Fig. 4, the paths in a syntactic network correspond to the vertical tree structures in treebanks and therefore somewhat reflect the layers of the language structures.

The more layers the language structures have, the more complicated they are to process for humans and machines. Since closeness, in-closeness, and out-closeness are related to the paths, they may be able to describe the constructive complexity of language units: higher closeness probably indicate smaller syntactic constructive complexity of language units including a specific word.

'Betweenness' is short for *betweenness centrality*. It is a measure of a node's centrality in a network. It is measured by the number of shortest paths that pass through

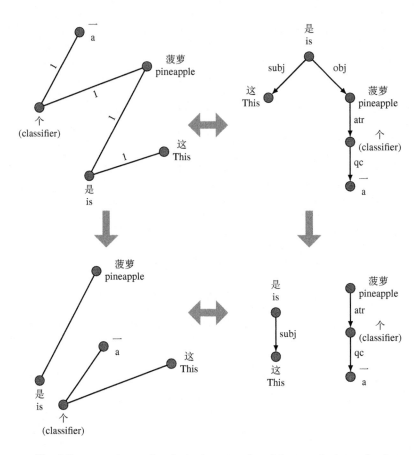

Fig. 4 Correspondence of paths in the network and the trees in the treebank

that node. Linguistically, the betweenness reflects the importance and unavoidability of a word in the whole language network or system.

These network properties of the three function words are shown in Table 3.

In our study, although the sizes of the original texts of XWLB and SHSS are similar in terms of word tokens, the sizes of the XWLB and SHSS networks are quite different due to the difference in the lexical richness (word types). In order to make the data more comparable, we standardized the original data. The results are shown in Table 3. The table clearly shows that:

- *le* has a zero out-degree, out-closeness, and betweenness because it cannot govern other words in our analysis of Chinese while *zai* and *de* have both in-degree and out-degree.
- *de* has the maximum value, that is, 1, for degree, in-degree, closeness, in-closeness, and betweenness for both XWLB and SHSS. These parameters show that *de* is the most important word, the center, in the syntactic networks.

Table 3 The frequency information of the three function words *le*, *de* and *zai*. R: rank, F_1: frequency, W: function word, F_2: frequency in 10,000, SD: standardized degree, SOD: standardized out-degree, SID: standardized in-degree, SOC: standardized closeness, SOC: standardized out-degree, SIC: standardized in-degree, SB: standardized betweenness, MCFD: Modern Chinese Frequency Dictionary*.

Features	le		zai		de	
	XWLB	SHSS	XWLB	SHSS	XWLB	SHSS
degree	133	234	222	131	964	830
SD	0.14	0.28	0.23	0.16	1	1
out-degree	0	0	88	61	504	405
SOD	0	0	0.17	0.12	1	0.82
in-degree	133	234	134	70	460	425
SID	0.29	0.55	0.29	0.16	1	1
closeness	0.35	0.44	0.41	0.44	0.50	0.56
SC	0.70	0.79	0.82	0.79	1	1
out-closeness	0	0	0.26	0.28	0.37	0.40
SOC	0	0	0.65	0.60	0.92	0.85
in-closeness	0.16	0.23	0.18	0.21	0.22	0.28
SIC	0.72	0.83	0.83	0.78	1	1
betweenness	0	0	0.03	0.01	0.32	0.27
SB	0	0	0.09	0.05	1	1

* In Chinese, *le* and *zai* also can be content words even though these phenomena are not common. The Modern Chinese Frequency Dictionary doesn't distinguish these difference but we believe the deviation of the data won't change the fact that these two function words are among the most common Chinese words.

- *le* has zero betweenness, which means that it is not a global central node of the networks. However, it does have a high degree which indicates it as a local central node.
- In many ways, *zai* is a word somewhere between *de* and *le*. It has a high degree which means that it is a local central node. At the same time, its closeness and betweenness are higher than that of *de* and lower than that of *le*. So its importance for the global structure is also between these two words.
- *le*'s degree is higher in SHSS than in XWLB, which shows that the combinatory possibilities of *le* are more diverse in spontaneous speech. On the contrary, words that *zai* can connect with are more diverse in written style, especially when it comes to the in-degree, or rather, the words can *zai* syntactically depend on.

4.2 Network Manipulation

One advantage of the language network model is that it views a language as a connected system. Without the language network approach, describing the language system is more like talking about an unspecified abstract structure. The language network model gives a more specific structure to the language system and provides different computational tools that have proven to be successful in sociology and computer science, able to describe the different elements of, as in our case, a language system. So we tried to manipulate the XWLB and SHSS networks to find out the roles of the three chosen function words in the language networks systems. The way we tried follows a very simple logic. If you want to know the function of one element in a system, the simplest way is to remove it from the system and see what the consequences are. We respectively removed the vertices representing *zai*, *de*, and *le* from XWLB and SHSS language networks to see what consequences these removals may bring about with regard to some important network features, including *the number of vertices, the number of isolated vertices, average degree, the average path length, and the diameter* before and after removing the vertex. The results are shown in Table 4.

Table 4 The network data before and after removing the function words. *Num: Numbers of vertices, IV: Isolated Vertices, AD: Average Degree.*

	network	Num	IV	AD	APL	DR
XWLB	original	4,011	0	6.15	3.58	12
	de removed	4,010	42	5.67	3.93	12
	le removed	4,010	0	6.09	4.56	20
	zai removed	4,010	17	6.04	4.59	20
SHSS	original	2,601	0	8.56	3.05	9
	de removed	2,600	57	7.92	3.25	10
	le removed	2,600	0	8.38	3.95	13
	zai removed	2,600	5	8.46	3.96	13

The numbers of vertices are the numbers of word types in the treebank. Although the sizes of XWLB and SHSS are similar, the numbers of vertices of XWLB and SHSS networks, or the sizes of the networks, are obviously different due to the difference in lexical richness.

The isolated vertices are those vertices without any neighbors. This is particularly interesting. According to the data, removing the most frequent word *de* caused the most isolated vertices in both XWLB and SHSS but there are no isolated vertices after removing *le*. All the remained vertices are still fully connected. So, if we see the network as the model of the syntactic structure of the language as revealed in the treebank, then removing *le* seems to cause no significant trouble here. The whole structure didn't suffer from a systematic collapse, even though *le* was a high frequency word with very high degrees. At the same time, removing *zai* caused isolated vertices in XWLB and SHSS networks, though the *zai* has lower frequency than *le* in the treebank and lower degrees in the network. In other words, removing this word node from the network led to systematic collapse. The reason is simple: *le* can only be a dependent. As Fig. 5 shows: In the simple full connected network there is a vertex A that only has in-degree and no out-degree. Therefore, vertex A is dependent and requires a governor vertex. Furthermore, since you can't reach any other vertex via A, A doesn't convey any unique information between its neighbors. In other words, removing A from the network won't render any vertex isolated. In most of network-based linguistic studies, language networks were treated as undirected so it is considered that removing *le* from undirected networks may give a more abstract picture of the overall importance of this word. But even if we treated the network in Fig. 5 as undirected, removing vertex A still does not create any isolate vertex. It seems that whether removing a word node would cause a systematic collapse is irrelevant with the types of links, directed or undirected, in a network.

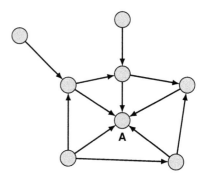

Fig. 5 A simple network example

This result fits a common sense in syntax that the governors are somehow more important than dependents when it comes to the structural completeness of sentences. But it is very difficult to quantify this syntactic importance. We can see that these three function words, though all having high frequency and high degrees, play very different roles in the system. As a result, it seems safe to claim that *de* is the most important function word and *zai* is more important than *le* in this network, or the syntactic system. In other words, the syntactic importance of specific words can be quantified in this way. Developing a numeric scale of a well-defined notion of *syntactic importance* is left for future research.

Since *le*, *de*, and *zai* are all frequent words with high degrees, removing them from the network reduces the average degree of the networks, which is correlated with the degree of the removing word.

Other interesting parameters are the average path length and the diameter. Our study shows that removing *de* didn't cause the increase of the diameter of the XWLB while removing *le* or *zai* did. Although removing *de* caused a slight increase of the diameter in SHSS, removing *le* or *zai* reached a higher number. The diameter is highly correlated with the size of the network. Since the size of the two networks are quite different, it is not so easy to define the real reason or reasons behind these phenomena but it may be of some help to compare the phenomena in the same network. It is obvious that removing *le* or *zai* would influence the diameter of the networks more than removing *de*. Since the diameter is defined by the biggest distance of any two vertexes in the network, removing *de* did not change the distance at all in XWLB and increase the diameter from 9 to 10 in SHSS while removing *le* or *zai* made the distance increased by 8, or, by 66.67%, in XWLB and 4, or, 44.44%, in SHSS.

Removing three function words all caused the increase of the average path length. Removing *de* caused the least increase, removing *le* cased more increase, and removing *zai* caused the most increase in both XWLB and SHSS.

Diameter and average path length are two parameters related to the distance between nodes. Linguistically, they can be seen as the index of 'how any two words are connected with each other?, how many intermediate words bridge these two words? And what there intermediate words are?'. As mentioned before, the data showed that removing *le* or *zai* would bring about more distance between words than removing *de* does, although *de* is the center of the whole networks. The reason is related to these two words' ability in creating short cuts for other words and the irreplaceability of this ability. As the center of networks, there are many paths passing the node *de* and according to the closeness of *de*, we can see that *de* has the shortest distance to all the other words. But removing *de* from the networks didn't create much trouble for other words to connect each other. The reason is either *de* doesn't have the ability of 'creating short cuts for other words' or this node has this ability but it is replaceable. Removing *le* and *zai* caused more trouble, because *le* and *zai* have this ability and they are irreplaceable. No other words can completely replace their connecting roles. It also explains why the function word *le* is only a local central node, but has a very high frequency: it can irreplaceably 'create short cuts'. The best explanation for this, to our mind, is that people tend to save time and energy when sentences are understandable and make use of such system-inherent properties of words we investigated here.

5 Conclusion

This paper addresses the importance of developing network techniques of treebank exploitation for syntactic research ranging from theorem verification to discovery of new relations invisible to the eye.

We attach much importance in particular to the network tools and show how a treebank can, and, in our view, should be seen as a unique network.

We have shown in more detail, by comparing the function words *le*, *de*, and *zai*, that the frequency of words is not equivalent to the word's importance in the syntactic structure, pointing to a notion that we may call the *syntactic centrality* of the word. The importance in the syntactic structure is still a vague notion that needs to be refined further, but simple network manipulations like removal of the word nodes in question can reveal properties of the words that seem to be closely related to the their structural roles. For example, a word A whose removal breaks the network in parts is clearly more important than a word B whose removal preserves the connectedness of the network (as the word only occupies marginal nodes). Since the results shown in this paper confirm well-known facts concerning these three function words, the same method can be applied to other function words, and perhaps content words as well. Ongoing research includes analysis of following words: *wo* 'I, me, myself', *shi* 'are, am, yes', *ge* 'individual, entries', *yi* 'one, single', *zhe* 'this, it, these', *bu* 'do not, need not', *ta* 'he, him', *shuo* 'speak, talk, say', *ren* 'person, people, human being', and *dao* 'arrive, reach, get to'.

We leave above words for further research, aiming to develop the notion of *syntactic centrality* into a quantifiable value that could allow quantitative comparison between any words.

This study shows that the language network approach can provide not only an easy and direct access to getting a graphic output but also a fresh perspective on language analyzing.

Acknowledgements. This work was supported in part by the National Social Science Fund of China (11& ZD188). We thank Prof. Alexander Mehler from the Goethe University Frankfurt am Main for inviting us to attend the Conference on Modeling Linguistic Networks, where we presented the preliminary results of this paper. We also thank Kim Gerdes for the valuable comments and Zhao Yiyi, Guan Runchi for the treebank annotation work.

References

Abramov, O., Mehler, A.: Automatic Language Classification by Means of Syntactic Dependency Networks. Journal of Quantitative Linguistics 18(4), 291–336 (2011)

Allee, V.: Spectral Methods Cluster Words of the Same Class in a Syntactic Dependency Network. International Journal of Bifurcation and Chaos 17(7), 2453–2463 (2007)

Amancio, D.R., Antiqueira, L., Pardo, T.A.S., da F. Costa, L., Oliveira Jr., O.N., Nunes, M.G.V.: Complex Networks Analysis of Manual and Machine Translations. International Journal of Modern Physics C 19(4), 583–598 (2008)

Amancio, D.R., Nunes, M.G.V., Oliveira Jr., O.N., Pardo, T.A.S., Antiqueira, L., da F. Costa, L.: Using Metrics from Complex Networks to Evaluate Machine Translation. Physica A 390(1), 131–142 (2011)

Arbesman, S., Strogatz, S.H., Vitevitch, M.S.: The Structure of Phonological Networks Across Multiple Languages. International Journal of Bifurcation and Chaos 20(3), 679–685 (2010)

Banisch, S., Araújo, T., Louçã, J.: Opinion Dynamics and Communication Networks. Advances in Complex Systems 13(1), 95–111 (2010)

Barabási, A.L., Bonabeau, E.: Scale-free Networks. Scientific American 288(5), 50–59 (2003)

Borgatti, S.P.: NetDraw: Graph Visualization Software. Analytic Technologies, Harvard (2002)

Borgatti, S.P., Everett, M.G., Freeman, L.C.: Ucinet for Windows: Software for Social Network Analysis. Analytic Technologies, Harvard (2002)

Borge-Holthoefer, J., Arenas, A.: Semantic Networks: Structure and Dynamics. Entropy 12(5), 1264–1302 (2010)

Čech, R., Mačutek, J.: Word Form and Lemma Syntactic Dependency Networks in Czech: A Comparative Study. Glottometrics 19, 85–98 (2009)

Čech, R., Mačutek, J., Žabokrtský, Z.: The Role of Syntax in Complex Networks: Local and Global Importance of Verbs in a Syntactic Dependency Network. Physica A 390(20), 3614–3623 (2011)

Chen, X., Liu, H.: Central Nodes of the Chinese Syntactic Networks. Chinese Science Bulletin 56(1), 735–740 (2011)

Chen, X., Xu, C., Li, W.: Extracting Valency Patterns of Word Classes from Syntactic Complex Networks. In: Gerdes, K., Hajicova, E., Wanner, L. (eds.) Proceedings of Depling 2011, Barcelona, pp. 165–172 (2011)

Christiano, S.T., Raphael, A.D.: Network-based Stochastic Competitive Learning Approach to Disambiguation in Collaborative Networks. Journal of Nonlinear Science 23(1), 013139 (2013)

De Saussure, F.: Course in General Linguistics. Columbia University Press, New York (1916)

Ferrer i Cancho, R.: The Structure of Syntactic Dependency Networks: Insights from Recent Advances in Network Theory. In: Problems of Quantitative Linguistics, pp. 60–75 (2005)

Ferrer i Cancho, R., Capocci, A., Caldarelli, G.: Spectral Methods Cluster Words of the Same Class in a Syntactic Dependency Network. International Journal of Bifurcation and Chaos 17(7), 2453–2463 (2007)

Ferrer i Cancho, R., Solé, R.V., Köhler, R.: Patterns in syntactic dependency networks. Physical Review E 69(5), 051915 (2004)

Gong, T., Baronchelli, A., Puglisi, A., Loreto, V.: Exploring the Role of Complex Networks in Linguistic Categorization. Artificial Life 18(1), 107–121 (2012)

Hudson, R.: Language Networks: The New Word Grammar. Oxford University Press, Oxford (2007)

Ke, J., Yao, Y.A.O.: Analysing Language Development from a Network Approach. Journal of Quantitative Linguistics 15(1), 70–99 (2008)

Klammer, T.P., Schulz, M.R., Della Volpe, A.: Analyzing English Grammar. Pearson Education, India (2000)

Kretzschmar, W.A.: The Linguistics of Speech. Cambridge University Press, New York (2009)

Lamb, S.M., Newell, L.E.: Outline of Stratificational grammar. ASUC Store, University of California, San Francisco (1962)

Levy, R., Manning, C.: Is it harder to parse Chinese, or the Chinese Treebank? In: Proceedings of the 41st Annual Meeting on Association for Computational Linguistics, vol. 1, pp. 439–446 (2003)

Li, J., Zhou, J.: Chinese Character Structure Analysis based on Complex Networks. Physica A 380, 629–638 (2007)

Li, Y., Wei, L., Li, W., Niu, Y., Luo, S.: Small-world Patterns in Chinese Phrase Networks. Chinese Science Bulletin 50(3), 287–289 (2005)

Liu, H.: Syntactic Parsing based on Dependency Relations. Grkg/Humankybernetik 47, 124–135 (2006)

Liu, H.: The Complexity of Chinese Dependency Syntactic Networks. Physica A 387(12), 3048–3058 (2008)

Liu, H.: Statistical Properties of Chinese Semantic Networks. Chinese Science Bulletin 54(16), 2781–2785 (2009)

Liu, H.: Language Clusters based on Linguistic Complex Networks. Chinese Science Bulletin 55(30), 3458–3465 (2010)
Liu, H., Cong, J.: Language Clustering with Word Cooccurrence Networks based on Parallel Texts. Chinese Science Bulletin 58(10), 1139–1144 (2013)
Liu, H., Xu, C.: Can Syntactic Networks Indicate Morphological Complexity of a Language? Europhysics Letters 93(2), 28005 (2011)
Liu, Z.Y., Sun, M.S.: Chinese Word Co-occurrence Network: Its Small World Effect and Scale-free Property. Journal of Chinese Information Processing 21(6), 52–58 (2007)
Mehler, A., Diewald, N., Waltinger, U., Gleim, R., Esch, D., Job, B., Küchelmann, T., Pustylnikov, O., Blanchard, P.: Evolution of Romance language in written communication: Network analysis of late Latin and early Romance corpora. Leonardo 44(3), 244–245 (2011)
Mehler, A., Lücking, A., Menke, P.: Assessing Cognitive Alignment in Different Types of Dialog by Means of a Network Model. Neural Networks 32, 159–164 (2012)
Mehler, A., Lücking, A., Weiß, P.: A Network Model of Interpersonal Alignment in Dialog. Entropy 12(6), 1440–1483 (2010)
Mukherjee, A., Choudhury, M., Ganguly, N., Basu, A.: Language Dynamics in the Framework of Complex Networks: A Case Study on Self-organization of the Consonant Inventories. In: Villavicencio, A., Poibeau, T., Korhonen, A., Alishahi, A. (eds.) Cognitive Aspects of Computational Language Acquisition, pp. 51–78. Springer, Netherlands (2013)
Nooy, W., Mrvar, A., Batagelj, V.: Exploratory Network Analysis with Pajek. Cambridge University Press, New York (2005)
Peng, G., Minett, J.W., Wang, W.S.Y.: The Networks of Syllables and Characters in Chinese. Journal of Quantitative Linguistics 15(3), 243–255 (2008)
Shannon, P., Markiel, A., Ozier, O., Baliga, N.S., Wang, J.T., Ramage, D., Amin, N., Schwikowski, B., Ideker, T.: Cytoscape: A Software Environment for Integrated Models of Biomolecular Interaction Networks. Genome Research 13(11), 2498–2504 (2003)
Watts, D.J., Strogatz, S.H.: Collective Dynamics of 'Small-world' Networks. Nature 393(6684), 440–442 (1998)
Yu, S.: Modern Chinese Grammatical Information Dictionary Explanation. Tsinghua University, Beijing (1998)
Yu, S., Liu, H., Xu, C.: Statistical Properties of Chinese Phonemic Networks. Physica A 390(7), 1370–1380 (2011)

Non-crossing Dependencies: Least Effort, Not Grammar

Ramon Ferrer-i-Cancho

Abstract. The use of null hypotheses (in a statistical sense) is common in hard sciences but not in theoretical linguistics. Here the null hypothesis that the low frequency of syntactic dependency crossings is expected by an arbitrary ordering of words is rejected. It is shown that this would require star dependency structures, which are both unrealistic and too restrictive. The hypothesis of the limited resources of the human brain is revisited. Stronger null hypotheses taking into account actual dependency lengths for the likelihood of crossings are presented. Those hypotheses suggests that crossings are likely to reduce when dependencies are shortened. A hypothesis based on pressure to reduce dependency lengths is more parsimonious than a principle of minimization of crossings or a grammatical ban that is totally dissociated from the general and non-linguistic principle of economy.

1 Introduction

A substantial subset of theoretical frameworks under the general umbrella of "generative grammar" or "generative linguistics" have been kidnapped by the idea that a deep theory of syntax requires that one neglects the statistical properties of the system (Miller and Chomsky 1963) and abstracts away from functional factors such as the limited resources of the brain (Chomsky 1965).

This radical assumption disguised as intelligent abstraction led to the distinction between competence and performance (see Jackendoff 1999, Sect. 2.4 for a historical perspective from generative grammar), a dichotomy that is sometimes regarded as a soft methodological division (Jackendoff 1999, p. 34) or as theoretically unmotivated (Newmeyer 2001). A sister radical dichotomy is the division between

Ramon Ferrer-i-Cancho
Complexity and Quantitative Linguistics Lab, LARCA Research Group,
Departament de Ciències de la Computació, Universitat Politècnica de Catalunya (UPC),
Campus Nord, Edifici Omega, Jordi Girona Salgado 1-3, 08034 Barcelona, Catalonia, Spain
e-mail: rferrericancho@cs.upc.edu

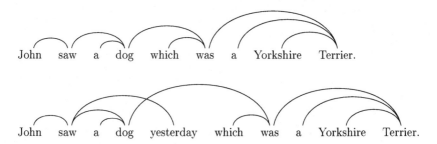

Fig. 1 Top an English sentence without crossings. **Bottom** a variant of the previous sentence with one dependency crossing (the dependency between "saw" and "yesterday" crosses the dependency between "dog" and "was" and vice versa). Adapted from McDonald et al. (2005).

grammar and usage (Newmeyer 2003). A revision of those views has led to the proposal of competence-plus, *"a package consisting of the familiar recursive statements of grammatical rules, plus a set of numerical constraints on the products of those rules"* (Hurford 2012). Interestingly, certain approaches reconcile competence with performance by regarding grammar as a store of "frozen" or "fixed" performance preferences (Hawkins 2004, p. 3) or by opening the set of numerical constraints of competence-plus to performance factors (Hurford 2012). Other examples of approaches that reject the dichotomous view of language are emergent grammar (Hopper 1998), synergetic linguistics (Köhler 2005) or probabilistic syntax (Manning 2002). The challenges of the competence/performance are not specific to generative linguistics. For instance, *"the competence/performance distinction is also embodied in many symbolic models of language processing"* (Christiansen and Chater 1999) and integrated with some refinements in language evolution research (Hurford 2012).

Again from the perspective of standard model selection (Burnham and Anderson 2002), the competence/performance dichotomy, even in soft versions, has a serious risk: if a more parsimonious theory exists based on performance, one that has the same or even superior explanatory power, it may not be discovered and if so, it will not be sufficiently endorsed. Astonishingly, linguistic theories that belittle the role of the limited resources of the human brain for structural constraints of syntax are presented as minimalistic (e.g. Chomsky 1995). In contrast, standard model selection favors theories with a good compromise between simplicity (often coming from a suitable abstraction or idealization) and explanatory power (Burnham and Anderson 2002).

A follower of the competence-performance split may consider that the opponents are unable to think in sufficiently abstract terms: opponents are being sidetracked by actual language use and limited computational resources and do not focus on the essence of syntax (in those views the essence of syntax is grammar (Newmeyer 2003) or certain features of such as recursion (Hauser et al. 2002);

in other approaches the essence of syntax is not grammar but dependencies (Hudson 2007; Frank et al. 2012)). The proponents of the split doctrine have not hesitated either to advertise functional approaches to linguistic theory as wrong (Miller 1968) or to attempt to dismantle attempts to turn research on language or communication more quantitative (e.g. Miller and Chomsky 1963; Niyogi and Berwick 1995; Suzuki et al. 2005). A real scientist however, will ask for the quality of a theory or a hypothesis in terms of the accuracy of its definitions, its testability, the statistical analyzes that have been performed to support it, the null hypotheses, the trade-off between explanatory power and parsimony of the theory, and so on.

If the limited resources of the brain are denied, one might be forced to blame grammar for the occurrence of certain patterns. Using standard model selection terms (Burnham and Anderson 2002), forwarding the responsibility to grammar implies the addition of more parameters to the model, indeed unnecessary parameters, as it will be shown here through a concrete phenomenon. The focus of the current article is a striking pattern of syntactic dependency trees of sentences that was reported in the 1960s: dependencies between words normally do not cross when drawn over the sentence (Lecerf 1960; Hays 1964) – e.g., Fig. 1. The problem of dependency crossings looks purely linguistic but it goes beyond human language: crossings have also been investigated in dependency networks where vertices are occurrences of nucleotides A, G, U, and C and edges are U-G and Watson-Crick base pairs, i.e. A-U, G-C (Chen et al. 2009). Having in mind various domains of application helps a researcher to apply the right level of abstraction. Becoming a specialist in human language or certain linguistic phenomena helps to find locally optimal theories, causes the illusion of scientific success when becoming the world expert of a certain topic but does not necessarily produce compact, coherent, general and elegant theories.

Here new light is shed on the origins of non-crossing dependencies by means of two fundamental tools of the scientific method: null hypotheses and unrestricted parsimony (unrestricted parsimony in the sense of being a priori open to favor theories that make fewer assumptions; not in the sense that parsimony has to be favored neglecting explanatory power). Unfortunately, the definition of null hypotheses (in a statistical sense) is rare in theoretical linguistics (although it is fundamental in biology or medicine). Even in the context of quantitative linguistics research, clearly defined null hypotheses or baselines are present in certain investigations (e.g. Ferrer-i-Cancho, Hernández-Fernández, et al. 2013; Ferrer-i-Cancho 2004) but are missing in others (e.g. Ferrer-i-Cancho and Moscoso del Prado Martín 2011; Moscoso del Prado 2013). When present, they are not always tested rigorously (Ferrer-i-Cancho and Elvevåg 2009). In the context of quantitative research, claims about the efficiency of language have been made lacking a measure of cost and evidence that such a cost is below chance (Piantadosi et al. 2011). A deep theory of language requires (at least) metrics of efficiency, tests of their significance and an understanding of the relationship between the minimization of the costs that they define and the emergence of the target patterns, e.g., Zipf's law of abbreviation (Ferrer-i-Cancho, Hernández-Fernández, et al. 2013).

To our knowledge, claims for the existence of a universal grammar have never been defended by means of a null hypothesis in a statistical sense (e.g. Jackendoff 1999; Uriagereka 1998), and a baseline is missing in research where grammar is seen as a conventionalization of performance constraints (Hawkins 2004) or in research where competence is complemented with quantitative constraints (Hurford 2012). As for the latter, baselines would help one to determine which of those constraints must be stored by grammar or competence-plus.

The first question that a syntactician should ask as a scientist when investigating the origins of a syntactic property X is: could X happen by chance? The question is equivalent to asking if grammar (in the sense of some extra knowledge) or specific genes are needed to explain property X. Accordingly, the major question that this article aims to answer is: could the low frequency of crossings in syntactic dependency trees be explained by chance, maybe involving general computational constraints of the human brain?

The remainder of the article is organized as follows. Section 2 reviews our minimalistic approach to the syntactic dependency structure of sentences. Section 3 considers the null hypothesis of a random ordering of the words of the sentence and shows that keeping the expected number of crossings small requires unrealistic constraints on the ensemble of possible dependency trees (only star trees would be possible). Section 4 considers alternative hypotheses, discarding the vague or heavy hypothesis of grammar and focusing on two major hypotheses: a principle of minimization of crossings and a principle of minimization of the sum of dependency lengths. The analysis suggests that the number of crossing and the sum of dependency lengths are not perfectly correlated but their correlation is strong. Of the two principles, dependency length minimization offers a more parsimonious account of many more linguistic phenomena. Interestingly, that principle is motivated by the need of minimizing cognitive effort. A challenge for the hypothesis that the rather small number of crossings of real sentence is a side-effect of minimization of dependency lengths is (a) determining the degree of that minimization that the real number of crossings requires and (b) if that degree is realistic. Section 5 presents a stronger null hypothesis that addresses the challenge with knowledge about edge lengths. That null hypothesis allows one to predict the number of crossings when the length of one of the edges potentially involved in a crossing is known but words are arranged at random. Thus, that predictor uses the actual dependency lengths to estimate the number of crossings. Interestingly, that predictor provides further support for a strong correlation between crossings and dependency lengths: analytical arguments suggest that it is likely that a reduction of dependency lengths causes a drop in the number of crossings. Section 6 considers another predictor based on a stronger null hypothesis where the sum of dependency lengths is given but words are arranged at random. Preliminary numerical results indicate a strong correlation between the mean number of crossings and the sum of dependency lengths over all the possible orderings of the words of real sentences. Interestingly, that null hypothesis leads to a predictor that requires less information about a real sentence than the previous predictor (only the sum of dependency lengths is needed) and paves the way to understanding the rather low number of crossings in real sentences as a consequence of global cognitive constraints on dependency lengths. Section 7 compares the predictions of the three predictors on a small set of sentences. The results suggest that the predictor based on the sum of dependency lengths is the best candidate. There it is also demonstrated that p-value testing can be used to investigate the adequacy of the best candidate. Interestingly, the best candidate was not rejected in that sample of sentences. Finally, Section 8 reviews and discusses the idea that least effort, not grammar, is the reason for the small number of crossings of real sentences.

2 The Syntactic Dependency Structure of Sentences

This article borrows the minimalistic approach to the syntactic dependency structure of sentences of dependency grammar (Hudson 1984; Mel'čuk 1988) and recent progress in cognitive sciences (Frank et al. 2012):

- No hierarchical phrase structure is assumed in the sense that the structure of a sentence is defined simply as a tree where vertices are words and edges are syntactic dependencies. This is a fundamental assumption of our approach: tentatively, the network defining the dependencies between words might be a disconnected forest or a graph with cycles – these are possibilities that have not been sufficiently investigated (Hudson 1984). A general theory of crossings in nature cannot obviate the fact that RNA structures cannot be modeled with trees but can be modeled with forests (Chen et al. 2009).[1] Although the choice of a tree of words as the reference model for sentence structure (e.g. McDonald et al. 2005) is to some extent arbitrary, a tree is optimal for being the kind of network that is able to connect all words with the smallest amount of edges (Ferrer i Cancho and Solé 2003).
- Words establish direct relationships that are not necessarily mediated by syntactic categories (non-terminals in the phrase structure formalism and generative grammar evolutions). This skepticism about syntactic categories (as entities by its own, not epiphenomena) goes beyond dependency grammar, e.g., construction grammar (Goldberg 2003).

Along the lines of Frank et al. (2012), link direction is irrelevant for the arguments in this article. Even within the dependency grammar formalism, dependencies are believed to be directed (from heads to modifiers/complements; Mel'čuk 1988; Hudson 1984). A minimalistic approach to dependency syntax should not obviate the fact that the accuracy of dependency parsing improves if link direction is neglected (Gómez-Rodríguez et al. 2014).

---

[1] In those RNA structures, vertex degrees do not exceed one (Chen et al. 2009) and thus cycles are not possible but connectedness is not either (the handshaking lemma (Bollobás 1998, p. 4) indicates that such a graph cannot have more than $n/2$ edges, being n the number of vertices, and thus cannot be connected because that needs at least $n-1$ edges).

3 The Null Hypothesis

Let n be the number of vertices of a tree. Let k_i be the degree of the i-th vertex of a tree and $k_1, ..., k_i, ..., k_n$ its degree. By K_α, we denote

$$K_\alpha = \sum_{i=1}^{n} k_i^\alpha, \qquad (1)$$

where α is a natural number. In a tree, K_1 only depends on n, i.e. (Noy 1998),

$$K_1 = 2(n-1) \qquad (2)$$

and thus the 1st moment of degree is

$$\langle k \rangle = \frac{K_1}{n} = 2 - \frac{2}{n}. \qquad (3)$$

Let $E_0[C]$ be the expected number of crossings in a random linear arrangement of a dependency tree with a given degree sequence, i.e. (Ferrer-i-Cancho 2013b)

$$E_0[C] = \frac{n}{6}\left(n - 1 - \langle k^2 \rangle\right), \qquad (4)$$

where $\langle k^2 \rangle$ is the 2nd moment of degree, i.e.

$$\langle k^2 \rangle = \frac{K_2}{n}. \qquad (5)$$

Thus, the expected number of crossings depends on the number of vertices (n) and the 2nd moment of degree ($\langle k^2 \rangle$). The higher the hubiness (the higher $\langle k^2 \rangle$) the lower the expected number of crossings.

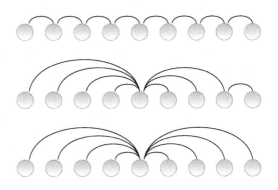

Fig. 2 Linear arrangements of trees of nine vertices. **Top** a linear tree. **Center** quasi-star tree. **Bottom** star tree.

A star tree is a tree with a vertex of degree $n - 1$ while a linear tree is a tree where no vertex degree exceeds two (Fig. 2; Ferrer-i-Cancho 2013b). For a given number of vertices, $E_0[C]$ is minimized by star trees, for which $E_0[C] = 0$, whereas $E_0[C]$

is maximized by linear trees, for which (Ferrer-i-Cancho 2013a; Ferrer-i-Cancho 2013b)

$$E_0[C] = \frac{1}{6}n(n-5) + 1. \tag{6}$$

As $E_0[C]$ depends on the $\langle k^2 \rangle$ of the tree, the null hypothesis that the tree structures are chosen uniformly at random among all possible labeled trees is considered next. The Aldous-Brother algorithm allows one to generate uniformly random labeled spanning trees from a graph (Aldous 1990; Broder 1989). Here a complete graph is assumed to be the source for the spanning trees. A low number of crossings cannot be attributed to grammar if $E_0[C]$ is low.

$E_0[C]$ is the expectation of C given a degree sequence. Indeed, that expectation can be obtained just from knowledge about $\langle k^2 \rangle$ and n (Eq. (4)). The expectation of C for uniformly random labeled trees is

Proposition 3.1

$$E[E_0[C]] = \frac{1}{6}(n-1)\left(n-5+\frac{6}{n}\right)$$
$$= \frac{n^2}{6} - n + \frac{11}{6} - \frac{1}{n}. \tag{7}$$

Proof. On the one hand, the degree variance for uniformly random labeled trees is (Moon 1970; Noy 1998)

$$V[k] = \langle k^2 \rangle - \langle k \rangle^2 = \left(1 - \frac{1}{n}\right)\left(1 - \frac{2}{n}\right). \tag{8}$$

Applying Eq. (3), it is obtained

$$\langle k^2 \rangle = \left(1 - \frac{1}{n}\right)\left(5 - \frac{6}{n}\right). \tag{9}$$

On the other hand,

$$E[E_0[C]] = E\left[\frac{n}{6}(n-1-\langle k^2 \rangle)\right] \quad \text{applying Eq. (4)}$$
$$= \frac{n}{6}(n-1-E[\langle k^2 \rangle])$$
$$= \frac{n}{6}\left(n-1-\left(1-\frac{1}{n}\right)\left(5-\frac{6}{n}\right)\right) \quad \text{applying Eq. (9)}$$
$$= \frac{1}{6}(n-1)\left(n-5+\frac{6}{n}\right).$$

□

Figure 3 shows that uniformly random labeled trees exhibit a high value of $E_0[C]$ that is near the upper bound defined by linear trees. Thus, it is unlikely that the rather low frequency of crossings in real syntactic dependency trees (Mel'čuk 1988; Liu 2010) is due to uniform sampling of the space of labeled trees. However, one cannot exclude the possibility that real dependency trees belong to a subclass of random trees for which $E_0[C]$ is low (e.g., the uniformly random trees may not be spanning trees of a complete graph). This possibility is explored next.

A quasi-star tree is defined as a tree with one vertex of degree $n-2$, one vertex of degree 2 and the remainder of vertices of degree 1 (Fig. 2). A quasi-star tree needs $n \geq 3$ to exist (Appendix). The sum of squared degrees of such a tree is (Appendix)

$$K_2^{\text{quasi}} = n^2 - 3n + 6 \tag{10}$$

and thus the degree 2nd moment of a quasi-star tree is

$$\langle k^2 \rangle^{\text{quasi}} = \frac{K_2^{\text{quasi}}}{n} = n - 3 + \frac{6}{n}. \tag{11}$$

Eq. (4) and Eq. (11) allow one to infer that

$$E_0[C] = \frac{n}{3} - 1 \tag{12}$$

for a quasi-star tree. Figure 3 shows the linear growth of the expected number of crossings as a function of the number of vertices in quasi-star trees. Interestingly, if a tree has a value of $\langle k^2 \rangle$ that exceeds $\langle k^2 \rangle^{\text{quasi}}$, it has to be a star tree (see Appendix). For this reason, Fig. 3 suggests that star trees are the only option to obtain a small constant number of crossings. A detailed mathematical argument will be presented next.

If it is required that the expected number of crossings does not exceed a, i.e. $E_0[C] \leq a$, Eq. (4) gives

$$\langle k^2 \rangle \geq n - 1 - \frac{6a}{n}. \tag{13}$$

Notice that the preceding result has been derived making no assumption about the tree topology. We aim to investigate when a $E_0[C] \leq a$ implies a star tree.

As a tree whose value of $\langle k^2 \rangle$ exceeds $\langle k^2 \rangle^{\text{quasi}}$ must be a star tree (Appendix), Eq. (13) indicates that if

$$n - 1 - \frac{6a}{n} > \langle k^2 \rangle^{\text{quasi}} \tag{14}$$

then $E_0[C] \leq a$ requires a star tree. Applying the definition of $\langle k^2 \rangle^{\text{quasi}}$ in Eq. (11) to Eq. (14), we obtain that a star tree is needed to expect at most a crossings if

$$n > 3a + 3. \tag{15}$$

Thus, Eq. (15) implies that a hub tree is needed to expect at most one crossings by chance ($a = 1$) for $n > 6$ (this can be checked with the help of Fig. 3). In order to have at most one crossing by chance, the structural diversity must be minimum because star trees are the only possible labeled trees. To understand the heavy constraints

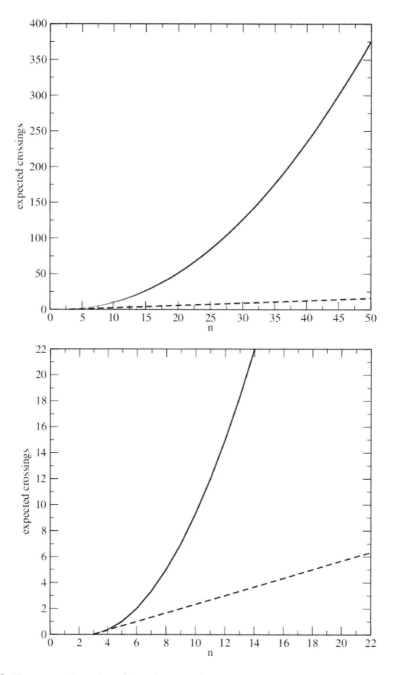

Fig. 3 The expected number of crossings as a function of the number of vertices of the tree (n) in random linear arrangements of vertices for linear trees (*black solid line*), uniformly random labeled trees (*gray line*) and quasi-star trees (*dashed line*). **Top** the whole picture up to $n = 50$. **Bottom** a zoom of the left-bottom corner.

imposed by $a = 1$ on the possible trees, consider $t(n)$, the number of unlabeled trees of n that can be formed (Table 1). When $n = 4$, the only trees than can be formed are a star tree and a linear tree, which gives $t(4) = 2$. In contrast, the star tree is only one out of 19320 possible unlabeled trees when $n = 16$. The decrease in diversity is more radical as n increases (Table 1). The choice of $a = 2$ does not change the scenario so much: Equation (15) predicts that sentences of length $n > 9$ should have a star tree structure if no more than two crossings are to be expected. Real syntactic dependency trees from sufficiently longer sentences are far from star trees, e.g., Fig. 1 (Mel'čuk 1988; Hays 1964; Lecerf 1960).

Table 1 $t(n)$, the number of unlabeled trees of n vertices (Seoane 2013)

n	$t(n)$	n	$t(n)$	n	$t(n)$
1	1	8	23	15	7741
2	1	9	47	16	19320
3	1	10	106	17	48629
4	2	11	235	18	123867
5	3	12	551	19	317955
6	6	13	1301		
7	11	14	3159		

4 Alternative Hypotheses

It has been shown that the low frequency of crossings is unexpected by chance in random linear arrangements of real syntactic dependency trees. As scientists, the next step is exploring the implications of this test and evaluating alternative hypotheses. Vertex degrees ($\langle k^2 \rangle$), which are an aspect of sentence structure, have been discarded as the only origin for the low frequency of crossings. This is relevant for some views where competence or grammar concern the structure of a sentence (Chomsky 1965; Jackendoff 1999; Newmeyer 2003). Discussing what competence or grammar is or should be is beyond the scope of this article but it is worth examining common reactions of language researchers when encountering a pattern:

- For statistical patterns such as Zipf's law for word frequencies and Menzerath's law, it was concluded that the patterns are inevitable (Miller and Chomsky 1963; Solé 2010) – see Ferrer-i-Cancho and Elvevåg (2009) and Ferrer-i-Cancho, Forns, et al. (2013) for a review of the weaknesses of such conclusions.
- Concerning syntactic regularities in general, a naive but widely adopted approach is blaming (universal) grammar, the faculty of language or similar concepts (Newmeyer 2003; Hauser et al. 2002). The fact that one is unable to explain a certain phenomenon through usage is considered as a justification for grammar (e.g. Newmeyer 2003). However, a rigorous justification requires a proof of the impossibility of usage to account for the phenomenon. To our knowledge, that proof is never provided.

- In the context of dependency grammar, crossings dependencies have been banned (Hudson 1984) or it has been argued that most phrases cannot have crossings or that crossings turn sentences ungrammatical (Hudson 2007, p. 130). It is worrying that the statement is not the conclusion of a proof of the impossibility of a functional explanation. Furthermore, the argument of "ungrammaticality" is circular and sweeps processing difficulties under the carpet.

In the traditional view of grammar or the faculty of language, the limited resources of the human brain are secondary or anecdotal (Hauser et al. 2002; Hudson 2007). Recurring to grammar or a language faculty implies more assumptions, e.g. grammar would be the only reason why dependencies do not cross so often, and an explanation about the origins of the property would be left open (the explanation would be potentially incomplete). The property might have originated in grammar as a kind of inevitable logical or mathematical property, or might be supported by genetic information of our species, or it might also have been transferred to grammar (culturally or genetically) and so on. Thus, a grammar that is responsible for non-crossing dependencies would not be truly minimalistic (parsimonious) at all if the phenomenon could be explained by a universal principle of economy (universal in the sense of concerning the human brain not necessarily exclusively). This is likely the case of current approaches to dependency grammar at least (e.g. Hudson 2007; Mel'čuk 1988; Hudson 1984).

Fig. 4 A linear arrangements of the vertices of a linear tree that maximizes D (the sum of dependency lengths) when edge crossings are not allowed.

4.1 A Principle of Minimization of Dependency Crossings

A tempting hypothesis is a principle of minimization of dependency crossings (e.g. Liu 2008) which can be seen as a quantitative implementation of the ban of crossings (Hudson 1984; Hudson 2007). This minimization can be understood as a purely formal principle (a principle of grammar detached from performance constraints but then problematic for the reasons explained above) or a principle related to performance. A principle of the minimization of crossings (or similar ones) is potentially problematic for at least three reasons:

- It is an inductive solution that may overfit the data.

- It is naive and superficial because it does not distinguish the consequences (e.g., uncrossing dependencies) from the causes. A deep theory of language requires distinguishing underlying principles from products: the principle of compression (Ferrer-i-Cancho, Hernández-Fernández, et al. 2013) from the law of abbreviation (a product), the principle of mutual information maximization from vocabulary learning biases (another product, Ferrer-i-Cancho 2013c), and so on.
- If patterns are directly translated into principles, the risk is that of constructing a fat theory of language when merging the tentative theories from every domain. When integrating the principle of minimization of crossings with a general theory of syntax, one may get two principles: a principle of minimization of crossings and a principle of dependency length minimization. In contrast, a theory where the minimization of crossings is seen as a side-effect of a principle of dependency length minimization (Ferrer-i-Cancho 2006; Liu 2008; Morrill et al. 2009; Ferrer-i-Cancho 2013a) might solve the problem in one shot through a single principle of dependency length minimization. However, it has cleverly been argued that a principle of minimization of crossings might imply a principle of dependency length minimization (Liu 2008) and thus a principle of minimization of crossings might not imply any redundancy.

Thus, it is important to review the hypothesis of the minimization of dependency length and the logical and statistical relationship with the minimization of C.

4.2 A Principle of Minimization of Dependency Lengths

The length of a dependency is usually defined as the absolute difference between the positions involved (the 1st word of the sentence has position 1, the 2nd has position 2 and so on). In Fig. 1, the length of the dependency between "John" and "saw" is $2 - 1 = 1$ and the length of the dependency between "saw" and "dog" is $4 - 2 = 2$. In this definition, the units of length are word tokens (it might be more precise if defined in phonemes, for instance). If d_i is the length of the i-th dependency of a tree (there are $n - 1$ dependencies in a tree) and $g(d)$ is the cost of a dependency of length d, the total cost of a linguistic sequence from the perspective of dependency length is the total sum of dependency costs (Ferrer-i-Cancho 2014; Ferrer-i-Cancho 2015), which can defined as

$$D = \sum_{i=1}^{n} g(d_i), \qquad (16)$$

where $g(d)$ is assumed to be a strictly monotonically increasing function of d (Ferrer-i-Cancho 2015). The mean cost of a tree structure is defined as $\langle d \rangle = D/(n-1)$. If g is the identity function ($g(d) = d$) then D is the sum of dependency lengths (and $\langle d \rangle$ is the mean dependency length). It has been hypothesized that D or equivalently $\langle d \rangle$ is minimized by sentences (see Ferrer-i-Cancho (2015) for a review). The hypothesis does not imply that the actual value of D has to be the minimum in absolute terms. Hereafter we assume that $g(d)$ is the identity function ($g(d) = d$).

The minimum D that can be obtained without altering the structure of the tree is the solution of the minimum linear arrangement problem in computer science (Chung 1984). Another baseline is provided by the expected value of D in a random arrangement of the vertices, which is (Ferrer-i-Cancho 2004)

$$E_0[D] = (n-1)(n+1)/3. \tag{17}$$

Statistical analyzes of D in real syntactic dependency trees have revealed that D is systematically below chance (below $E_0[D]$) for sufficiently long sentences but above the value of a minimum linear arrangement on average (Ferrer-i-Cancho 2004; Ferrer-i-Cancho and Liu 2014).

D is one example of a metric or score to evaluate the efficiency of a sentence from a certain dimension (see Morrill (2000) and Hawkins (1998) for similar metrics on syntactic structures). Stating clearly the metric that is being optimized is a requirement for a rigorous claim about efficiency of language. For instance, consider the sentence on top of Fig. 5 and the version below that arises from the right-extraposition of the clause "who I knew". Notice that the dependency tree is the same in both cases (only word order varies). It has been argued that theories of processing based on the distance between dependents *"predict that an extraposed relative clause would be more difficult to process than an in situ, adjacent relative clause"* (Levy et al. 2012). However, that does not grant one to conclude that the sentence on top of Fig. 5 should be easier to process than the sentence below from that perspective: one has $D = 3 \times 1 + 2 + 4 + 6 = 15$ for the sentence without the extraposition and $D = 3 \times 1 + 2 \times 2 + 3 = 10$ for the one with right-extraposition suggesting that the easier sentence is precisely the sentence with right-extraposition. The prediction about the cost of extraposition in Levy et al. (2012) is an incomplete argument. The ultimate conclusion about the cost of extraposition requires considering all the dependency lengths, i.e. a true efficiency score. A score of sentence locality is needed to not rule out prematurely accounts of the processing difficulty of non-projective orderings that are based purely on *"dependency locality in terms of linear positioning"* (Levy et al. 2012). The issue is tricky even for studies where a quantitative metric of dependency length such as D is employed: it is important to not mix values of the metric coming from sentences of different length to draw solid conclusions about a corpus or a language (Ferrer-i-Cancho and Liu 2014). The need of strengthening quantitative standards (Gibson and Fedorenko 2010) and also the need of appropriate controls (Culicover and Jackendoff 2010; Ferrer-i-Cancho and Moscoso del Prado Martín 2011; Moscoso del Prado 2013) in linguistic research are challenges that require the serious commitment of each of us.

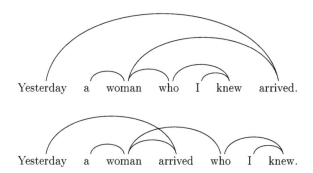

Fig. 5 Top an English sentence with a relative clause ("who I knew"). **Bottom** the same sentence with a right extraposition of the relative clause. Adapted from Levy et al. (2012).

4.3 The Relationship between Minimization of Crossings and Minimization of Dependency Lengths

Let us examine the logical relationship between the two principles above from the perspective of their global minima. On the one hand, the minimum value of C is 0 (Ferrer-i-Cancho 2013a) and the minimum value of D is obtained by solving the minimum linear arrangement problem (Baronchelli et al. 2013) or a generalization (Ferrer-i-Cancho 2015), which yields D_{min}. At constant n and $\langle k^2 \rangle$, there are two facts:

- $C = 0$ does not imply $D = D_{min}$ in general. This can be shown by means of two extreme configurations, a star tree, which maximizes $\langle k^2 \rangle$ and a linear tree, which minimizes $\langle k^2 \rangle$ (Ferrer-i-Cancho 2013b):

  - A star tree implies $C = 0$ (Ferrer-i-Cancho 2013a). In that tree, $D = D_{min}$ holds only when the hub is placed at the center (Ferrer-i-Cancho 2015). If the hub is placed at one of the extremes of the sequence, D is maximized for that tree (Ferrer-i-Cancho 2015). Those results still hold when $g(d)$ is not the identity function but a strictly monotonically increasing function of d (Ferrer-i-Cancho 2015). Furthermore, the placement of the hub in one extreme implies the maximum D that a non-crossing tree (not necessarily a star) can achieve, which is $D = n(n-1)/2$ (Ferrer-i-Cancho 2013a).
  - A linear tree can be arranged linearly with $C = 0$ and $D = D_{min} = n - 1$ (as in Fig. 2), which has no crossings and coincides with the smallest D than an unrestricted tree can achieve (as $d_i \geq 1$ and a tree has $n - 1$ edges). In contrast, a linear arrangement of the kind of Fig. 4 has $C = 0$ but yields $D = n(n-1)/2$, i.e. the maximum value of D that a non-crossing tree can achieve (Ferrer-i-Cancho 2013a).

- $D = D_{\min}$ does not imply $C = 0$ in general. It has been shown that a linear arrangement of vertices with crossings can achieve a smaller value of D than that of a minimum linear arrangement that minimizes D when no crossings are allowed (Hochberg and Stallmann 2003) – Fig. 6.

Thus, there is not a clear relationship between the minima of D and C when one abstracts from the structural properties of real syntactic dependency trees. The impact of the real properties of dependency structures for the arguments should be investigated in the future.

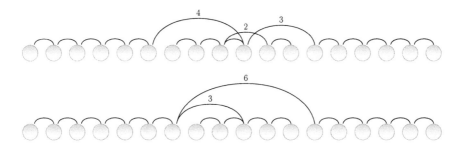

Fig. 6 Minimum linear arrangements of the same tree (only the length of edges that are longer than unity is indicated). **Top** a minimum linear arrangement of a tree. The total sum of dependency lengths is $D = 4 + 2 + 3 + 14 = 23$. **Bottom** a minimum linear arrangement of the same tree when crossings are disallowed. The total sum of dependency lengths is $D = 6 + 3 + 15 = 24$. Adapted from Hochberg and Stallmann (2003).

As for the statistical relationship between C and D, statistical analyzes support the hypothesis of a positive correlation between both at least in the domain between $n - 1$, the minimum value of D, and $D = E_0[D]$ (Ferrer-i-Cancho 2006; Ferrer-i-Cancho 2013b). For instance, crossings practically disappear if the vertices of a random tree are ordered to minimize D. The relationship between C and D in random permutations of vertices of the dependency trees is illustrated in Fig. 7: C tends to increase as D increases from $D = D_{\min}$ onwards. Results obtained with similar metrics (Liu 2008; Liu 2007) are consistent with such a correlation. For instance, a measure of dependency length reduces in random trees when crossings are disallowed (Liu 2007).

This opens the problem of causality, namely if the minimization of D may cause a minimization of C, or a minimization of C may cause a minimization of D, or both principles cannot be disentangled or simply both principles are epiphenomena (correlation does not imply causality). Solving the problem of causality is beyond the scope of this article but we can however attempt to determine rationally which of the two forces, minimization of D or minimization of C, might be the primary force by means of qualitative version of information theoretic model selection (Burnham and Anderson 2002). The apparent tie between these two principles will be broken by the more limited explanatory power of the minimization of C. The point

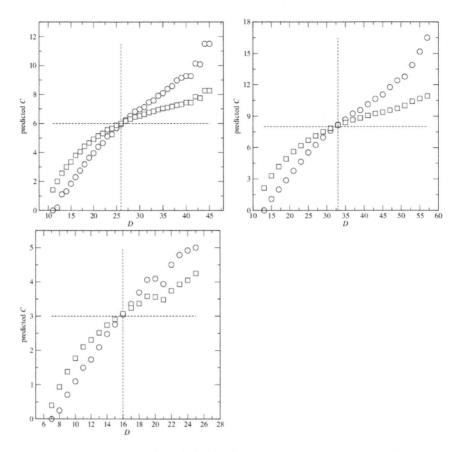

Fig. 7 Predictions about number of dependency crossings (C) as a function of the sum of dependency lengths (D) for dependency trees of real sentences. $E[C|D]$, the average C in all the possible permutations with a given value of D (*circles*), is compared against the average $E_1[C]$ in those permutations (*squares*). $E_1[C]$ is a prediction about C based on information on the distance of just one of the vertices potentially involved in a crossing. The *vertical dashed line* indicates the value of $E_0[D]$ and the *horizontal line* indicates the value of $E_0[C]$ (given a tree, those values are easy to compute with the help of Eqs. (17) and (4), respectively). The value of $E[C|D]$ and its prediction is only shown for values of D achieved by at least one permutation because certain values of D cannot be achieved by a given tree. For a given tree, $D_{\min}$ and $D_{\max}$ are, respectively, the minimum and the maximum value of D that can be reached ($D_{\min} \leq D \leq D_{\max}$). **Top-Left** sentence on top Fig. 1 with $D_{\min} = 11$, $D_{\max} = 45$, $E_0[D] = 26$ and $E_0[C] = 6$. **Top-Right** sentence at the bottom of Fig. 1 with $D_{\min} = 13$, $D_{\max} = 57$, $E_0[D] = 33$ and $E_0[C] = 8$. Even values of D are not found. **Bottom-Left** results for the two sentences in Fig. 5 with $D_{\min} = 7$, $D_{\max} = 25$, $E_0[D] = 16$ and $E_0[C] = 3$. Notice that the results are valid for the couple of sentences in Fig. 5 because they have the same structure in a different order. For this reason it is not surprising that $D_{\min}$, $D_{\max}$, $E_0[D]$, $E_0[C]$ and $E[C|D]$ coincide. Interestingly, $E_1[C]$ turns out to be the same for both, too.

is focusing on the phenomena that a principle of minimization of C cannot illuminate. A challenge for that principle is the ordering of subject (S), object (O) and verb (V). The dependency structure of that triple is a star tree with the verb as the hub (Ferrer-i-Cancho 2015). A tree of less than four vertices cannot have crossings (Ferrer-i-Cancho 2013a). Thus, $C = 0$ regardless of the ordering of the elements of the triple. Interestingly, the principle of minimization of C cannot explain why languages abandon SOV in favor of SVO (Gell-Mann and Ruhlen 2011). In contrast, the attraction of the verb towards the center combined with the structure of the word order permutation space can explain it (Ferrer-i-Cancho 2015). Another challenge for a principle of the minimization of C are the relative ordering of dependents of nominal heads in SVO orders, that have been argued to preclude regression to SOV (Ferrer-i-Cancho 2015). To sum up, a single principle of minimization of C would compromise explanatory power and if its limitations were complemented with an additional principle of dependency length minimization then parsimony would be compromised.

The reminder of the article is aimed at investigating a more parsimonious explanation for the ubiquity of non-crossing dependencies based on the minimization of D as the primary force (Ferrer-i-Cancho 2006; Liu 2008; Morrill et al. 2009; Ferrer-i-Cancho 2013a). The minimization of D would be a consequence of the minimization of cognitive effort: longer dependencies are cognitively more expensive (Liu 2008; Morrill 2000; Gibson 2000; Hawkins 1994). We will investigate two null hypotheses that allow one to predict the number of crossings as function of dependency lengths, which are determined by cognitive pressures.

5 A Stronger Null Hypothesis

Here we consider a predictor for the number of crossings when some information about the length of the arcs is known. The predictor guesses that number by considering, for every pair of edges that may potentially cross (pairs of edges sharing vertices cannot cross), the probability that they cross knowing the length of one of the edges and assuming that the vertices of the other edge have been arranged linearly at random. The null hypothesis in Sect. 3 predicts the number of crossings in the same fashion but replacing that probability by the probability that two edges cross when both are arranged linearly at random (no arc length is given).

The null hypothesis in Sect. 3 and the null hypothesis that will be explored in the current section, are reminiscent of two null hypotheses that are used in networks theory: random binomial graphs and random graphs with an arbitrary degree sequence or degree distribution (Molloy and Reed 1995; Molloy and Reed 1998; Newman et al. 2001). In our case, information about dependency length plays an equivalent role to vertex degree in those models.

5.1 The Probability That Two Edges Cross

Let $\pi(v)$ be the position of the vertex v in the linear arrangement ($\pi(v) = 1$ if v is the 1st vertex of the sequence, $\pi(v) = 2$ it is the 2nd vertex and so on). $u \sim v$ is used to indicate an edge between vertices u and v, namely that u and v are connected. A vertex position q is covered by the edge $u \sim v$ if and only if $min(\pi(u), \pi(v)) < q < max(\pi(u), \pi(v))$. A position q is external to the edge $u \sim v$ if and only if $q < min(\pi(u), \pi(v))$ or $q > max(\pi(u), \pi(v))$. $s \sim t$ crosses $u \sim v$ if and only if

- either $\pi(s)$ is covered by $u \sim v$ and $\pi(t)$ is external to $u \sim v$
- or $\pi(t)$ is covered by $u \sim v$ and $\pi(s)$ is external to $u \sim v$.

Notice that edges that share vertices cannot cross.

Let us consider first that no information is known about the length of two arcs. The probability that two edges, $s \sim t$ and $u \sim v$, cross when arranged linearly at random is $p(cross) = 1/3$ if the edges do not share any vertex and $p(cross) = 0$ otherwise (Ferrer-i-Cancho 2013b). We will investigate $p(cross|d)$, the probability that two edges, $s \sim t$ and $u \sim v$, cross when arranged linearly at random knowing that (a) one of the edges has length d, e.g., $|\pi(u) - \pi(v)| = d$ and (b) the edges do not share any vertex.

If $s \sim t$ and $u \sim v$ share a vertex, then $p(cross|d) = 0$. If not,

Proposition 5.1

$$p(cross|d) = 2\frac{(d-1)(n-d-1)}{(n-2)(n-3)}$$
$$= \frac{2(-d^2 + nd - n + 1)}{(n-2)(n-3)}. \tag{18}$$

Proof. To see this notice that $d - 1$ is the number of vertex positions covered by the edge of length d and $n - d - 1$ is the number of vertices that are external to that edge. Once the edge of length d has been arranged linearly, there are

$$\binom{n-2}{2} = \frac{(n-2)(n-3)}{2} \tag{19}$$

possible placements for the two vertices of the other edge of which only $(d-1)(n-d-1)$ involve a position covered by the edge of length d and another one that is external to that edge. □

$p(cross|d)$ and $p(cross) = 1/3$ are related, i.e.

$$\sum_{d=1}^{n-1} p(cross|d) p(d) = \frac{1}{3}, \tag{20}$$

where

$$p(d) = \frac{2(n-d)}{n(n-1)} \tag{21}$$

is the probability that the linear arrangement of the two vertices of an edge yields a dependency of length d (Ferrer-i-Cancho 2004). Equation (20) is easy to prove applying the definition of conditional probability ($p(\text{cross}|d) = p(\text{cross},d)/p(d)$), which gives

$$\sum_{d=1}^{n-1} p(\text{cross}|d)p(d) = \sum_{d=1}^{n-1} p(\text{cross},d) = p(\text{cross}) = \frac{1}{3}. \quad (22)$$

When n takes the smallest value needed for the possibility of crossings, i.e. $n = 4$ (Ferrer-i-Cancho 2013b), Eq. (18) yields $p(\text{cross}|1) = p(\text{cross}|3) = 0$ and $p(\text{cross}|2) = 1$. It is easy to show that

- $p(\text{cross}|d)$ is symmetric, i.e. $p(\text{cross}|d) = p(\text{cross}|n-d)$,
- $p(\text{cross}|d)$ has two minima ($p(\text{cross}|d) = 0$), at $d = 1$ and $d = n-1$.
-
$$p(\text{cross}|d) \leq p_{\max}(\text{cross}|d), \quad (23)$$

where

$$p_{\max}(\text{cross}|d) = \frac{\frac{n^2}{2} - 2(n-1)}{(n-2)(n-3)}. \quad (24)$$

To see this notice that $p(\text{cross}|d)$ is a function (Eq. (18)) that has a maximum at $d = d^* = n/2$. Applying $d = d^*$, Eq. (18) gives Eq. (24). As $p(\text{cross}|d)$ is not defined for non-integer values of d, equality in Eq. (23) needs that d^* is integer, namely that n is even. In the limit of large n, one has that $p_{\max}(\text{cross}|d) = 1/2$.
- Accordingly, $p(\text{cross}|d)$ has either a maximum at $d = n/2$ if n is even or two maxima, at $d = \lfloor n/2 \rfloor$ and $d = \lceil n/2 \rceil$ when n is even because d is a natural number.

5.2 The Expected Number of Edge Crossings

Imagine that the structure of the tree is defined by an adjacency matrix $A = \{a_{uv}\}$ such that $a_{uv} = 1$ if the vertices u and v are linked and $a_{uv} = 0$ otherwise. Let C be the number of edge crossings and $C(u,v)$ be the number of crossings where the edge formed by u and v is involved ($C(u,v) = 0$ if u and v are unlinked), i.e.

$$C = \frac{1}{4} \sum_{u=1}^{n} \sum_{v=1}^{n} a_{uv} C(u,v) \quad (25)$$

and

$$C(u,v) = \frac{1}{2} \sum_{s=1, s \neq u,v}^{n} \sum_{t=1, t \neq u,v}^{n} a_{st} C(u,v;s,t), \quad (26)$$

where $C(u,v;s,t)$ indicates if u,v and s,t define a couple of edges that cross ($C(u,v;s,t) = 1$ if they cross, $C(u,v;s,t) = 0$ otherwise). Thus, the expectation of C is

$$E[C] = \frac{1}{4} \sum_{u=1}^{n} \sum_{v=1}^{n} a_{uv} E[C(u,v)]. \tag{27}$$

In turn, the expectation of $C(u,v)$ is

$$E[C(u,v)] = \frac{1}{2} \sum_{s=1, s \neq u,v}^{n} \sum_{t=1, t \neq u,v}^{n} a_{st} E[C(u,v;s,t)]. \tag{28}$$

As $C(u,v;s,t)$ is an indicator variable,

$$E[C(u,v;s,t)] = p(s \sim t \ \& \ u \sim v \ \text{cross}), \tag{29}$$

namely $E[C(u,v;s,t)]$ is the probability that the edges $s \sim t$ and $u \sim v$ cross knowing that they do not share any vertex.

The number of edges that can cross the edge $u \sim v$ is $n - k_u - k_v$ (edges that share a vertex with $u \sim v$ cannot cross), which gives (Ferrer-i-Cancho 2013b)

$$E[C(u,v)] = (n - k_u - k_v) p(\text{cross}|u \sim v) \tag{30}$$

assuming that the probability that the edge $u \sim v$ crosses another edge depends at most on $u \sim v$. Applying the last result to Eq. (27), we obtain a general predictor of the number of crossings under a null of hypothesis x, i.e.

$$E_x[C] = \frac{1}{4} \sum_{u=1}^{n} \sum_{v=1}^{n} a_{uv} (n - k_u - k_v) p_x(\text{cross}|u \sim v), \tag{31}$$

where x indicates if the identity of one of the edges potentially involved a crossing, i.e. $u \sim v$ is actually known ($x = 1$ if it is known; $x = 0$ otherwise). Equation (31) defines a family of predictors where x is the parameter. $x = 0$ corresponds to the null hypothesis in Sect. 3. To see it, notice that $x = 0$, i.e.

$$p_x(\text{cross}|u \sim v) = p(\text{cross}) = 1/3 \tag{32}$$

transforms Eq. (31) into Eq. (4) (Ferrer-i-Cancho 2013b). $E_x[C]$ can be seen as a simple approximation to the expected number of crossings in a linear random arrangement of vertices when all edge lengths are given. While $E_0[C]$ is a true expectation, $E_1[C]$ is not for not conditioning on the same lengths for every pair of edges that may cross.

A potentially more accurate prediction of C with regard to Eq. (4) is obtained when $x = 1$. For simplicity, let us reduce the knowledge of an edge to the knowledge of its length. Let us define $d_{uv} = |\pi(u) - \pi(v)|$ as the distance between the vertices u

and v. If $x = 1$, the substitution of $p_x(\text{cross}|u \sim v)$ by $p(\text{cross}|d_{uv})$ in Eq. (31) yields $E_1[C]$, a prediction of C where, for ever pair of edges that many potentially cross, the length of one edge is given and a random placement is assumed for the other edge, i.e.

$$E_1[C] = \frac{1}{4} \sum_{u=1}^{n} \sum_{v=1}^{n} a_{uv}(n - k_u - k_v) p(\text{cross}|d_{uv}). \tag{33}$$

Figure 7 shows that $E_1[C]$ is positively correlated with D on permutations of the vertices of the trees of a real sentences. Interestingly, the values of $E_1[C]$ overestimate the average C when $D < E_0[D]$ and underestimate it when $D > E_0[D]$.

Variations in dependency lengths can alter $E_1[C]$. A drop in $p(\text{cross}|d_{uv})$ leads to a drop in $E_1[C]$. It has been shown above that $p(\text{cross}|d_{uv})$ is minimized by $d = 1$ and $d = n-1$ and maximized by $d_{uv} \approx n/2$. When $d_{uv} < n/2$, a decrease in d_{uv} decreases $p(\text{cross}|d_{uv})$. In contrast, a decrease in d_{uv} when $d_{uv} > n/2$, increases $p(\text{cross}|d_{uv})$. However, the shortening of edges is more likely to decrease $p(\text{cross}|d)$ because

- Tentatively, the linear arrangement of a tree can only have $n - d$ edges of length d (Ferrer-i-Cancho 2004).
- Under the null hypothesis that edges are arranged linearly at random, short edges are more likely than long edges. In that case, $p(d)$, the probability that an edge has length d satisfies $p(d) \propto n - d$ (Eq. (21)).
- The potential ordering of a sentence is unlikely to have many dependencies of length greater than about $n/2$ because the cost of a dependency is positively correlated with its length (Morrill 2000; Gibson and Warren 2004).
- From an evolutionary perspective, initial states are unlikely to involve many long dependencies (Ferrer-i-Cancho 2014).

Thus, it is unlikely that the shortening of edges increases $E_1[C]$. Instead, this is likely to decrease $E_1[C]$. Figure 7 shows that

- the average $E_1[C]$ is tends to increase as D increases
- the mean C is bounded above by $E_1[C]$ (in the domain $D \leq E_0[D]$)

in concrete sentences.

Applying the definition of $p(\text{cross}|d)$, given in Eq. (18), to Eq. (33), yields

$$E_1[C] = \frac{B_2 - B_1}{(n-2)(n-3)}, \tag{34}$$

where

$$B_2 = \frac{1}{2} \sum_{u=1}^{n} \sum_{v=1}^{n} a_{uv}(n - k_u - k_v)(n - d_{uv}) d_{uv} \tag{35}$$

and

$$B_1 = \frac{n-1}{2} \sum_{u=1}^{n} \sum_{v=1}^{n} a_{uv}(n - k_u - k_v)$$

$$= \frac{n-1}{2} \left(n \sum_{u=1}^{n} \sum_{v=1}^{n} a_{uv} - 2 \sum_{u=1}^{n} k_{uv} \sum_{v=1}^{n} a_{uv} \right)$$

$$= \frac{n-1}{2} \left(n \sum_{u=1}^{n} k_u - 2 \sum_{u=1}^{n} k_u^2 \right)$$

$$= \frac{n-1}{2} \left(2n(n-1) - 2n \langle k^2 \rangle \right) \qquad \text{applying Eqs. (2) and (5)}$$

$$= n(n-1)\left(n - 1 - \langle k^2 \rangle\right). \tag{36}$$

On the one hand, notice that $B_1 \geq 0$ because $n \geq 1$ and $\langle k^2 \rangle \leq n-1$ (Ferrer-i-Cancho 2013a). On the other hand, notice that $B_2 \geq 0$ because

- The fact that $C(u,v) < 0$ is impossible by definition and also the fact that $C(u,v) \leq n - k_u - k_v$ (Ferrer-i-Cancho 2013a) yields $n - k_u - k_v \geq 0$.
- $d_{uv} \leq n - 1$ by definition (Ferrer-i-Cancho 2013a).

Thus, $E_1[C]$ is proportional to the difference between two terms: a term B_2 that depends on dependency lengths and a term B_1 that depends exclusively on vertex degrees and sentence length (in words). Interestingly, Eq. (4) allows one to see $E_1[C]$ as a function of $E_0[C]$ since $B_1 = 6(n-1)E_0[C]$.

6 Another Stronger Null Hypothesis

Another prediction from a stronger null hypothesis is $E[C|D]$, the expected value of C when one of the orderings (permutations) preserving the actual value of D is chosen uniformly at random. Figure 7 shows that knowing the exact value of D, a positive correlation between $E[C|D]$ and D follows for a concrete sentence (this is what circles are showing) and, interestingly, $E[C|D]$ predicts a smaller number of crossings than the average $E_1[C]$ as a function of D.

7 Predictions, Testing and Selection

Table 2 compares $E_0[C]$, $E_1[C]$ and $E[C|D]$ for the example sentences that have appeared so far. Table 2 shows that, according to the normalized error,

- $E_0[C]$ makes the worst predictions.
- The predictions of $E[C|D]$ are the best except in one sentence where $E_1[C]$ wins.

The take home message here is that it is possible to make quantitative predictions about the number of crossings, an aspect that theoretical approaches to the problem of crossings dependencies have missed (Morrill et al. 2009; Vries et al. 2012; Levy et al. 2012). This quantitative requirement should not be regarded as a mathematical *divertimento*: science is about predictions (Bunge 2013).

Table 2 The predicted crossings by each of the three null hypotheses, i.e. (1) random linear arrangement of vertices, (2) random linear arrangement with knowledge of the length of one of the edges that can potentially cross, and (3) random linear arrangement of vertices at constant sum of dependency lengths, for the example sentences employed in this article. Results of p-value testing for the 3rd null hypothesis are also included. n is the number of vertices of the tree, $\langle k^2 \rangle$ is the degree second moment about zero, D is the sum of dependency lengths, C is the actual number of crossings, C_{max} is the potential number of crossings, i.e. $C_{max} = n(n-1-\langle k^2 \rangle)/2$ (Ferrer-i-Cancho 2013a). $E_0[C], E_1[C]$ and $E[C|D]$ are the predicted number of crossings according to the 1st, 2nd and 3rd hypotheses, respectively. $\varepsilon_0[C]$, $\varepsilon_1[C]$ and $\varepsilon[C|D]$ are the normalized error of the 1st, 2nd and 3rd hypotheses, respectively, i.e. $\varepsilon_x[C] = |C - E_x[C]|/C_{max}$ and $\varepsilon[C|D] = |C - E[C|D]|/C_{max}$. Left and right p-values are provided for two statistics, C and $|C - E[C|D]|$ under the 3rd null hypothesis. R is the number of permutations of the vertex sequence where D coincides with the original value.

	Figure 1, top	Figure 1, bot.	Figure 5, top	Figure 5, bot.
n	9	10	7	7
$\langle k^2 \rangle$	4	4.2	3.4	3.4
D	13	17	15	10
C	0	1	0	1
C_{max}	18	24	9	9
$E_0[C]$	6	8	3	3
$\varepsilon_0[C]$	0.33	0.29	0.33	0.22
$E_1[C]$	2.4	4.2	1.2	2.2
$\varepsilon_1[C]$	0.13	0.14	0.13	0.13
$E[C|D]$	1.1	2	2.8	1.1
$\varepsilon[C|D]$	0.062	0.041	0.31	0.011
Left p-value of C	0.28	0.37	0.058	0.69
Right p-value of C	1	0.88	1	0.75
Left p-value of $|C - E[C|D]|$	0.94	0.54	0.97	0.43
Right p-value of $|C - E[C|D]|$	0.33	0.71	0.088	1
R	288	6664	548	102

Our exploration of some sentences suggest that $E[C|D]$ makes better predictions in general. Notwithstanding we cannot rush to claim that $E[C|D]$ is the best model among those that we have considered only for that reason. According to standard model selection, the best model is one with the best compromise between the quality of its fit (the error of its predictions) and parsimony (the number of parameters, Burnham and Anderson 2002). Applying a rigorous framework for model selection is beyond the scope of this article but we can examine the parsimony of each model qualitatively to shed light on the best model. Interestingly, the three predictors vary concerning the amount of information that suffices to make a prediction. A comparison of the minimal information that each needs to make a prediction would be a better approach but that is beyond the scope of the current article. The value of $E_0[C]$ can be computed knowing only n and $\langle k^2 \rangle$ (Eq. (4)). The value of $E_1[C]$ can be computed knowing only n, the length of every edge and the degrees of the nodes forming every edge (Eqs. (18) and (33)). The calculation of $E[C|D]$ is less demanding than that of $E_1[C]$ concerning edge lengths. i.e. the total sum of the their lengths suffices (the length of individual edges is irrelevant), but still employs some information about edges, e.g., the nodes forming every edge (Sect. 6). Our preliminary results (Table 2) and our analysis of parsimony from the perspective of sufficient information suggests that $E[C|D]$ has a better trade-off between the quality of its predictions and parsimony with respect to $E_1[C]$. Interestingly, none of the models has free parameters if the ordering of the vertices and the syntactic dependency structure of the sentence is known as we have assumed so far.

It is possible to perform traditional p-value testing of our models. For simplicity we focus on the null hypothesis that is defined in Sect. 6, i.e. permutations of vertices where the sum of edge lengths coincides with the true value (Table 2 shows the number of permutations of this kind for each sentence). We consider two statistics for test test. First, C, the actual number of crossings of a dependency tree. Then one can define the left p-value as the proportion of those permutations with a number of crossings at most as large as the true value. Similarly, one can define the right p-value as the proportion of those permutations with a number of crossings at least as large as the true value. One has to choose a significance level α. The significance level must be such that there can be p-values bounded above by α a priori. Otherwise, one is condemned to make type II errors. The smallest possible p-value is $1/R$, where R is the number of permutations yielding the original value of D (there is at least one permutation giving the same value of the statistic, i.e. the one that coincides with the original linear arrangement). Thus, α must exceed $1/R_{\min}$, being $R_{\min}$ the smallest value of R in Table 2. $R_{\min} = 102$ yields $\alpha \geq 1/102 \approx 0.01$. Thus we can safely choose a significance level of $\alpha = 0.05$. Table 2 shows that the left and right p-values are always below α, suggesting that the real numbers of crossings are compatible with those of this null hypothesis. However, notice that the p-value is borderline for one sentence (Fig. 5, top). Second, we consider $|C - E[C|D]|$, the absolute value of the difference between the actual number of crossings and the expected value, as another statistic to perform p-value testing. Accordingly, we define left and right p-values for the latter statistic following the same rationale used for the p-values of C. Table 2 shows $|C - E[C|D]|$ is neither significantly low nor significantly large, thus providing support for the hypothesis that the number of crossings is determined by global constraints on dependency lengths.

Our p-value tests should be seen as preliminary statistical attempts. Future research should involve more languages and large dependency treebanks. Moreover, information theoretic models selection offers a much powerful approach over traditional p-value testing (Burnham and Anderson 2002) and should be explored in the future.

8 Discussion

In this article, the possibility that the low number of crossings in dependency trees is a mere consequence of chance has been considered. As the expected number of crossings (when edge lengths are not given) decreases as the degree 2nd moment increases, a high hubiness could lead to a small number of crossings by chance, when dependency lengths are unconstrained. However, it has been shown that the hubiness required to have a small number of crossings in that circumstance would imply star trees, which are problematic for at least two reasons: real syntactic dependency trees of a sufficient length are not star trees and the diversity of possible syntactic dependency structures would be seriously compromised (Sect. 3). One cannot exclude that hubiness has some role in decreasing the potential number of crossings in sentences (Ferrer-i-Cancho 2013a; Liu 2007) but it cannot be the only reason. "Grammar" has been examined as an explanation for the rather low frequency of crossings and the more parsimonious hypothesis of the minimization of syntactic dependency lengths (Ferrer-i-Cancho 2006; Liu 2008; Morrill et al. 2009; Ferrer-i-Cancho 2013a) has been revisited (Sect. 4). Stronger null hypotheses involving partial information about dependency lengths suggest that the shortening of the dependencies is likely to imply a reduction of crossings (recall empirical evidence in Fig. 7 and general mathematical arguments based on one of those stronger null hypotheses in Sect. 5.2). Moreover, it has been shown that a null hypothesis incorporating global information about dependency lengths, i.e. the sum of dependency lengths (Sect. 6), allows one to make specially accurate predictions about the actual number of crossings. The error of those predictions is neither surprisingly low nor surprisingly high (Sect. 7). Our findings provide support for the hypothesis that uncrossing dependencies could be a side-effect of dependency length minimization, a principle that derives from the limited computational resources of the human brain (Liu 2008; Morrill 2000; Gibson 2000; Hawkins 1994). A universal grammar, a faculty of language or a competence-plus limiting the number of crossings might not be necessary. Just a version of Zipf's least effort might suffice (Zipf 1949).

Upon a superficial analysis of facts, it is tempting to conclude that crossings cause processing difficulties and thus should be reduced. That follows easily from the correlation between crossings and dependency lengths that has been found in real and artificial sentence structures (e.g. Ferrer-i-Cancho 2013b; Liu 2007). However, the cognitive cost of crossings does not need to be a direct consequence of the crossing (Liu 2008) but a side effect of the longer dependencies that crossings are likely to involve (Ferrer-i-Cancho 2006; Morrill et al. 2009). It has been argued that a single principle of minimization of dependency crossings would compromise seriously parsimony and explanatory power (Sect. 4).

Although Fig. 7 shows that solving the minimum linear arrangement problem yields zero crossings for concrete sentences ($E[C|D] = 0$ and thus $C = 0$ for $D = D_{\min}$), it is important to bear in mind that dependency length minimization cannot promise to reduce crossings to zero in all cases: the minimum linear arrangement of a tree can involve crossings (Fig. 6). Interestingly, this means that the presence of some crossings in a sentence does not contradict a priori pressure for dependency length minimization or pressure for efficiency. Recall also the examples of real English in Fig. 5, showing that an ordering without crossings can have a higher sum of dependency lengths than one with crossings. This is specially important for Dutch, where crossing structures abound and have been shown to be easier to process that parallel non-crossing structures in German (Bach et al. 1986). We believe that our theoretical framework might help to illuminate experiments suggesting that orderings with crossings tax working memory less than orderings with nesting (Vries et al. 2012).

In this article, we have addressed the problem of non-crossing dependencies from a really theoretical perspective. The arguments are a priori valid for any language and any linguistic phenomenon. This is a totally different approach from the investigation of non-crossing dependencies in a given language with a specific phenomenon (e.g., extraposition of relative clauses in English as in Levy et al. (2012), see also Bach et al. (1986) for other languages). With such a narrow focus, the development of a general theory of word order is difficult. Generativists have been criticized for not having developed a general theory of language but a theory of English (Evans and Levinson 2009). If one takes seriously recent concerns about the limits of building a theory from a sample of languages (Piantadosi and Gibson 2014), it follows that hypotheses about non-crossing dependencies that abstract from linguistic details like ours (see also Vries et al. 2012) should receive more attention in the future.

Although we believe that least effort is the main reason why crossings dependencies do not occur very often in languages, we do not believe that pressure for dependency length minimization is the only factor involved in word order phenomena. The maximization of predictability or the structure of word order permutation space are crucial ingredients for a non-reductionist theory of word order (Ferrer-i-Cancho 2015; Ferrer-i-Cancho 2014). Word order is a multiconstraint engineering problem (Ferrer-i-Cancho 2014).

Tentatively, a deep theory of syntax does not imply grammar or a language faculty exclusively. A well-known example is the case of sentence acceptability that may derive in some cases from processing constraints (Morrill 2000; Hawkins 2004). Our findings suggest that grammar may not be an autonomous entity but a series of phenomena emerging from physical or psychological constraints. Grammar might simply be an epiphenomenon (Hopper 1998). This is a more parsimonious hypothesis than grammar as a conventionalization of processing constraints (Hawkins 2004). Grammar may require fewer parameters than commonly believed. This is consistent with the idea that it would be desirable that the quantitative constraints of competence-plus are replaced by a theory of processing complexity or that the content of the "plus" derives from memory and integration costs (Hurford 2012). Being the "plus" part non-empty, the point is elucidating whether the "competence" part is indeed empty or at least lighter than commonly believed.

A deep theory of language cannot be circumscribed to language: a deep physical theory for the fall of inanimate objects is also valid for animate objects (with certain refinements). A deep theory of syntactic structures and their linear arrangement does not need to be valid only for human language but also for other natural systems producing and processing sequences and operating under limited resources. For these reasons, our results should be extended to non-tree structures to investigate crossings in RNA structures (Chen et al. 2009).

Acknowledgements. The deep and comprehensive review of Steffen Eger has been crucial for the preparation of the final version of this article. We owe him the current proof of Proposition 8.2. We are also grateful to D. Blasi, R. Čech, M. Christiansen, J. M. Fontana, S. Frank, E. Gibson, H. Liu and G. Morrill for helpful comments or discussions. All remaining errors are entirely ours. This work was supported by the grant *Iniciació i reincorporació a la recerca* from the Universitat Politècnica de Catalunya and the grants BASMATI (TIN2011-27479-C04-03) and OpenMT-2 (TIN2009-14675-C03) from the Spanish Ministry of Science and Innovation.

Appendix

Tree Reduction

As any tree of at least two vertices has at least two leaves (Bollobás 1998, p. 11), a tree of $n+1$ vertices (with $n > 1$) yields a reduced tree of n vertices by removing one of its leaves. Notice that such a reduction from a tree of $n+1$ to a tree of n nodes will never produce a disconnected graph.

Consider that a tree has a leaf that is attached to a vertex of degree k. Then, the sum of squared degrees of a tree of $n+1$ vertices, i.e. $K_2(n+1)$, can be expressed as a function of $K_2(n)$, the sum of squared degrees of a reduced tree of n vertices, i.e.

$$\begin{aligned} K_2(n+1) &= K_2(n) + k^2 - (k-1)^2 + 1 \\ &= K_2(n) + 2k \end{aligned} \tag{37}$$

with $n \geq 0$.

The Only Tree That Has Degree Second Moment Greater Than That of a Quasi-star Tree Is a Star Tree.

A quasi-star tree is a tree with one vertex of degree $n-2$, one vertex of degree 2 and the remainder of vertices of degree 1. As a tree must be connected, that tree needs $n > 2$. The sum of squared degrees of a quasi-star tree is

$$K_2^{\text{quasi}}(n) = (n-2)^2 + 4 + n - 2 = n^2 - 3n + 6. \tag{38}$$

The sum of squared degrees of a star tree is (Ferrer-i-Cancho 2013a)

$$K_2^{\text{star}}(n) = n(n-1). \tag{39}$$

$K_2^{star}(n)$ is an upper bound of $K_2^{quasi}(n)$, more precisely,

Proposition 8.1. *For all $n \geq 3$,*

$$K_2^{quasi}(n) \leq K_2^{star}(n) \tag{40}$$

with equality if and only if $n = 3$.

Proof. Applying the definitions in Eqs. (38) and (39) to Eq. (40), we obtain

$$n^2 - 3n + 6 \leq n(n-1), \tag{41}$$

that is $n \geq 3$. □

$K_2(n)$ is maximized by star trees (Ferrer-i-Cancho 2013a). Quasi-star trees yield the second largest possible value of $K_2(n)$, more precisely

Proposition 8.2. *For all $n \geq 3$, it holds that*

$$K_2(n) > K_2^{quasi}(n) \implies K_2(n) = K_2^{star}(n) \tag{42}$$

for any tree with n vertices.

Proof. Denote the antecedent of the implication in Eq. (42) by $L(n)$ and the consequent by $R(n)$. We show by induction that, for all $n \geq 3$, $L(n) \implies R(n)$.

For $n = 3$, the fact that the only possible tree is both a star and a quasi-star tree implies that $L(n)$ is false. Thus, Eq. (42) holds trivially.

Let $n > 3$. For the induction step, assume that $L(n) \implies R(n)$, and also assume that $L(n+1)$ holds. We must show that then also $R(n+1)$ holds.

Consider an arbitrary tree T_{n+1} with $n+1$ vertices and consider the tree T_n, on n vertices, with a leaf l removed from T_{n+1}.

If $L(n)$ holds, then, by the induction hypothesis, $R(n)$ holds, i.e., $K_2(n) = K_2^{star}(n)$. A tree with n vertices for which $R(n)$ holds must be a star tree (Ferrer-i-Cancho 2013b); thus, T_n is a star tree. Then, the leaf vertex l is

- either attached to the hub of the star tree, in which case the resulting tree T_{n+1} is also a star tree, so that $K_2(n+1) = K_2^{star}(n+1)$, i.e., $R(n+1)$, holds.
- or attached to a leaf of T_n, in which case T_{n+1} is a quasi-star tree, contradicting that $L(n+1)$ holds.

Conversely, if $L(n)$ does not hold, then $K_2(n) \leq K_2^{quasi}(n)$. Accordingly, a tree T_{n+1} with a leaf of degree k satisfies (for any k)

$$K_2(n+1) = K_2(n) + 2k \leq K_2^{quasi}(n) + 2k = n^2 - 3n + 6 + 2k, \tag{43}$$

being $K_2(n)$ the sum of squared degrees of the reduced tree. Now, we have assumed that $L(n+1)$ holds, i.e., that

$$K_2(n+1) > K_2^{quasi}(n+1) = n^2 - n + 4, \tag{44}$$

thanks to Eq. (38). Combining Eqs. (43) and (44), it is obtained

$$n^2 - n + 4 < K_2(n+1) \leq n^2 - 3n + 6 + 2k,$$

which implies

$$2n < 2 + 2k.$$

But this would require that $k > n - 1$, that is, $k = n$, since the maximum degree of a vertex in a tree with $n+1$ vertices is n. In other words, T_{n+1} would be forced to have a vertex of degree $k = n$, whence T_{n+1} is a star tree, so that T_n also is a star tree (removing a leaf from a star tree yields a star tree). But, this would contradict that $L(n)$ does not hold since $K_2^{\text{star}}(n) > K_2^{\text{quasi}}(n)$ for $n > 3$ (Proposition 8.1). □

References

Aldous, D.: The Random Walk Construction of Uniform Spanning Trees and Uniform Labelled Trees. SIAM J. Disc. Math. 3, 450–465 (1990), doi:10.1137/0403039

Bach, E., Brown, C., Marslen-Wilson, W.: Crossed and Nested Dependencies in German and Dutch: A Psycholinguistic Study. Language and Cognitive Processes 1(4), 249–262 (1986), doi:10.1080/01690968608404677

Baronchelli, A., Ferrer-i-Cancho, R., Pastor-Satorras, R., Chatter, N., Christiansen, M.: Networks in Cognitive Science. Trends in Cognitive Sciences 17, 348–360 (2013), doi:10.1016/j.tics.2013.04.010

Bollobás, B.: Modern Graph Theory. Graduate Texts in Mathematics. Springer, New York (1998)

Broder, A.: Generating Random Spanning Trees. In: Symp. Foundations of Computer Sci., pp. 442–447. IEEE, New York (1989)

Bunge, M.: La ciencia. Su método y su filosofía. Laetoli, Pamplona (2013)

Burnham, K.P., Anderson, D.R.: Model Selection and Multimodel Inference. A Practical Information-theoretic Approach, 2nd edn. Springer, New York (2002)

Chen, W.Y.C., Han, H.S.W., Reidys, C.M.: Random k-noncrossing RNA Structures. Proceedings of the National Academy of Sciences 106(52), 22061–22066 (2009), doi:10.1073/pnas.0907269106

Chomsky, N.: Aspects of the Theory of Syntax. MIT Press, Cambridge (1965)

Chomsky, N.: The Minimalist Program. MIT Press (1995)

Christiansen, M.H., Chater, N.: Toward a Connectionist Model of Recursion in Human Linguistic Performance. Cognitive Science 23(2), 157–205 (1999), doi:10.1207/s15516709cog2302_2

Chung, F.R.K.: On Optimal Linear Arrangements of Trees. Comp. & Maths. with Appls. 10(1), 43–60 (1984), doi:10.1016/0898-1221(84)90085-3

Culicover, P.W., Jackendoff, R.: Quantitative Methods Alone Are not Enough: Response to Gibson and Fedorenko. Trends in Cognitive Sciences 14(6), 234–235 (2010), doi:10.1016/j.tics.2010.03.012

Evans, N., Levinson, S.C.: The Myth of Language Universals: Language Diversity and its Importance for Cognitive Science. Behavioral and Brain Sciences 32, 429–492 (2009), doi:10.1017/S0140525X0999094X

Ferrer i Cancho, R., Solé, R.V.: Optimization in Complex Networks. In: Pastor-Satorras, R., Rubí, J.M., Díaz-Guilera, A. (eds.) Statistical Mechanics of Complex Networks. Lecture Notes in Physics, vol. 625, pp. 114–125. Springer, Berlin (2003), doi:10.1007/b12331

Ferrer-i-Cancho, R.: Euclidean Distance between Syntactically Linked Words. Physical Review E 70, 056135 (2004), doi:10.1103/PhysRevE.70.056135

Ferrer-i-Cancho, R.: Why Do Syntactic Links not Cross? Europhysics Letters 76(6), 1228–1235 (2006), doi:10.1209/epl/i2006-10406-0

Ferrer-i-Cancho, R.: Hubiness, Length, Crossings and Their Relationships in Dependency Trees. Glottometrics 25, 1–21 (2013a)

Ferrer-i-Cancho, R.: Random Crossings in Dependency Trees (2013b), http://arxiv.org/abs/1305.4561

Ferrer-i-Cancho, R.: The Optimality of Attaching Unlinked Labels to Unlinked Objects (2013c), http://arxiv.org/abs/1310.5884

Ferrer-i-Cancho, R.: Why Might SOV Be Initially Preferred and Then Lost or Recovered? A Theoretical Framework. In: Cartmill, E.A., Roberts, S., Lyn, H., Cornish, H. (eds.) The Evolution of Language – Proceedings of the 10th International Conference (EVOLANG10). Evolution of Language Conference (Evolang 2014), April 14-17, pp. 66–73. Wiley, Vienna (2014), doi:10.1142/9789814603638_0007

Ferrer-i-Cancho, R.: The Placement of the Head that Minimizes Online Memory: a Complex Systems Approach. Language Dynamics and Change 4(2) (in press, 2015)

Ferrer-i-Cancho, R., Elvevåg, B.: Random Texts Do not Exhibit the Real Zipf's-law-like Rank Distribution. PLoS ONE 5(4), e9411 (2009), doi:10.1371/journal.pone.0009411

Ferrer-i-Cancho, R., Forns, N., Hernández-Fernández, A., Bel-Enguix, G., Baixeries, J.: The Challenges of Statistical Patterns of Language: the Case of Menzerath's law in Genomes. Complexity 18(3), 11–17 (2013), doi:10.1002/cplx.21429

Ferrer-i-Cancho, R., Hernández-Fernández, A., Lusseau, D., Agoramoorthy, G., Hsu, M.J., Semple, S.: Compression as a Universal Principle of Animal Behavior. Cognitive Science 37(8), 1565–1578 (2013), doi:10.1111/cogs.12061

Ferrer-i-Cancho, R., Liu, H.: The Risks of Mixing Dependency Lengths from Sequences of Different Length. Glottotheory 5(2), 143–155 (2014), doi:10.1515/glot-2014-0014

Ferrer-i-Cancho, R., Moscoso del Prado Martín, F.: Information Content Versus Word Length in Random Typing. Journal of Statistical Mechanics, L12002 (2011), doi:10.1088/1742-5468/2013/07/L07001

Frank, S.L., Bod, R., Christiansen, M.H.: How hierarchical is language use? Proceedings of the Royal Society B: Biological Sciences 279, 4522–4531 (2012), doi:10.1098/rspb.2012.1741

Gell-Mann, M., Ruhlen, M.: The Origin and Evolution of Word Order. Proceedings of the National Academy of Sciences USA 108(42), 17290–17295 (2011), doi:10.1073/pnas.1113716108

Gibson, E.: The Dependency Locality Theory: a Distance-based Theory of Linguistic Complexity. In: Image, Language, Brain, pp. 95–126. The MIT Press, Cambridge (2000)

Gibson, E., Fedorenko, E.: Weak Quantitative Standards in Linguistics Research. Trends in Cognitive Sciences 14(6), 233–234 (2010), doi:10.1016/j.tics.2010.03.005

Gibson, E., Warren, T.: Reading Time Evidence for Intermediate Linguistic Structure in Long-Distance Dependencies. Syntax 7, 55–78 (2004), doi:10.1111/j.1368-0005.2004.00065.x

Goldberg, A.E.: Constructions: a New Theoretical Approach to Language. Trends in Cognitive Sciences 7(5), 219–224 (2003), doi:10.1016/S1364-6613(03)00080-9

Gómez-Rodríguez, C., Fernández-González, D., Bilbao, V.M.D.: Undirected Dependency Parsing. Computational Intelligence (2014), doi:10.1111/coin.12027

Hauser, M.D., Chomsky, N., Fitch, W.T.: The Faculty of Language: What Is It, Who Has It and How Did It Evolve? Science 298, 1569–1579 (2002), doi:10.1126/science.298.5598.1569

Hawkins, J.A.: A Performance Theory of Order and Constituency. Cambridge University Press, New York (1994)

Hawkins, J.A.: Some Issues in a Performance Theory of Word Order. In: Siewierska, A. (ed.) Constituent Order in the Languages of Europe. Mouton de Gruyter, Berlin (1998)

Hawkins, J.A.: Efficiency and Complexity in Grammars. Oxford University Press, Oxford (2004)

Hays, D.: Dependency Theory: a Formalism and Some Observations. Language 40, 511–525 (1964)

Hochberg, R.A., Stallmann, M.F.: Optimal One-page Tree Embeddings in Linear Time. Information Processing Letters 87, 59–66 (2003), doi:10.1016/S0020-0190(03)00261-8

Hopper, P.J.: Emergent Grammar. In: Tomasello, M. (ed.) The New Psychology of Language: Cognitive and Functional Approaches to Language Structure, pp. 155–175. Lawrence Erlbaum, Mahwah (1998)

Hudson, R.: Word Grammar. Blackwell, Oxford (1984)

Hudson, R.: Language Networks. The New Word Grammar. Oxford University Press, Oxford (2007)

Hurford, J.R.: Chapter 3. Syntax in the Light of Evolution. In: The Origins of Grammar. Language in the Light of Evolution II, pp. 175–258. Oxford University Press, Oxford (2012)

Jackendoff, R.: Foundations of Language. Oxford University Press, Oxford (1999)

Köhler, R.: Synergetic Linguistics. In: Quantitative Linguistik. Ein internationals Handbuch. Quantitative Linguistics: An International Handbook, pp. 760–775. Walter de Gruyter, Berlin (2005)

Lecerf, Y.: Programme des Conflits - Modèle des Conflits. Rapport CETIS No. 4. Euratom, pp. 1–24 (1960)

Levy, R., Fedorenko, E., Breen, M., Gibson, E.: The Processing of Extraposed Structures in English. Cognition 122(1), 12–36 (2012), doi:10.1016/j.cognition.2011.07.012

Liu, H.: Probability Distribution of Dependency Distance. Glottometrics 15, 1–12 (2007)

Liu, H.: Dependency Distance as a Metric of Language Comprehension Difficulty. Journal of Cognitive Science 9, 159–191 (2008)

Liu, H.: Dependency Direction as a Means of Word-order Typology: A Method Based on Dependency Treebanks. Lingua 120(6), 1567–1578 (2010), doi:10.1016/j.lingua.2009.10.001

Manning, C.D.: Probabilistic Syntax. In: Bod, R., Hay, J., Jannedy, S. (eds.) Probabilistic Linguistics, pp. 289–341. MIT Press, Cambridge (2002)

McDonald, R., Pereira, F., Ribarov, K., Hajič, J.: Non-projective Dependency Parsing Using Spanning Tree Algorithms. In: Proceedings of the Conference on Human Language Technology and Empirical Methods in Natural Language Processing, HLT 2005, pp. 523–530. Association for Computational Linguistics, Vancouver (2005), doi:10.3115/1220575.1220641

Mel'čuk, I.: Dependency Syntax: Theory and Practice. State of New York University Press, Albany (1988)

Miller, G.A.: Introduction. In: Zipf, G.K. (ed.) The Psycho-Biology of Language: an Introduction to Dynamic Psychology, pp. v–x. MIT Press, Cambridge (1968)

Miller, G.A., Chomsky, N.: Finitary Models of Language Users. In: Luce, R.D., Bush, R., Galanter, E. (eds.) Handbook of Mathematical Psychology, vol. 2, pp. 419–491. Wiley, New York (1963)

Molloy, M., Reed, B.: A Critical Point for Random Graphs with a Given Degree Sequence. Random Structures and Algorithms 6, 161–180 (1995), doi:10.1002/rsa.3240060204

Molloy, M., Reed, B.: The Size of the Largest Component of a Random Graph on a Fixed Degree Sequence. Combinatorics, Probability and Computing 7, 295–306 (1998), doi:10.1017/S0963548398003526

Moon, J.: Counting Labelled Trees. Canadian Math. Cong. (1970)

Morrill, G.: Incremental Processing and Acceptability. Computational Linguistics 25(3), 319–338 (2000), doi:10.1162/089120100561728

Morrill, G., Valentín, O., Fadda, M.: Dutch Grammar and Processing: A Case Study in TLG. In: Bosch, P., Gabelaia, D., Lang, J. (eds.) TbiLLC 2007. LNCS, vol. 5422, pp. 272–286. Springer, Heidelberg (2009)

Moscoso del Prado, F.: The Missing Baselines in Arguments for the Optimal Efficiency of Languages. In: Knauff, M., Pauen, M., Sebanz, N., Wachsmuth, I. (eds.) Proceedings of the 35th Annual Conference of the Cognitive Science Society, pp. 1032–1037. Cognitive Science Society, Austin (2013)

Newman, M.E.J., Strogatz, S.H., Watts, D.J.: Random Graphs with Arbitrary Degree Distribution and Their Applications. Phys. Rev. E 64, 026118 (2001), doi:10.1103/PhysRevE.64.026118

Newmeyer, F.J.: The Prague School and North American Functionalist Approaches to Syntax. Journal of Linguistics 37, 101–126 (2001), doi:10.1017/S0022226701008593

Newmeyer, F.J.: Grammar is Grammar and Usage is Usage. Language 79, 682–707 (2003), doi:10.1353/lan.2003.0260

Niyogi, P., Berwick, R.C.: A Note on Zipf's Law, Natural Languages, and Noncoding DNA Regions. A.I. Memo No. 1530/C.B.C.L. Paper No.118 (1995)

Noy, M.: Enumeration of Noncrossing Trees on a Circle. Discrete Mathematics 180, 301–313 (1998), doi:10.1016/S0012-365X(97)00121-0

Piantadosi, S.T., Tily, H., Gibson, E.: Word Lengths are Optimized for Efficient Communication. Proceedings of the National Academy of Sciences 108(9), 3526–3529 (2011)

Piantadosi, S.T., Gibson, E.: Quantitative Standards for Absolute Linguistic Universals. Cognitive Science 38(4), 736–756 (2014), doi:10.1111/cogs.12088

Seoane, N.J.A.: Number of Trees with n Unlabeled Nodes. In: The On-Line Encyclopedia of Integer Sequences (2013), http://oeis.org/A000055

Solé, R.V.: Genome Size, Self-organization and DNA's Dark Matter. Complexity 16(1), 20–23 (2010), doi:10.1002/cplx.20326

Suzuki, R., Tyack, P.L., Buck, J.: The Use of Zipf's Law in Animal Communication Analysis. Anim. Behav. 69, 9–17 (2005), doi:10.1016/j.anbehav.2004.08.004

Uriagereka, J.: Rhyme and Reason. An Introduction to Minimalist Syntax. The MIT Press, Cambridge (1998)

de Vries, M.H., Petersson, K.M., Geukes, S., Zwitserlood, P., Christiansen, M.H.: Processing Multiple Non-adjacent Dependencies: Evidence from Sequence Learning. Philosophical Transactions of the Royal Society B: Biological Sciences 367(1598), 2065–2076 (2012), doi:10.1098/rstb.2011.0414

Zipf, G.K.: Human Behaviour and the Principle of Least Effort. Addison-Wesley, Cambridge (1949)

Part IV
Dynamics

Simulating the Effects of Cross-Generational Cultural Transmission on Language Change

Tao Gong and Lan Shuai

Abstract. Language evolves in a socio-cultural environment. Apart from biological evolution and individual learning, cultural transmission also casts important influence on many aspects of language evolution. In this paper, based on the lexicon-syntax coevolution model, we extend the acquisition framework in our previous work to examine the roles of three forms of cultural transmission spanning the offspring, parent, and grandparent generations in language change. These transmissions are: those between the parent and offspring generations (PO), those within the offspring generation (OO), and those between the grandparent and offspring generations (GO). The simulation results of the considered model and relevant analyses illustrate not only the necessity of PO and OO transmissions for language change, thus echoing our previous findings, but also the importance of GO transmission, a form of cross-generational cultural transmission, on preserving the mutual understandability of the communal language across generations of individuals.

1 Introduction

As a socio-cultural phenomenon, language is acquired and exchanged in a population of individuals during cultural transmission. Generally speaking, *cultural transmission* is the process by which information is passed from individual to individual via social learning mechanisms such as imitation, teaching, or language (Mesoudi and Whiten 2008). In linguistics, *cultural transmission* refers to the process of language adaptation in a community via communications of individuals from the same or different generations (Christiansen and Kirby 2003b). Many traditional frameworks of language evolution focused primarily on biological evolution and individual learning. For example, Chomsky's single-speaker-single-listener framework

---

Tao Gong · Lan Shuai
Haskins Laboratories, New Haven, CT, USA
e-mail: {gtojty,susan.shuai}@gmail.com

(Chomsky 1972) ignored completely the socio-cultural factors during language interactions, and many studies incorporating this framework examined primarily the biological capacities for language and the nature of language learning abilities (Hauser et al. 2002; Jackendoff 2002). However, the roles of cultural transmission in the origin of human communication system and the evolution of human language have been revealed in recent laboratory experiments (e.g., Galantucci (2005), Kirby et al. (2002), Cornish (2010), and Scott-Phillips and Kirby (2002)), and typological studies have suggested that cultural transmission might play important roles in linguistic diversity (Evans and Levinson 2009) and shaping particular language structures (Dunn et al. 2011). Noting these, evolutionary linguists start to reconsider the roles of cultural transmission in language evolution (e.g., Christiansen and Kirby (2003b), Labov (2001), Mufwene (2001)).

Apart from empirical explorations, computer models of language acquisition and/or interactions (e.g., Cangelosi and Parisi (2002), Christiansen and Kirby (2003a), Bickerton and Szathmáry (2009), and Steels (2012)) have long taken socio-cultural factors into account (e.g., Ke et al. (2008), Gong et al. (2008), and Gong et al. (2012)), and different models have incorporated various forms of cultural transmission. For example, Axelrod adopted *horizontal transmission* (communications among individuals of the same generation) into his model studying the effect of this transmission on cultural dissemination (Axelrod 1997). Horizontal transmission was also involved in many lexical or grammatical evolution models (e.g., Steels (2012) and Ke et al. (2008)). Kirby simulated *vertical transmission* (a member of one generation talking to a biologically-related member of a later generation) in his iterated learning model tracing the origin of compositionality via this transmission across a series of single-member generations (Kirby 1999). Vertical transmission was also set up in many laboratory experiments of language evolution (e.g., Kirby et al. (2002) and Cornish (2010)). Together with horizontal and vertical transmissions, *oblique transmission* (a member of one generation talking to a non-biologically-related member of a later generation) was also implemented in some modeling studies (e.g., Smith and Hurford (2003) and Lenaerts et al. (2005)).

In order to simulate a realistic cultural environment and systematically analyze the effect of cultural transmission on language evolution, we proposed an acquisition framework that unified the above three forms of cultural transmission between consecutive generations of language learners (Gong et al. 2008; Gong 2010). Compared with other studies that also involved more than one form of cultural transmission (e.g., Vogt (2005)), this acquisition framework defined explicit parameters to respectively control the ratios of these forms of transmission in the total number of transmissions taking place during language acquisition. In this way, this framework allows analyzing the respective and collective effects of those transmissions on the origin and change of communal language across generations of individuals. Our simulations not only revealed an integrated role of oblique transmission that combined the roles of horizontal and vertical transmissions in preserving linguistic understandability within and across generations, but also showed that both horizontal and oblique transmissions were more necessary than vertical transmission for language evolution in a multi-individual cultural environment.

Despite of these insightful findings, this framework was restricted to two consecutive generations, within which all three forms of transmission took place. In reality, cultural transmission and language acquisition may span more than two generations. For example, apart from parents, grandparents may also care for grandchildren whose parents are unable or unwilling to do so, either as "kinship care-givers" where arrangements are made by social services (Farmer and Moyers 2008) or in households where grandparents and grandchildren live together without a parent present (Nandy and Selwyn 2011). The Childcare and Early Years Survey of Parents 2009 in UK showed that around 26 percent of the 6,700 surveyed parents having children under age 14 had received help with childcare from grandparents (Smith et al. 2010). Analysis of care arrangements for children of different ages also revealed that 44 percent of children were regularly cared for by grandparents (Fergusson et al. 2008). Apart from UK, in other communities such as China and Russia, such grandparent-grandchild interactions are more frequent and childcare performed by grand-parents becomes more prominent. Psychological research has suggested that early child-care performed by grandparents could play important roles in developing child's cognition, task-related behaviors, and language (Sylva et al. 2011). Inspired by these studies, research in evolutionary linguistics needs to pay attention to the role of such grandparent-grandchild transmissions in language evolution, and examine whether the role of such cross-generational transmissions is similar to or distinct from those of vertical, oblique, or horizontal transmissions that take place between consecutive generations of individuals.

In order to discuss these issues, we extend our acquisition framework to incorporate cross-generational cultural transmission. Instead of both language origin and change as in our previous work, we focus on language change, since the relatively shorter life span of our ancestors at the early stage of language development might not allow grandparent-grandchild transmissions (the late life survival, reaching the 4th decade of life as seen today, seldom occurred in early human populations before the advent of settled agriculture (Crews and Gerber 2003), whereas a mature human language must have already emerged during the stage of hunter-gatherers).

In the following sections, we describe our modified acquisition framework (Sect. 2); illustrate the simulation results accordingly (Sect. 3); and finally, summarize the roles of these forms of transmission on language change and discuss other relevant issues concerning cultural transmission and the acquisition framework (Sect. 4). In this study, we adopt the lexicon-syntax coevolution model (Gong 2009; Gong 2011) as the language evolution model. The details of this model are briefly introduced in Appendix.

2 Modified Acquisition Framework

This acquisition framework (see Fig. 1) was modified from our previous work (Gong 2010). Instead of two consecutive generations, this framework consists of the grandparent (G), parent (P), and offspring (O) generations. In each round of generation turnover (see Fig. 1(a)), first, half of the individuals in the P generation are chosen to reproduce, each producing an offspring having no linguistic knowledge. These offspring comprise the offspring generation. Then, a learning stage begins, during which the offspring communicates either with another offspring, or with an individual from the P or G generation (note that in the first round of generation turnover, there is not yet a G generation, learning occurs either within the O generation or between the P and O generations). After that, the individuals in the P generation replace those in the G generation, and those in the O generation replace their parents in the P generation. After replacement, a new round of reproduction, learning, and replacement begins.

This framework incorporates three forms of cultural transmission that span over the G, P, and O generations (see Fig. 1): *transmissions between two offspring* (*OO*, offspring can be either speakers or listeners), *transmissions between parents and offspring* (*PO*, parents are speakers and offspring are listeners), and *transmissions between grandparents and offspring* (*GO*, grandparents are speakers and offspring are listeners). Similar to the previous work, we define three parameters, *OOrate*, *POrate*, and *GOrate*, respectively denoting the proportions of these three forms of cultural transmission in the total transmissions taking place during the learning stage, $OOrate + POrate + GOrate = 1.0$. After excluding some cases (e.g., $GOrate = 0.0$, which makes the framework identical to the original one considering only the P and O generations; $OOrate = 1.0$, $POrate = GOrate = 0.0$, which is the unrealistic case of purely horizontal transmissions), we set up 54 cases based on different combinations of *OOrate*, *POrate*, and *GOrate* (see Table 1). In each of these cases, we conduct 20 simulations for statistical analysis, and in each simulation, the population of the first generation contains 10 individuals and there are in total 100 rounds of generation turnover. In order to illustrate the simulation results, we also borrow the surface ternary plot used in our previous work (see Fig. 1(c) for how to read such plot).

In the language change simulations based on the lexicon-syntax coevolution model (see Appendix), the individuals in the first P generation share a complete set of lexical and syntactic rules that are respectively associated to the corresponding syntactic categories. This shared linguistic knowledge enables all individuals to accurately produce and comprehend all integrated meanings in the semantic space, and the mutual understandability of this initial communal language is 1.0. Due to different combinations of the OO, PO, and GO transmissions, this communal language may or may not be sufficiently transmitted across generations. Similar to our previous work, we adopt two indices to evaluate the mutual understandability of the communal language in later generations. These indices are: *Understanding Rate* (*UR*) between individuals from two consecutive P generations (UR_{con}), and *UR* between individuals from the very first P generation and those from the later

Cross-Generational Transmission on Language Change

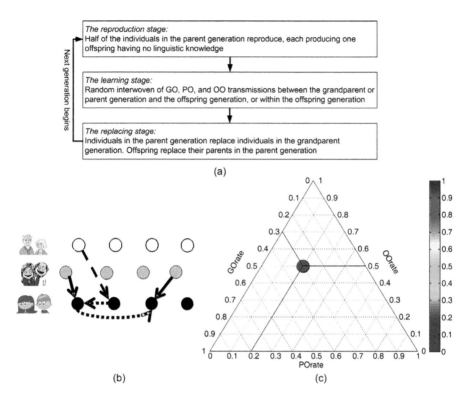

Fig. 1 The modified acquisition framework: (**a**) the reproduction, learning, and replacement stages during generation turnover; (**b**) the three forms of cultural transmission in the learning stage, where different types of balls represent individuals from the G (white balls), P (grey balls), and O (black balls) generations, and different types of arrows denote OO (dotted arrows), PO (solid arrows), and GO (dashed arrows) transmissions; (**c**) an example of the surface ternary plot showing the index value in the case $OOrate = 0.5$, $POrate = 0.2$, $GOrate = 0.3$, the value can be read from the color map on the side.

P generation (UR_{ini}) (see Eq. (1) and Eq. (2), where *SemSize* is the size of the semantic space shared by all individuals, and *PopSize* is the population size). When calculating these indices, we let each individual in the first P generation talk to each individual in the second P generation, and record the ratio of the meanings that are successfully produced by the speaker and accurately understood by the listener. High UR_{con} indicates that a communal language is well understood by individuals from consecutive generations, and high UR_{ini} indicates that individuals in a later generation can well understand the language of the first generation. By analyzing the average UR_{con} and UR_{ini} throughout 100 generations in different cases, we discuss the necessity of these forms of transmissions and identify in which cases an initial communal language can be largely preserved across many generations.

Table 1 The 54 cases formed by different combinations of *OOrate*, *POrate*, and *GOrate*

Cases	OOrate	POrate	GOrate
1	0.0	0.1	0.9
2	0.0	0.2	0.8
...	...	...	...
9	0.0	0.9	0.1
10	0.1	0.0	0.9
...	...	...	...
18	0.1	0.8	0.1
19	0.2	0.0	0.8
...	...	...	...
52	0.8	0.0	0.2
53	0.8	0.1	0.1
54	0.9	0.0	0.1

$$UR_{con} = \frac{\sum \text{Correctly Understood Meanings between Generations i-1 and i}}{SemSize \times PopSize \times (PopSize - 1)} \quad (1)$$

$$UR_{ini} = \frac{\sum \text{Correctly Understood Meanings between Generations 0 and i}}{SemSize \times PopSize \times (PopSize - 1)} \quad (2)$$

In this modified acquisition framework, we do not distinguish vertical and oblique transmissions; both GO and PO transmissions can be either vertical (biological parents or grandparents talk to their offspring) or oblique (non-biological parents or grandparents talk to their offspring). Our previous work has already shown that purely vertical transmissions failed to maintain good mutual understandability of the communal language across multi-individual generations. In addition, we assume that only offspring update their linguistic knowledge as listeners during transmissions, whereas individuals from the P or G generations do not. Such learning constraint was also adopted in other studies (Hurford and Kirby 1999) (but also see Sect. 4 for more discussions). Furthermore, instead of actual number of transmissions a particular offspring participants, the proportion parameters release the dependence of simulation results on certain parameters, such as the number of individuals chosen for reproduction, the number of offspring, the number of individuals in different generations, and the total number of transmissions during the learning stage. Once the values of these parameters are reasonably assigned (excluding extreme or unrealistic cases, such as the cases of single offspring or too few transmissions, etc.), the simulation results show similar tendencies under identical settings of *OOrate*, *POrate*, and *GOrate*.

3 Simulation Results

Based on the evaluating indices, UR_{con} and UR_{ini}, and the surface ternary plot, we illustrate the simulation results in those 54 cases. These results are analyzed from

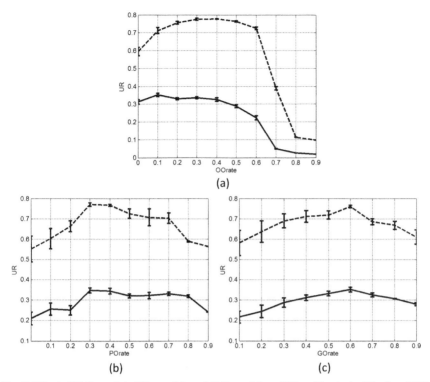

Fig. 2 Effect of *OOrate* (**a**), *POrate* (**b**), and *GOrate* (**c**) on UR_{con} (the dashed line) and UR_{ini} (the solid line). The error bars denote standard deviations.

two aspects: the respective effect of each form of transmissions and the collective effect of all three forms of transmissions.

On the first aspect, we show respectively in Fig. 2(a)-(c) the average UR_{con} and the average UR_{ini} of 20 simulations across 100 generations under cases with fixed *OOrate*, *POrate*, and *GOrate*. Due to the correlation (*OOrate* + *POrate* + *GOrate* = 1.0), the number of cases having a particular *OOrate* (*POrate* or *GOrate*) is not equal, and we take the mean values in cases all having a particular *OOrate* (*POrate* or *GOrate*). This manipulation helps neutralize the effects of the other forms of transmission to a certain extent, and allows focusing on the effect of a particular form of transmission.

As shown in Fig. 2, UR_{con} is always higher than UR_{ini} throughout cases. This echoes the dynamic equilibrium of language evolution shown in our previous work: in the short run, individuals from consecutive generations can largely understand each other, whereas in the long run, individuals of later generations may no longer understand well those of early generations, i.e., language change is inevitable. In addition, both higher UR_{ini} and higher UR_{con} tend to occur in cases having intermediate *OOrate*, *POrate* and *GOrate*. This indicates that all these forms of transmissions are needed for preserving a communal language across generations.

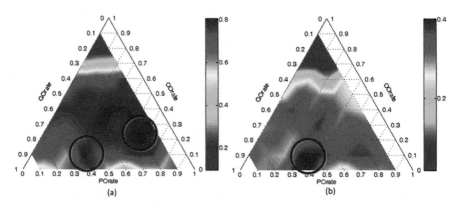

Fig. 3 Collective effects of *OOrate*, *POrate*, and *GOrate* on UR_{con} (**a**) and UR_{ini} (**b**). Circles point out the regions having relatively high UR_{con} and UR_{ini}.

Furthermore, when *OOrate* is below 0.5, UR_{con} increases, but UR_{ini} does not change much. This suggests that though helping maintain mutual understandability across consecutive generations, OO transmission is not that efficient in preserving a communal language across many generations. Finally, when OO_{rate} is too high (around 0.7 to 0.9), there is a notable drop in UR_{ini} and UR_{con} (see Fig. 2(a)), and the values in these cases are even lower than those in cases *OOrate*=0.0. This also shows that too many horizontal transmissions fail to maintain a communal language across generations.

On the second aspect, we show respectively in Fig. 3(a) and Fig. 3(b) the average UR_{ini} and the average UR_{con} in all 54 cases. As shown in these figures, the regions having higher *OOrate* (around the top angle) have much lower UR_{con} and UR_{ini}. This is consistent with the results shown in Fig. 2(a). However, in the regions having either high *POrate* (around the right angle) or high *GOrate* (around the left angle), the mutual understandability of the communal language is not low. This reveals the important roles of PO and GO transmissions in preserving a communal language across generations. As for UR_{con}, there are two regions having relatively higher UR_{con} than others (as circled out in Fig. 3(a)), both of which have smaller *OOrate*, but either high *POrate* or high *GOrate*. This reflects the similar roles of PO and GO transmissions in maintaining the mutual linguistic understandability in consecutive generations. As for UR_{ini}, there is only one region having the highest UR_{ini} (as circled out in Fig. 3(b)), which has low *OOrate* (around 0.1), relatively low *POrate* (around 0.3), but high *GOrate* (around 0.6 to 0.7). This suggests that compared with PO transmission GO transmission is more efficient in keeping a communal language with a relatively high mutual understandability in later generations.

4 Discussions and Conclusions

Based on the particular language model, our simulations of language change suggest that all the three forms of cultural transmission are necessary for preserving the mutual understandability of the communal language. To be specific, PO and GO transmissions are efficient at preserving the mutual understandability in both the short (across consecutive generations) and long (across many generations) runs, whereas OO transmission is only good at maintaining the mutual understandability within generations. These roles of OO and PO transmissions in language change are in line with those identified in our previous work.

As for the newly added cross-generational transmission (GO transmission), despite of its similar roles to PO transmissions, we further show that GO transmission is more efficient in keeping the mutual understandability in the long run; although language change is inevitable, given fixed $OOrate$, similar proportions of GO transmissions lead to higher UR_{ini} compared with similar proportions of PO transmissions (see Fig. 3(a)). To our knowledge, this is the first attempt that simulates cross-generational cultural transmission and analyzes its roles in language evolution.

Language maintenance across generations relies upon cultural transmission between individuals from different generations. PO transmission offers fundamental contacts between individuals of generation i and those of generation $i+1$, and GO transmission offers opportunities for individuals of generation i to contact those of generation $i+2$. Even if offspring fail to learn sufficient linguistic knowledge from their parents, offspring can still obtain some shared linguistic knowledge from their grandparents. In a cultural setting involving both PO and GO transmissions, the language state of the new generation is collectively determined by the states of previous two generations. Although different types of linguistic knowledge may have different efficiencies to be transferred during cultural transmission (in the adopted lexicon-syntax coevolution model, learning lexical items based on detection of recurrent patterns is faster than learning syntactic rules, since the latter have to be obtained based on similar usage of lexical items in exchanged sentences, see Appendix for details), in terms of general information transmission, such setting is more reliable than the one in which the state of the new generation is dependent solely on the state of the immediately previous generation. In analogy, this finding also dovetails with many theories of system control or prediction. For example, it has been accepted that second or higher order Markov chain models (the state at time step i is determined by previous states at $i-1, i-2, \ldots$) could be more reliable than a first-order Markov chain model (the state at time step i is determined solely by the state at $i-1$) in predicting future states of a system that involves correlations or mutual influences between states at previous time steps and those at later ones, such as weather forecast (Spoof and Pryor 2008), speech recognition and machine translation (Jurafsky and Martin 2009), etc. In other words, apart from language evolution, the findings of this study are also insightful to other socio-cultural phenomena that involve information transmission in a chain of multi-individual groups.

One may argue that in the current setting, the O generation at each time step contains only 5 offspring, who will replace half of the adults in the P generation. In other words, during generation turnover, some adults in the P generation will not be replaced and become individuals in the G generation. Then, some GO transmissions involving such individuals at the current time step may be identical to some PO transmissions at the previous time step, both of which involve identical speakers

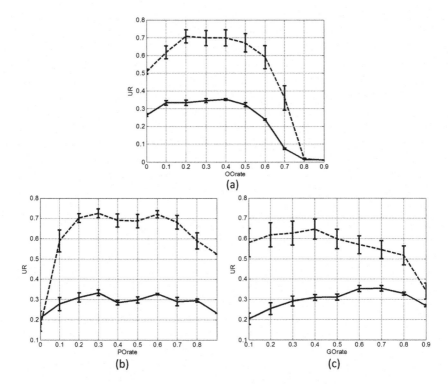

Fig. 4 Effect of *OOrate* (**a**), *POrate* (**b**), and *GOrate* (**c**) on UR_{con} (the dashed line) and UR_{ini} (the solid line) under the setting in which all adults produce offspring and are replaced by those offspring at each round of generation turnover. The error bars denote standard deviations.

and naive offspring. In order to show that such mixed transmissions do not affect the simulation results, we further conduct simulations in which during each round of generation turnover all individuals in the P generation produce offspring, and all these individuals are replaced by these offspring after the learning stage. In this setting, PO and GO transmissions in each generation involve totally different individuals. Since the number of offspring is doubled, we also double the total number of cultural transmission in the learning stage. The simulation results under this setting are shown in Fig. 4 and Fig. 5.

Due to the different number of offspring and the complete replacement of the P generation in each new generation, these simulations report different results from the previous setting. For example, when *OOrate* is low, UR_{con} becomes lower than that in the previous setting, which indicates that due to complete replacement OO transmission helps contribute to the mutual understandability of the communal language. Similarly, in this new setting, a certain proportion of PO transmission is also needed to maintain a relatively high UR_{con}. In addition, the regions having the best UR_{con} (circled in Fig. 5(a)) or the best UR_{ini} (circled in Fig. 5(b)) do not overlap as

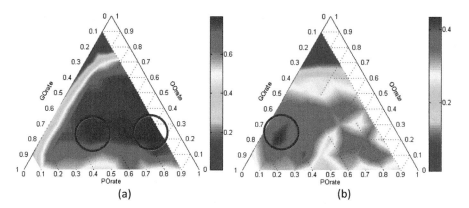

Fig. 5 Collective effects of *OOrate*, *POrate*, and *GOrate* on UR_{con} (**a**) and UR_{ini} (**b**) under the setting in which all adults produce offspring and are replaced by those offspring at each round of generation turnover. Circles point out the regions having relatively high UR_{con} and UR_{ini}.

in the previous setting, which clearly separate the roles of PO and GO transmissions in maintaining the communal language in the short run (UR_{con}) and in the long run (UR_{ini}). Finally, due to the different number of offspring, the regions having better UR_{con} and UR_{ini} become different from those in the previous setting. Despite of these interesting differences, we can still summarize some similar effects of OO, PO and GO transmissions on language change in this new setting. This reflects the fact that the general findings concerning the effect of OO, PO and GO transmissions on language change are less dependent on particular settings of certain parameters.

The whole study is based upon the acquisition framework that explicitly defines individuals from different generations and various forms of cultural transmission within and across these generations. Though involving many arbitrary settings (e.g., it simulates only punctuated generation turnover, horizontal transmissions occur only among individuals in the O generation, each individual in the P generation only reproduces one offspring, etc.), this framework offers an efficient conceptual tool to clarify the cultural environment in which language is learned and exchanged, to quantify various forms of cultural transmission, and to evaluate their respective and collective roles in language evolution. Apart from computer simulations, this framework can also help set the cultural transmission environment in laboratory experiments of language evolution. Moreover, based on this framework, we can easily incorporate other forms of cultural transmission, which pay the way for the future work exploring the effect of cultural transmission on language evolution. For example, by allowing children to talk to adults and also allowing adults to occasionally update their linguistic knowledge, we can incorporate children-talking-to-parents/grandparents transmission in our study. Then, by evaluating the effect of such transmissions on diffusion of linguistic innovation introduced by children, we can evaluate some widely-discussed questions in evolutionary linguistics, such as who, children or adults, are the primary driving forces for language change (Gívon 1998; Mufwene 2008).

Finally, note that although the simulations in this paper are based only on this particular language model tracing the evolution of both lexical items and simple syntax, some of the conclusions are informative and insightful to simulation studies based on other language models and general discussions on the roles of cultural transmission in language evolution. For example, as shown in our previous work (Gong 2010), apart from the lexicon-syntax coevolution model, simulations based on the category game model (Baronchelli et al. 2010) report similar results. However, since the category game model cannot simulate language change, we cannot use it to further verify the generality of the results in this paper. In addition, it is worth noting some apparent differences between the simulation results and the reality. For example, although the model seems complex in terms of involved behavioral dynamics (these aspects of dynamics have been examined in detail in (Gong 2009; Gong 2011), it only takes account of most fundamental semantic structures in language and general learning mechanisms particulary for acquiring simple word orders. These simplifications make the simulation results only conceptually informative to empirical research. For example, our simulations only touch upon change in fundamental expressions, whereas language change takes place in many aspects and appears faster than what is simulated here. Nonetheless, compared with mathematical models that transform acquisition and communication processes into abstract equations, our behavioral model implements detailed learning processes and enables tracking the origin and evolution of specific linguistic features. Therefore, to better understand the reality, we suggest that mathematical models, behavioral models, as well as generalizable frameworks are all necessary. With gradual increase in the complexity of these models and frameworks, we may eventually understand and predict many evolutionary phenomena involving language, individual learners, and the socio-cultural environment.

Acknowledgements. Gong acknowledges support from the US NIH Fund (R01HD071988-01A1). We thank Prof. Alexander Mehler from the Goethe University Frankfurt am Main for inviting us to attend the Conference on Modeling Linguistic Networks, where we presented the preliminary results of this paper.

Appendix

The lexicon-syntax coevolution model encodes language as *meaning-utterance mappings* (*M-U mappings*).

On the meaning side, individuals share a semantic space containing a fixed number of *integrated meanings*, each having a predicate-argument structure, e.g., "predicate ⟨ agent ⟩" or "predicate ⟨ agent, patient ⟩", where predicate, agent, and patient are thematic notations. *Predicates* refer to actions that individuals conceptualize (e.g., "run" or "chase"), and *arguments* entities on or by which those actions are performed (e.g., "fox" or "tiger"). Some predicates each take a single argument,

Lexical rules
  Holistic rules: **Compositional rules:**
  (a) "chase<wolf, bear>"←→/a b/ (0.5)   (c) "wolf"←→/f/ (0.6)
  (b) "hop<deer>"←→/c/ (0.4)   (d) "chase<#, bear>"←→/e h * g/ (0.7)
Syntactic rules
  (1) Category 1 (S) << Category 2 (V) (SV) (0.8)
  (2) Category 3 (O) >> Category 2 (V) (VO) (0.4)
Categories
Category 1 (S): *List of lexical rules:*
           {"wolf"←→/b c/ (0.7)} [0.5]
           List of syntactic rules:
           Category 1 (S) << Category 2 (V) (0.8)
           Category 3 (O) >> Category 1 (S) (0.4)
Category 2 (V): *List of lexical rules:*
           {"fight<#,#>"←→/e/ (0.6)} [0.5]
           List of syntactic rules:
           Category 1 (S) << Category 2 (V) (0.8)
Category 3 (O): *List of lexical rules:*
           {"fox"←→/a/ (0.5)} [0.7]
           List of syntactic rules:
           Category 3 (O) >> Category 1 (S) (0.4)

Fig. A1 Examples of lexical rules, syntactic rules, and syntactic categories. "♯" denotes unspecified semantic item, and "⋆" unspecified syllable(s). S, V, and O are syntactic roles of categories. Numbers enclosed by ( ) denote rule strengths, and those by [ ] association weights. "≪" denotes the local order *before*, and "≫" *after*. Compositional rules can combine, if specifying each constituent in an integrated meaning exactly once, e.g., rules (c) and (d) can combine to form "chase ⟨ wolf, bear ⟩", and the corresponding utterance is /ehfg/.

e.g., "run ⟨ tiger ⟩" ("a tiger is running"); others each take two, e.g., "chase ⟨ tiger, fox ⟩" ("a tiger is chasing a fox"), where the first constituent within ⟨⟩, "tiger", denotes the *agent* (the action instigator) of the predicate "chase", and the second, "fox", the *patient* (the entity that undergoes the action). Meanings having identical agent and patient constituents (e.g., "fight ⟨ fox, fox ⟩") are excluded.

On the utterance side, integrated meanings are encoded by *utterances*, each comprising a string of syllables chosen from a signaling space. An utterance encoding an integrated meaning can be segmented into subparts, each mapping one or two semantic constituents; and meanwhile, subparts can combine to encode an integrated meaning. Using predicate-argument structures to denote semantics and combinable syllables to form utterances has been adopted in many *structured simulations* (Wagner et al. 2003).

Individuals are simulated as *artificial agents*. Via learning mechanisms, individuals can: (i) acquire linguistic knowledge from M-U mappings obtained in previous communications; and (ii) produce utterances to encode integrated meanings and comprehend utterances in communications.

Linguistic knowledge is encoded by *lexicon*, *syntax*, and *syntactic categories* (some examples are shown in Fig. A1).

An individual's lexicon consists of a number of lexical rules. Some of these rules are *holistic*, each mapping an integrated meaning onto an utterance (sentence), e.g., "run ⟨ tiger ⟩" ↔ /abcd/ indicates that the meaning "run ⟨ tiger ⟩" can be encoded by the utterance /abcd/, and also that /abcd/ can be decoded as "run ⟨ tiger ⟩"; others are *compositional*, each mapping particular semantic constituent(s) onto a subpart of an utterance, e.g., "fox" ↔ /ef/ (a word rule) or "chase ⟨ wolf, ♯ ⟩" ↔ /gh ⋆ i/ (a phrase rule, "↔" denotes an unspecified constituent, and "⋆" unspecified syllable(s)).

In order to form a sentence using compositional rules, the utterances of these rules need to be ordered. A syntactic rule specifies an order between two lexical items, e.g., "tiger" ↔ /ef/ ≪ "fox" ↔ /abc/ denotes that the utterance /ef/ encoding the constituent "tiger" lies *before* - but not necessarily immediately before - /abc/ encoding "fox". Based on one such local order, "predicate ⟨ agent ⟩" meanings can be expressed; based on combination of two or three local orders, "predicate ⟨ agent, patient ⟩" meanings can be expressed.

Syntactic categories are formed in order for syntactic rules acquired from some lexical items to be applied productively to other lexical items with the same thematic notation. A syntactic category comprises a set of lexical rules and a set of syntactic rules that specify the orders in utterance between these lexical rules and those from other categories. For the sake of simplicity, we simulate a nominative-accusative language and exclude the passive voice. A category associating lexical rules encoding the thematic notation of agent can be denoted as a subject (S) category, since that thematic notation usually corresponds to the syntactic role S. Similarly, patient corresponds to object (O), and predicate to verb (V). A local order between two categories can be denoted by their syntactic roles, e.g., an order before between an S and a V category can be denoted by S ≪ V, or simply SV.

Lexical and syntactic knowledge collectively encode integrated meanings. As shown in Fig. A1, to express "fight ⟨ wolf, fox ⟩" using the lexical rules respectively from the S, V, and O categories and the syntactic rules SV and SO, the resulting sentence can be /bcea/ or /bcae/, following SVO or SOV.

Every lexical or syntactic rule has a *strength*, indicating the probability of successfully using its M-U mapping or local order. A lexical rule also has an *association weight* to the category that contains it, indicating the probability of successfully applying the syntactic rules of this category to the utterance of that lexical rule. Both strengths and association weights lie in [0.0 1.0]. A newly-acquired rule has strength 0.5, the same as the weight of a new association of a lexical rule to a category. These numeral parameters realize a strength-based competition during communications and a gradual forgetting of linguistic knowledge. *Forgetting* occurs regularly after a number (scaled to the population size) of communications. During forgetting, all individuals deduct a fixed amount from each of their rules' strengths and association weights, after that, lexical or syntactic rules with negative strengths are discarded, lexical rules with negative association weights to some categories are removed from those categories, and categories having no lexical rules are discarded, together with their syntactic rules.

Individuals use general learning mechanisms to acquire linguistic knowledge. Lexical rules are acquired by detecting *recurrent patterns* (meanings and syllables appearing recurrently in at least two M-U mappings). Each individual has a buffer storing M-U mappings obtained in previous communications. New mappings, before being inserted into the buffer, are compared with those already in the buffer. As shown in Fig. A2(a), by comparing "hop 〈 fox 〉" ↔ /ab/ with "run 〈 fox 〉" ↔ /acd/, an individual can detect the recurrent pattern "fox" and /a/, and if there is no lexical rule recording such mapping in the individual's rule list, a lexical rule "fox" ↔ /a/ with initial strength 0.5 will be acquired.

Syntactic categories and syntactic rules are acquired according to thematic notations of lexical rules and local order relations of their utterances in M-U mappings. As shown in Fig. A2(b), evident in M-U mappings (1) and (2), syllables /d/ of rule (i) and /ac/ of rule (iii) precede /m/ of rule (ii). Since both "wolf" and "fox" are agents in these meanings, rules (i) and (iii) are associated into an S category (Category 1), and the order *before* between these two rules and rule (ii) is acquired as a syntactic rule. Similarly, in M-U mappings (1) and (3), /m/ of rule (ii) and /b/ of rule (iv) follow /d/ of rule (i), thus leading to a V category (Category 2) associating rules (ii) and (iv) and a syntactic rule *after*. Now, since Categories 1 and 2 respectively associate rules (i) and (iii), and (ii) and (iv), the two syntactic rules are updated as "Category 1 (S) ≪ Category 2 (V)", indicating that the syllables of lexical rules in the S category should precede those in the V category.

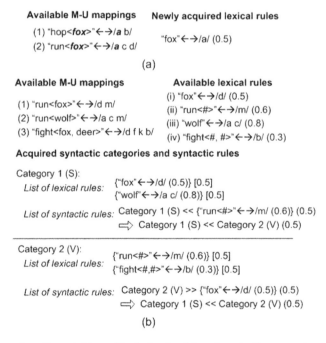

Fig. A2 Examples of acquisition of lexical rules (**a**) and syntactic categories and syntactic rules (**b**). M-U mappings are itemized by Arabic numbers, and lexical rules by Roman numbers.

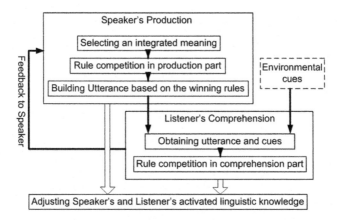

Fig. A3 Diagram of an utterance exchange. The dotted block indicates unreliable cues.

Such item-based learning mechanisms have been traced in empirical studies of language acquisition (Mellow 2008), and the categorization process resembles *the verb-island hypothesis on language acquisition* (Tomasello 2003).

A communication involves two individuals (a *speaker* and a *listener*), who perform a number of utterance exchange, each proceeding as follows (see Fig. A3).

In production, the speaker (hereafter as "she") randomly selects an integrated meaning from the semantic space to produce. She activates: (i) lexical rules encoding some (compositional rules) or all (holistic rules) constituents in this meaning; (ii) categories associating those lexical rules and having appropriate syntactic roles; and (iii) syntactic rules in those categories that can regulate those lexical rules in sentences. These rules form candidate sets for production. She calculates the *combined strength* (CS_{prod}) of each set (see Eq. (1), where *Avg* means taking average, *aso* association weights, *str* rule strengths, *Lex* lexical rules, *Cat* categories, and *Syn* syntactic rules):

$$CS_{prod} = Avg(str(Lex(s))) + Avg(aso(Cat) \times str(Syn(s))) \quad (A1)$$

$CS_{production}$ is the sum of the lexical contribution (the average strength of the lexical rules in this set) and the syntactic contribution (the average product of (i) the strengths of the syntactic rules regulating the lexical rules and (ii) the association weights of those lexical rules to the categories in this set). As shown in Fig. A1, the three categories, three lexical rules encoding "wolf", "fight $\langle \sharp, \sharp \rangle$" and "fox", and two syntactic rules SV and SO form a candidate set to encode "fight$\langle$wolf, fox$\rangle$". Its CS_{prod} is 0.98, where the lexical contribution is 0.6 $((0.7+0.6+0.5)/3)$ and the syntactic contribution is 0.38 $((0.8(0.7+0.6)/2 + 0.4 \times (0.7+0.5)/2)/2)$.

After calculation, she selects the set of *winning rules* having the highest CS_{prod}, builds up the sentence accordingly, and transmits it to the listener. If lacking sufficient rules to encode the integrated meaning, she occasionally (based on *a random creation rate*) creates a holistic rule to encode the whole meaning.

In comprehension, the listener (hereafter as "he") receives the sentence from the speaker and an environmental cue (its meaning is set based on the reliability of cues (*RC*), which is the probability for the cue contains the speaker's intended meaning). Based on his linguistic knowledge, he activates (i) lexical rules whose syllables fully/partially match the heard sentence, (ii) categories associating these lexical rules, and (iii) syntactic rules in those categories whose orders match those of the lexical rules in the heard sentence. Then, he compares the cue's meaning with the one comprehended using the activated rules. According to the three conditions discussed in the main texts, he forms candidate sets for comprehension, and calculates the combined strength of each set (see Eq. (2)). For a set without a cue, the calculation is identical to CS_{prod}; for a set with a cue, cue strength is added; and for a set with only a cue, cue strength is CS_{comp}:

$$CS_{comp} = Avg(str(Lex(s))) + Avg(aso(Cat) \times str(Syn(s))) + str(Cue) \quad (A2)$$

After calculation, the listener selects the set of winning rules having the highest CS_{comp} to interpret the heard sentence. If CS_{comp} of the winning rules exceeds a confidence threshold, he adds the perceived M-U mapping to his buffer, and transmits a positive feedback to the speaker. Then, both individuals reward their winning rules by adding a fixed amount to their strengths and association weights, and penalize other competing ones by deducting the same amount from their strengths and association weights. Otherwise, he discards the perceived mapping and sends a negative feedback to the speaker. Then, both individuals penalize their winning rules. Such linear inhibition mechanism has been used in many models (e.g., Steels (2005)), though its details may be slightly different. For activated rules having initial strength and association weight values, the linguistic (lexical and syntactic) contribution is 0.75 ($0.5 + 0.5 \times 0.5$). In order to equally treat linguistic and non-linguistic information, we set cue strength and confidence threshold as 0.75. Together with forgetting, this strength-based competition leads to conventionalization of linguistic knowledge among individuals.

Throughout the whole utterance exchange, there is no check whether the speaker's encoded meaning matches the listener's decoded one. Equations (1) and (2) illustrate how non-linguistic information aids linguistic comprehension, by clarifying unspecified constituent(s) and enhancing rules helping achieve similar interpretations. Recent neuroscience studies (e.g., Esterman et al. (2009)) have shown that certain brain regions act as a domain-general cognitive control on categorization, stimulus-response mapping, and voluntary attention shift among perceptual inputs and memory representations. This provides the neural basis of the coordination of various types of information.

Table A1 lists the values of the parameters defining the semantic and signaling spaces, and acquisition and communication mechanisms. In language change simulations, individuals of the first generation share 12 compositional rules each encoding a semantic constituent, and 3 syntactic rules that can form a consistent global word order. These compositional and syntactic rules are respectively assigned to an S, a V and an O category. Based on these rules, individuals can accurately encode

Table A1 Parameter setting (see Gong (2009) for discussions on their effects)

Parameters	Values
Size of semantic space	64
Size of signaling space	30
Size of individual buffer	40
Size of rule list	60
No. utterance exchange per transmission	20
Random creation rate of holistic rules	0.25
Amount of adjustment on strengths/association weights in competition	0.1
Amount of adjustment on strengths/association weights in forgetting	0.01
Size of population	10
Forgetting frequency (scaled to population size)	10

all 64 integrated meanings in the semantic space, and offspring having no linguistic knowledge try learning these rules via different forms of cultural transmission.

In summary, this model presumes that: (i) individuals can conceptualize semantic constituents and integrated meanings having simple predicate-argument structures; (ii) they can use general learning mechanisms to handle lexical and syntactic rules, and compositional relationship; (iii) they can optimize their linguistic knowledge during communications. It implements concrete instance-based, learning mechanisms and major behaviors during linguistic communications, and traces a simultaneous acquisition and close interaction of lexical and syntactic knowledge.

References

Axelrod, R.: The Dissemination of Culture: A Model with Local Convergence and Global Polarization. J. Conflict Res. 41(2), 203–226 (1997)

Baronchelli, A., Gong, T., Puglisi, A., Loreto, V.: Modeling the emergence of universality in color naming patterns. Proc. Natl. Acad. Sci. USA 107, 2403–2407 (2010)

Bickerton, D., Szathmáry, E.: Biological Foundations and Origin of Syntax. MIT Press, Cambridge (2009)

Cangelosi, A., Parisi, D.: Simulating the Evolution of Language. Springer, London (2002)

Chomsky, N.: Language and Mind. Harcourt Brace Jovanovich, New York (1972)

Christiansen, M.H., Kirby, S.: Language Evolution. Oxford University Press, Oxford (2003a)

Christiansen, M.H., Kirby, S.: Language Evolution: Consensus and Controversies. Cog. Sci. 7(7), 300–307 (2003b)

Cornish, H.: Investigating How Cultural Transmission Leads to the Appearance of Design Without A Designer in Human Communication Systems. Inter. Stud. 11(1), 112–137 (2010)

Crews, D.E., Gerber, L.M.: Effects of Early Child-Care on Cognition, Language and Task-Related Behaviors at 18 Months: An English Study. Coll. Antropol. 27(1), 7–22 (2003)

Dunn, M., Greenhill, S.J., Levinson, S.C., Gray, R.D.: Language Evolution in the Laboratory. Nature 473, 79–82 (2011)

Esterman, M., Chiu, Y.-C., Tamber-Rosenau, B.J., Yantis, S.: Decoding Cognitive Control in Human Parietal Cortex. Proc. Natl. Acad. Sci. USA 106(42), 17974–17979 (2009)

Evans, N., Levinson, S.C.: The Myth of Language Universals: Language Diversity and Its Importance for Cognitive Science. Behav. Brain Sci. 32(5), 429–448 (2009)

Farmer, E., Moyers, S.: Fostering Effective Family and Friends Placements. Jessica Kingsley Publishers, London (2008)

Fergusson, E., Maughan, B., Golding, J.: Which Children Receive Grandparental Care and What Effect Does It Have? J. Child Psychol. Psyc. 49(2), 161–169 (2008)

Galantucci, B.: An Experimental Study of the Emergence of Human Communication Systems. Cog. Sci. 29(5), 737–767 (2005)

Gívon, T.: On the Coevolution of Language, Mind, and Brain. Evo. Com. 2, 45–116 (1998)

Gong, T.: Computational Simulation in Evolutionary Linguistics: A Study on Language Emergence. Institute of Linguistics, Academia Sinica, Taipei (2009)

Gong, T.: Exploring the Roles of Horizontal, Vertical, and Oblique Transmissions in Language Evolution. Adap. Behav. 18(3-4), 356–376 (2010)

Gong, T.: Simulating the Coevolution of Compositionality and Word Order Regularity. Inter. Stud. 12(1), 63–106 (2011)

Gong, T., Minett, J.W., Wang, W.S.-Y.: Exploring Social Structure Effect on Language Evolution Based on a Computational Model. Con. Sci. 20(2-3), 135–153 (2008)

Gong, T., Shuai, L., Tamariz, M., Jäger, G.: Studying Language Change Using Price Equation and Pólya-Urn Dynamics. PLoS ONE 7, e33171 (2012)

Hauser, M.D., Chomsky, N., Fitch, W.T.: The Faculty of Language: What Is It, Who Has It, and How Did It Evolve? Science 298, 1569–1579 (2002)

Hurford, J.R., Kirby, S.: Coevolution of Language Size and the Critical Period. In: Birdsong, D. (ed.) Second Language Acquisition and the Critical Period Hypothesis, pp. 39–63. Lawrence Erlbaum, Mahwah (1999)

Jackendoff, R.: Foundations of Language: Brain, Meaning, Grammar, Evolution. Oxford University Press, Oxford (2002)

Jurafsky, D., Martin, J.H.: Speech and Language Processing: An Introduction to Natural Language Processing, Computational Linguistics, and Speech Recognition. Pearson Prentice Hall, Upper Saddle River (2009)

Ke, J.-Y., Gong, T., Wang, W.S.-Y.: Language Change and Social Networks. Com. Comp. Phys. 3(4), 935–949 (2008)

Kirby, S.: Function, Selection and Innateness: The Emergence of Language Universals. Oxford University Press, Oxford (1999)

Kirby, S., Cornish, H., Smith, K.: Cumulative Cultural Evolution in the Laboratory: An Experimental Approach to the Origins of Structure in Human Language. Proc. Natl. Acad. Sci. USA 105(31), 10681–10686 (2002)

Labov, W.: Principles of Linguistic Change: Social Factors. Basil Blackwell, Oxford (2001)

Lenaerts, T., Jansen, B., Tuyls, K., De Vylder, B.: The Evolutionary Language Game: An Orthogonal Approach. J. Theor. Biol. 235(4), 566–582 (2005)

Mellow, J.D.: The Emergence of Complex Syntax: A Longitudinal Case Study of the ESL Development of Dependency Resolution. Lingua 118(4), 499–521 (2008)

Mesoudi, A., Whiten, A.: The Multiple Roles of Cultural Transmission Experiments in Understanding Human Cultural Evolution. Philo. Trans. R. Soc. B 363(1509), 3498–3501 (2008)

Mufwene, S.S.: The Ecology of Language Evolution. Cambridge University Press, Cambridge (2001)

Mufwene, S.S.: What Do Creoles and Pidgins Tell Us About the Evolution of Languages? In: Laks, B., Cleuziou, S., Demoule, J.-P., Encreve, P. (eds.) Origin and Evolution of Languages: Approaches, Models, Paradigms, pp. 272–297. Equinox Publishing, London (2008)

Nandy, S., Selwyn, J.: Spotlight on Kinship Care: Using Census Micro-data to Examine the Extent and Nature of Kinship Care in the UK. The Hadley Centre for Adoption and Fostering, Bristol (2011)

Scott-Phillips, T., Kirby, S.: Language Evolution in the Laboratory. Trends Cog. Sci. 14(9), 411–417 (2002)

Smith, K., Hurford, J.R.: Language Evolution in Populations: Extending the Iterated Learning Model. In: Banzhaf, W., Ziegler, J., Christaller, T., Dittrich, P., Kim, J.T. (eds.) ECAL 2003. LNCS (LNAI), vol. 2801, pp. 507–516. Springer, Heidelberg (2003)

Smith, R., Poole, E., Perry, J., Wollny, I., Reeves, A.: Childcare and Early Years Survey of Parents 2009. DFE-RR054, London (2010)

Spoof, J.T., Pryor, S.C.: On the Proper Order of Markov Chain Model for Daily Precipitation Occurrence in the Contiguous United States. J. Appl. Meteorol. 47, 2477–2486 (2008)

Steels, L.: The Emergence and Evolution of Linguistic Structure: From Lexical to Grammatical Communication Systems. Cog. Sci. 17(3-4), 213–230 (2005)

Steels, L.: Experiments in Cultural Language Evolution. John Benjamins, Amsterdam (2012)

Sylva, K., Stein, A., Leach, P., Bames, J., Malmberg, L.-E., the FCCC team: Effects of Early Child-Care on Cognition, Language and Task-Related Behaviors at 18 Months: An English Study. Brit. J. Dev. Psychol. 29(1), 18–45 (2011)

Tomasello, M.: Constructing a Language: A Usage-Based Theory of Language Acquisition. Harvard University Press, Cambridge (2003)

Vogt, P.: On the Acquisition and Evolution of Compositional Languages: Sparse Input and Productive Creativity of Children. Adap. Behav. 13(4), 325–346 (2005)

Wagner, K., Reggia, J.A., Uriagereka, J., Wilkinson, G.S.: Progress in the Simulation of Emergent Communication and Language. Adap. Behav. 11(1), 37–69 (2003)

Social Networks and Beyond in Language Change

Gareth J. Baxter

Abstract. We examine the effects of heterogeneous social interactions in a numerical model of language change based on the evolutionary utterance based theory developed by Croft. Two or more variants of a linguistic variable compete in the population. Social interactions can be separated into a symmetric weighted network of social contact probabilities, and asymmetric weightings given by speakers to each other's utterances, that is, social influence. Remarkably, when interactions are symmetric between speakers, the network structure has no effect on the mean time to consensus. On the other hand large disparities in social influence, even in rather homogeneous networks, can dramatically affect mean time to reach consensus (fixation). We explore a range of representative scenarios, to give a general picture of both aspects of social interactions, in the absence of explicit selection for any particular variant.

1 Introduction

The rich structures present in language are not static entities, but are fluid and changing. Over long time periods, every element of language can change, at every level of structural hierarchy. From pronunciation, spelling, vocabulary, to grammatical structures themselves. Yet, while language is constantly changing, it changes only a little at a time, a relative stability being necessary for communication. A language is useless if its speakers do not share a common idea of its meaning (McMahon 1994). Language is therefore in balance between change and the formation and maintenance of convention. Because of the social nature of language, it seems natural that social interactions are the driving force both of language change and of the maintenance of conventions, and the fixation of new conventions.

Gareth Baxter
University of Aveiro
e-mail: gjbaxter@ua.pt

In this Chapter we explore the effects of social networks and social influence in language change through numerical modeling. We take as our starting point Croft's 2000 usage-based account of language change. Croft's theory assumes that speakers learn and adapt to the usage patterns of those around them. The theory is formulated as a generalized evolutionary process (Hull 2001) in which tokens of linguistic structure are replicated by speakers when they take part in conversation, so that language change occurs through the accumulated effect of differential replication of variants during conversation. A numerical model of language change, based on Croft's theory, was developed in Baxter et al. (2006). Here we explore the limits of the effects of social network structure, and further social interaction effects, on the dynamics of this model. Many of the results described have previously been presented in Baxter et al. (2008) and Baxter et al. (2012). The advantage of creating numerical models of social phenomena is that it allows us to repeat experiments numerous times, to try out different parameters and assumptions, in a way that is simply impossible in the real world. It also allows us to check whether certain proposed mechanisms really do lead to the outcomes predicted.

Other models of language change have been proposed, based on different explanations, for example, that change occurs during learning of a language (Niyogi and Berwick 1997). Numerical modelling has been used by a number of groups in the broader area of language change and evolution, studying phenomena such as the emergence of universal grammars (Nowak et al. 2001), the emergence of language structure using iterated learning (Kirby 2002), effects of social networks on the establishment of a convention (Dall'Asta et al. 2006), the competition between languages for speakers (Kandler and Steele 2008). For a further review of various effort in modeling language evolution, see Vogt (2009). The model described in this chapter has many mathematical similarities with models of semiotic dynamics or opinion dynamics, see for example the review Castellano et al. (2009), biodiversity (Hubbell 2001) and population genetics (Crow and Kimura 1970).

In the following Section we outline Croft's theory and its assumptions in more detail, including a summary of the different mechanisms of propagation of a language change through a population. In Sect. 3 we describe the numerical model based on this theory. In Sect. 4 we summarise a method for mathematical analysis of this model. In Sects. 5 and 6 we use the model to explore the effects of social network structure and social influence on change propagation.

2 Utterance Selection Model of Language Change

Croft's framework treats language change as taking place through language use. Whenever a speaker produces an utterance, he or she replicates examples of linguistic structures (variables) that she or he has heard before, although the structures are often combined in novel ways (Croft 2000). This replication process is mediated by the speaker and her knowledge about her language, which is based in turn on the language use she has been exposed to in the past. This is an example of a usage-based theory, in that linguistic behavior is determined by language use in communicative interactions (see e.g. Bybee (2001) and Pierrehumbert (2003)).

A language variable may have two or more variants, essentially "different ways of saying the same thing", that coexist in the speech community. Examples include different pronunciations of a phoneme, different terms for the same object, different idiomatic expressions with the same meaning and so on. A language change then consists of several steps. Suppose that a particular variant is predominantly used by a given community. From time to time, a new variant may appear. Over time, this variant may be adopted by more and more members of the population, until finally it may become the new predominating variant. We can say that the old convention has been replaced by a new one. Of course this is not the final state of the language, as new variants are introduced in all parts of a language, and some of them go on to replace older variants, so that when viewing the language as a whole, language change is a continual process. In the present work, we wish to study the spreading process in a single language change. We will assume that two or more discrete variants already exist, and model their competition and spread through the population, until one variant becomes the new convention. In particular, we are interested in the effect of social network structure and social influence on this process.

Croft's account can be seen as a generalised evolutionary process (Hull 1988; Hull 2001). An evolutionary process requires a *replicator*, which is the entity that is replicated in some process, preserving most of its structure. For example, in biological evolution, the replicator may be the gene. In language change, the replicator is the *variable*, and the replication process is language use in face-to-face interaction. Replication must preserve much of the replicator's structure. For example, a speaker more or less accurately replicates the phonetic, grammatical, and semantic structure of the sounds, words, and constructions when she produces an utterance. The second required element is an *interactor*. The interactor interacts with its environment in such a way that it causes the differential replication of replicators. In biological evolution, the interactor would be the organism. It interacts with its environment in such a way that it causes differential replication of its replicators. This process is selection. In language change, the speaker is the interactor. The speaker interacts with her environment, in particular other speakers, causing differential replication of linguemes. The result is language change.

A variety of factors influencing the spreading of a variant through the population have been proposed. It is possible to classify these different mechanisms of propagation in evolutionary terms (Baxter et al. 2009; Blythe and Croft 2012). In a typical evolutionary selection model, each variant replicator has an associated fitness value, and differences in fitness result in differential replication. We call this selection mechanism *replicator selection*. In language change, replicator selection corresponds to the situation in which one variant may be inherently more favoured or more likely to be reproduced than another. This may be due to universal forces, such as articulatory or acoustic properties of sounds, or syntactic processing factors (Ohala 1983; Hawkins 2004). On the other hand, speakers may have a conscious or unconscious preference for one variant over another, due to for example association with certain subgroups of society and a speakers own identity (Labov 2001; LePage and Tabouret-Keller 1985). All of these mechanisms can be classified as replicator selection.

However, variant frequencies may change without any form of replicator selection being present, but simply by random fluctuations in the replication process. In population genetics, this process is called genetic drift (Crow and Kimura 1970), or *neutral evolution*. We prefer this second terminology, to avoid confusion with linguistic drift, which is an entirely unrelated phenomenon. In language change, structured social networks mean that a speaker is more likely to interact with certain speakers rather than others (Labov 2001). Because some speakers may happen to use some variants (replicators) more than others, it can bring about differential replication. We call this type of selection *neutral interactor selection*, in that the only factor that influences replication is the frequency of interaction with the interactor.

There is another type of interactor selection that is possible in language change. Interactors (interlocutors) may be preferred or dispreferred by a speaker no matter how frequently or infrequently she or he interacts with them, and their linguistic replications (utterances) are weighted accordingly. Thus, variants of a speaker whose productions are weighted more heavily will be differentially replicated. This type of interactor selection is unlike network structure, in which the weighting of variants is simply a consequence of frequency of interaction with different speakers producing different variants. We call this *weighted interactor selection*. Weighted interactor selection implies that a speaker's linguistic behavior is influenced not just by frequency of interaction but also a differential social valuation of particular speakers. These four mechanisms are summarized in Table 1.

Table 1 Types of selection present in language change

	Interaction Frequency	Interactor Behaviour	Replicator Behaviour
Neutral Evolution	equal	symmetric	symmetric
Neutral Interactor Selection	unequal	symmetric	symmetric
Weighted Interactor Selection	—	non symmetric	symmetric
Replicator Selection	—	—	non symmetric

3 Numerical Model

These different forms of selection can be explored through numerical modelling (Baxter et al. 2006). Here we are interested in the effects of social networks and influence, so we will focus on neutral evolution and on the different forms of interactor selection. We focus on a single linguistic variable, which we assume to be sufficiently independent of other language elements that it can be treated in isolation. Of course in reality language structures are interdependent, but for simplicity we consider only a single variable. This is similar to, in population genetics, the modeling the evolution of alleles of a single gene in early models (Crow and Kimura 1970). It means that more complex interactions such as chain shifts cannot be acoomodated in the present model.

Suppose that within a given speech community, one particular variant is reproduced much more than any other. This variant is the *convention* among that group of speakers. It may be that, over time, a different variant becomes more widely used amongst this group of speakers, eventually becoming dominant, i.e. becoming the new convention. This is an example of a language change, and it is the process we wish to model. We do not model the process of innovation, rather we assume the existence of multiple discrete *variants* of a linguistic variable, and track their frequency as they compete and propagate in the population through replication by the speakers.

A speaker of a language is able to track the frequencies with which she has heard particular variant forms used within speech community (Bybee 2001; Pierrehumbert 2003), and this knowledge in turn governs how she uses the variants. A speaker retains then a language model or linguistic representation which in turn governs the fruquency with which he or she uses the different variants.

Another important feature we wish to include is the observation that speakers may use unconventional variants if they have become entrenched (Croft 2000). For example, someone who has lived for a long time in one region may continue to use parts of the dialect of that region after moving to a completely new area. This is incorporated into our model in two ways. First, we shall assume that a given interaction (conversation) between two speakers has only a small effect on the established grammar. Second, speakers will reinforce their own way of using language by keeping a record of their own utterances.

The model is then constructed as follows. We consider a speech community consisting of N speakers, $i = 1, 2, ..., N$. We focus on this community's usage of a particular lingueme with a finite number $V \geq 2$ of variant forms. Each individual retains a model of the frequency with which the different variants are used in the speech community. The variable $x_{iv}(t)$ reflects speaker i's perception of the frequency with which lingueme variant v ($1 \leq v \leq V$) is used in the speech community at time t. These variables are normalized

$$\sum_{v=1}^{V} x_{iv}(t) = 1 \ \forall i, t \ . \tag{1}$$

In the following we will assume, for simplicity, that only two variants are present, a and b. We then need only a single variable to represent the grammar of each speaker, $x_{ia}(t) \equiv x_i(t)$, as the value of $x_{ib} \equiv y_i(t)$ is determined by the normalisation constraint, $x_i(t) + y_i(t) = 1$ at all times. The state of the system at time t is then the aggregation of grammars $\mathbf{x}(t) = \{x_1(t), ..., x_N(t)\}$.

In this Chapter, we wish to understand the effects of the network of social interactions. This is parametrised by the symmetric matrix **G**, which can be considered a weighted adjacency matrix. If speaker i can interact with speaker j, the relative frequency of their interactions is given by the matrix element G_{ij}. If they do not meet (that is, they do not know each other), this element is zero. The matrix is necessarily symmetric, because when i meets j, j always meets i. The elements G_{ij} are normalized such that the sum over distinct pairs $\sum_{\langle i,j \rangle} G_{ij} = 1$. Any particular network of interactions and relationships can be encoded in this matrix.

Social influence is more than just a network of contacts, however. The influence one speaker has on another is not necessarily the same as that which this other speaker has on the first. We encode a social influence **H**. Unlike **G**, **H** may be asymmetric. Its entries H_{ij} represent the relative value speaker i places on the utterances of j. This may originate in different social class, age, social group membership or simply personal preference. The entries in **H** do not need to be normalised. In principle they can take any value, but consideration of the dynamics of the model shows that values of a similar scale to λ, that is, much smaller than 1, have a significant effect. Generally we use values of this order, except in the cases of very broad power-law distributions (see below), in which case some values may be much larger.

As described in the previous section, these social factors, from an evolutionary standpoint, represent interactor selection. Neutral interactor selection can be encoded in the symmetric interaction network **G**, while weighted interactor selection arises from the combination of **G** with asymmetric interactions encoded in the social influence matrix **H**. The other main class of selection, replicator selection, acts as a function of (preference for) the variants themselves. In general, replicator selection is much stronger than interactor selection. Therefore, here we will generally assume that replicator selection is absent, and focus in interactor selection, which is where network effects are observed.

After choosing some initial values for speaker grammars x_i, and defining the network G_{ij} and influence H_{ij}, we allow the system to evolve by repeatedly iterating the following three steps in sequence, each iteration having duration δt:

1. Social interaction. A pair i, j of speakers is chosen with probability G_{ij}. See Fig. 1(a).

2. Reproduction. Each of the speakers selected in step 1 produces a set of w *tokens*, i.e., instances of variants. Each token is produced independently and at random, with the probability that speaker i utters variant v equal to the *production probability*, which for present purposes we set to be simply equal to the linguistic representation value x_{iv}. With two variants, the numbers of tokens $n_i(t)$ of variant a produced by speaker i is simply drawn from the binomial distribution

$$P(n_i | x_i) = \binom{w}{n_i} x_i^{n_i} (1 - x_i)^{w - n_i}. \tag{2}$$

Speaker j similarly produces a sequence of tokens according to his or her memory x_j. The randomness in this step is intended to model the observed variation in

Fig. 1 (Color online) Stylised representation of the model algorithm. (**a**) Speakers in the speech community interact with different probabilities according to a weighted social network (represented by different thicknesses of lines connecting them). Speakers i and j are chosen to interact with probability G_{ij}. (**b**) During the interaction, each speaker produces tokens of the linguistic variable according to their own internal linguistic representation. (**c**) Each speaker modifies their linguistic representation by a small amount, according to the frequencies with which the variants were reproduced in the conversation. A speaker records his or her own utterances as well as those of her interlocutor, weighted by a social influence factor H_{ij}. The update rule is given by Eq. (3). A new pair of speakers is then chosen, and the process repeats.

language use that was described in the previous section. This step is illustrated in Fig. 1(b).

3. Retention. The final step is to update each speaker's language model to reflect the actual language used in the course of the interaction. A small fraction ($\propto \lambda$) of existing speaker's memory is replaced by contributions which reflect both the tokens produced by herself and by her interlocutor. The social influence factor appears as a weighting factor H_{ij} applied to the tokens of his or her interlocutor, relative to her own. These considerations imply that, for speaker i,

$$x_i(t+\delta t) = \frac{1}{Z}\left\{x_i(t) + \lambda \left[\frac{n_i(t)}{w} + H_{ij}\frac{n_j(t)}{w}\right]\right\}. \qquad (3)$$

where Z is needed to ensures normalisation is maintained. The fraction of the memory occupied by the second variant, $y_i(t)$ obeys a similar equation $y_i(t+\delta t) = \frac{1}{Z}\left\{y_i(t) + \lambda \left[\frac{w-n_i(t)}{w} + H_{ij}\frac{w-n_j(t)}{w}\right]\right\}$ and summing the two equations we find that $x_i(t+\delta t) + y_i(t+\delta t) = \frac{1}{Z}[1+\lambda(1+H_{ij})]$. Normalisation requires that $x_i + y_i = 1$, so we must have $Z = 1 + \lambda(1+H_{ij})$.

The same rule applies for speaker j after exchanging all i and j indices. Figure 1(c) illustrates this step. The parameter λ, which affects how much the grammar changes as a result of the interaction is intended to be small, for reasons given in the previous section.

In simulations we repeatedly apply these three steps: choose two speakers according to **G**, each produces w tokens according to their current values of x_i, and then each speakers' x_i is updated using the update rule (3). The prevalence of the variants fluctuates in the population, until eventually one or another variant wins out, completely dominating the population, after which no further change occurs. (We do not model the process of introduction of new variants, which happens in real languages, and allows continual evolution of the language.) We call this *fixation*. The main quantities we may wish to understand, then, are the probability that a given variant is the one that fixes, and the mean time this takes to occur. These depend on the social structure and interactions, and on the initial conditions. In the following section we develop the mathematical analysis of the model, which allows the calculation of fixation probabilities and mean fixation times.

4 Analysis

It is possible to understand the behaviour of the model through analytic calculations. These are based on a so called continuous time approximation, where we define a new time scale such that each interaction step in the model corresponds to a very short time interval. The time can then be approximately treated as continuous, so that we can derive differential equations for the evolution of the model. The model is stochastic, that is, it contains random elements meaning that repeating a simulation does not necessarily produce the same outcome. The differential equations we write are therefore for the evolution over time of the probability distribution of the state of the system.

We define the time scale such that $\delta t = \lambda^2$, where λ is the small parameter controlling the amount by which a speaker's memory is updated after an interaction. We choose this relationship because, it turns out, the behaviour of the model in this new time scale is independent of λ (so long as it is small). Then taking the limit $\delta t \to 0$, the time becomes continuous. Using the process described in Baxter et al. (2006), which we recapitulate briefly in the Appendix here, we construct a Fokker-Planck equation for the process described in the previous section. This is an equation for the time evolution of the probability $P(\mathbf{x},t)$ that the system is in state $\mathbf{x}$ at time t. The equation reads

$$\frac{\partial P(\mathbf{x},t)}{\partial t} = \sum_i \sum_{j \neq i} m_{ij} \frac{\partial}{\partial x_i}[(x_i - x_j)P(\mathbf{x},t)] + \frac{1}{2} \sum_{i=1}^{N} G_i \frac{\partial^2}{\partial x_i^2}[x_i(1-x_i)P(\mathbf{x},t)] \quad (4)$$

where G_i is the total interaction probability for speaker i, $G_i \equiv \sum_{\langle ij \rangle} G_{ij}$ and $m_{ij} \equiv G_{ij} h_{ij}$. Here H_{ij} has been rescaled by λ, $h_{ij} = H_{ij}/\lambda$, which is necessary to maintain proper scaling as $\lambda \to 0$ as $\delta t \to 0$.

It is not possible to write an exact solution for this equation, which has N variables. However, as shown in Baxter et al. (2008) and Blythe (2010) remarkably, progress can be made. We use the observation that, after a relatively short period, the system relaxes to a quasi-stationary state, in which the change in the speakers grammars effectively become coupled. After this point, the dynamics can be largely described by a single collective variable

$$\xi(t) = \sum_{i=1}^{N} Q_i x_i(t), \quad (5)$$

where the coefficients Q_i depend on the structure of the matrices **G** and **H**, in a way that we will define shortly. The collective variable ξ does not have a direct physical interpretation, but it can be viewed as an effective prevalence of variant a in the population. The contribution from each speaker is weighted by the factor Q_i which captures the relative influence of that speaker in the population. See Blythe (2010) for a detailed discussion of the validity of this approximation. It is appropriate when the diameter of the network (the longest path in the network between two speakers) is small compared to the population size, as we would expect it to be in a social interaction network. As we will see, analytic results obtained are in excellent agreement with numerical simulations.

We introduce a matrix **M** by

$$M_{ij} = \begin{cases} m_{ij}, & \text{if } j \neq i \\ -\sum_{k \neq i} m_{ik}, & \text{if } j = i. \end{cases} \quad (6)$$

The center-of-mass weights Q_i then correspond to the left eigenvector of M_{ij}:

$$\sum_{i=1}^{N} Q_i M_{ij} = 0 \text{ with } \sum_{i=1}^{N} Q_i = 1. \quad (7)$$

The reason for this choice is that it implies that the average of $\xi(t)$ is conserved by the dynamics:

$$\frac{d\langle \xi(t) \rangle}{dt} = \sum_{i=1}^{N} Q_i \frac{d\langle x_i(t) \rangle}{dt} = \sum_{i,j} Q_i M_{ij} \langle x_j(t) \rangle = 0. \quad (8)$$

where angled brackets $\langle ... \rangle$ signify an expectation value, that is the value averaged over many realisations.

Since every instance of the process eventually reaches $x_i = 1$ or $x_i = 0 \forall i$, the probability that the final point is 1, i.e. fixation to variant a, is $\langle \xi(\infty) \rangle$. But the conservation of $\langle \xi(t) \rangle$ means that this is simply equal to its initial value $\xi(0)$.

We proceed by assuming that once the quasi-stationary state is reached, after a short initial relaxation time T_0, the dynamics depend almost entirely on this one variable. We can then write a Fokker-Planck equation for this single variable

$$\frac{\partial P(\xi,t)}{\partial t} = \frac{1}{2} \sum_{i=1}^{N} G_i Q_i^2 x_i(t)[1 - x_i(t)] \frac{d^2 P(\xi,t)}{d\xi^2}. \quad (9)$$

This equation applies from T_0, the time that the quasi-stationary state is reached, until fixation occurs at time T_f. Let $T = T_f - T_0$. The time T will be a good approximation to T_f as long as $T_0 \ll T_f$. Typically T_0 is of the order of N. If the overall time to fixation T_f is proportional to N^ν, then the approximation is valid as long as ν is larger than 1. As we will see, ν is normally equal to 2, but it may take smaller or larger values. Whenever it becomes close to 1, our method of approximation fails.

Taking T_0 as the new initial time, the time T obeys a reversed-time version of Eq. (9) (Gardiner 2004; Risken 1989):

$$-2 = \sum_{i=1}^{N} G_i Q_i^2 x_i(T_0) [1 - x_i(T_0)] \frac{d^2 T}{d\xi(0)^2}. \tag{10}$$

This equation still depends on the variables x_i at time T_0. We approximate $x_i(1 - x_i)$ in Eq. (10) by $\langle x_i(T_0) \rangle - \langle x_i(T_0)^2 \rangle$, their expected values in the quasi-stationary state. In this state, $\langle x_i \rangle \approx \xi(0)$. We estimate $\langle x_i(T_0)^2 \rangle$ by assuming that x_i and x_j are uncorrelated, and calculating the variance in the quasi-stationary state from the original Fokker-Planck equation, assuming that the rate of change of the variance of x_i is negligibly small in the quasi-stationary state. This approximation is discussed in more detail in Baxter et al. (2008) and Blythe (2010). The equation for T becomes

$$[\xi(0)(1 - \xi(0))] \frac{d^2 T}{d\xi(0)^2} = -2/r \tag{11}$$

where

$$r \approx \sum_i Q_i^2 G_i \frac{2\sum_{j \neq i} m_{ij}}{2\sum_{j \neq i} m_{ij} + G_i}. \tag{12}$$

The mean fixation time is obtained by integrating the equation for $T(\xi(0))$, with the boundary conditions $T(0) = T(1) = 0$ to give

$$T(\xi(0)) = -\frac{2}{r} [\xi(0) \ln \xi(0) + (1 - \xi(0)) \ln(1 - \xi(0))]. \tag{13}$$

This rather simple expression encapsulates the effects of arbitrary social network and social influence structures. The probability that variant a eventually reaches fixation can be read off as $\xi(0)$ (the probability that b fixes is $1 - \xi(0)$), while the mean time taken to reach consensus is given by (13). Given G_{ij} and H_{ij}, it remains only to find Q_i to complete the calculation.

In the next sections, we examine different social network structures **G** and social influence **H**, representing examples of the different forms of selection described in Sect. 2. Where possible the above formalism will be used to give analytic calculations, which we compare with numerical results. In more difficult cases, we give only numerical results, but we see that they are consistent with the patterns observed in the simple cases. To calculate mean fixation times, simulations were carried out until a state of fixation was reached, and the time at which this occurred was averaged over multiple runs. Unless otherwise stated, we used homogeneous initial conditions, that is, all $x_i(0)$ are initially set to the same value x_0.

Fig. 2 Fixation times within the socially-neutral utterance selection model (defined in the text) on various networks with N nodes. Networks are fully connected (square symbols), random regular graph with each possible edge connected with probability 0.2 (plusses) and hub-and-spoke network (stars) in which small groups of speakers are arranged in a star pattern, with speakers only interacting with speakers from the central group outside of their own group.

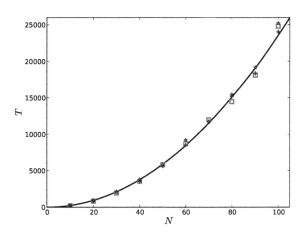

5 Social Networks in the Neutral Model

We begin with the simplest case, neutral interactor selection. That is, we examine the effect only of heterogeneous social contact frequencies, given by G_{ij}, while all other forms of selection are absent. This means that each speaker gives the same importance to every other speaker's utterances, and we set $H_{ij} = H$, a constant, for all i and j.

Setting these values in Eq. (12), and using the fact that the symmetry of m_{ij} implies that all Q_i are equal to $1/N$ (Baxter et al. 2008), we find that all dependence on G_{ij} cancels, leading to the remarkable result that the mean fixation time T doesn't depend on the social network structure at all (Baxter et al. 2008), and (when measured in terms of number of interactions) is proportional to N^2.

To confirm this results, in Fig. 2 we show numerical fixation times for a variety of network structures. We see that, indeed, there is little variation in the mean fixation times, across quite a diverse selection of network structures, and all are in agreement with the theoretical calculation. Furthermore, the probability of fixation also does not depend on network structure but simply on initial conditions.

These results enabled an evaluation of Trudgill's theory of new-dialect formation, as applied to the emergence of New Zealand English (Baxter et al. 2009). Trudgill has postulated (Trudgill 2004) that the large quantity of data that exist on the emergence on this English variant (Gordon et al. 2004) may be explained by assuming that social factors (which are modeled by H_{ij} in our model) are unimportant, and therefore this factor may be replaced by a constant. This meant that we did not need to know the details of the social network in order to apply our model to this situation. Incidentally, it turned out that the probability of certain variants emerging as the convention was indeed consistent with the neutral version of our model, but the time taken for this convention to arise would be much longer than was observed in New Zealand (Baxter et al. 2009).

6 Weighted Interactor Selection

We now move beyond the simple social network to consider weighted interactor selection. This is where the weight a speaker gives to the utterances of her interlocuter are not the same as those he gives to hers. This can accommodate a variety of social phenomena, from differential preference for the speech of certain social classes, groups, ages, or simply personal affinity.

Let us assume that H_{ij} is separable: $H_{ij} = \alpha_i \beta_j$. This is not an unreasonable assumption, it allows us to look at the case where speakers are influenced by (the α_i) or influence (the β_j) other speakers irrespective of the identity of their interlocutor. That is, that H_{ij} depends only on characteristics of individual speakers, and not on the social network.

Under this assumption, the solution for Q_i becomes (Baxter et al. 2012):

$$Q_i = \frac{\beta_i/\alpha_i}{\sum_j \beta_j/\alpha_j}, \qquad (14)$$

and then Eq. (12) for r becomes

$$r = \frac{1}{[\sum_j \beta_j/\alpha_j]^2} \sum_i \frac{\beta_i^2}{\alpha_i^2} G_i \frac{2\alpha_i \sum_{j \neq i} G_{ij}\beta_j}{2\alpha_i \sum_{j \neq i} G_{ij}\beta_j + G_i}. \qquad (15)$$

The fixation time is proportional to $1/r$, and so we will focus on the calculation of r. Note, however, that $\xi(0)$ will depend on the initial values of x_i. This has an effect on the fixation time, but only as a prefactor, and not on the scaling with population size. Unless otherwise stated, we will assume that $x_i(0) = x_0 \, \forall i$, then $\xi(0) = x_0$.

We further assume that the network of speaker interactions is large, random, and uncorrelated. The number of neighbours k_i that each speaker has, called their degree, is drawn at random from some distribution. In a large population, the social network is defined by this degree distribution. There is no correlation between the number of neighbors that different speakers have, so that $G_{ij} \propto k_i k_j$. With proper normalisation we have

$$G_{ij} \approx \frac{2k_i k_j}{(N\mu_1)^2} \quad \text{and} \quad G_i \approx \frac{2k_i}{N\mu_1}, \qquad (16)$$

where μ_1 is the mean of the degree distribution, that is, the mean number of neighbours a speaker has. Under these two assumptions we then obtain

$$r = \frac{1}{N\mu_1 [\sum_j \beta_j/\alpha_j]^2} \sum_i \frac{4\beta_i^2 k_i \sum_{j \neq i} k_j \beta_j}{2\alpha_i^2 \sum_{j \neq i} k_j \beta_j + \alpha_i(N\mu_1)}. \qquad (17)$$

Under these two approximations, we can try out different schemes for the interaction weightings. We are mainly interested in how the fixation time scales with N, and are in particular looking for significant deviations from the baseline result (found in Sect. 5) that T is proportional to N^2.

6.1 Asymmetry Independent of Network Structure

We first investigate the situation in which the social influence factors H_{ij} are assigned independently of the social network structure G_{ij}, so that there is no correlation between them. Consider first the case that the α_i are all equal, $\alpha_i = 1$, say, while the β_i take on arbitrary values. This means that different speakers' utterances carry different weights with their audience, but the importance given to them does not vary from listener to listener. Then

$$r = \frac{1}{N\mu_1[\sum_j \beta_j]^2} \sum_i \frac{4\beta_i^2 k_i \sum_{j \neq i} k_j \beta_j}{2\sum_{j \neq i} k_j \beta_j + N\mu_1}. \tag{18}$$

Suppose that the β_i are independently drawn from some distribution. Since they are selected independently from k_i,

$$\sum_{j \neq i} k_j \beta_j \approx N\mu_1 \langle \beta \rangle - k_i \beta_i. \tag{19}$$

For large N, we expand Eq. (18) in powers of $k_i \beta_i / N$ to obtain

$$r \approx \frac{4}{N^2 \mu_1 \langle \beta \rangle^2} \left\{ \frac{\mu_1 \langle \beta \rangle \langle \beta^2 \rangle}{[1 + 2\langle \beta \rangle]} - \frac{\mu_2 \langle \beta^3 \rangle}{N\mu_1 [1 + 2\langle \beta \rangle]^2} \right. \\ \left. - \frac{2\mu_3 \langle \beta^4 \rangle}{(N\mu_1)^2 [1 + 2\langle \beta \rangle]^3} - \cdots \right\}, \tag{20}$$

where μ_n is the nth moment of the degree distribution.

If the β_i are selected from a generic distribution, such as a Gaussian, the moments are well behaved, that is, they tend to a finite value for $N \to \infty$. This implies that $r \propto N^{-2}$ for large N and so the mean time to fixation grows as N^2 for large N. This is identical to the result obtained in the previous Section, for the case where H_{ij} had no structure at all, and suggests that if we are to look for deviations from this behavior then we must investigate distributions where the moments depend on N in some way. We therefore consider "fat tailed" distributions of β_i, in which some members of the community have a much larger influence than the norm. Consider a power law distribution

$$P(\beta) = A\beta^{-\gamma} \text{ for } \beta \geq \beta_0. \tag{21}$$

We find the dependence of r on N for different values of the exponent γ. In all ranges of γ, we find that the first term of Eq. (20) dominates, so that

$$r \approx \frac{4\langle \beta^2 \rangle}{N^2 \langle \beta \rangle [1 + 2\langle \beta \rangle]} \tag{22}$$

for large N.

Since for $\gamma > 3$ both $\langle \beta \rangle$ and $\langle \beta^2 \rangle$ have finite limits as $N \to \infty$, we recover the $T \propto N^2$ result found from more conventional distributions. For $1 < \gamma < 2$, Eq. (22)

Fig. 3 Mean fixation time as a function of population size for H_{ij} independent of degree, as described in Sect. 6.1. Results are for a fully connected network with β_i following a power-law distribution with decay exponent $\gamma = 2.0, 2.2, 2.6$ and constant α_i. Markers are average fixation times for 5000 numerical runs. Solid lines are expected scaling as given by Eq. (22), dashed lines are best fit curves of the form $T = AN^\zeta$.

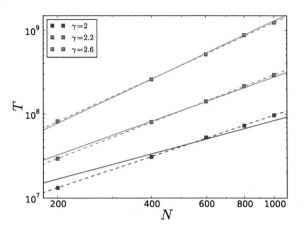

gives $T \propto N$ and for $2 < \gamma < 3$, $T \propto N^{(5-3\gamma)/(\gamma-1)}$, and so in this range the power of N varies from $3/2$ to 2, having the former value when $\gamma = 2$.

These calculations can be checked against numerical simulations of the model. The results shown in Fig. 3 are for a fully connected network of speakers, that is, each speaker is equally likely to speak with each of the other speakers, but similar results are obtained for a sparse interaction network in which each speaker had approximately an equal number of neighbors. The β_j were chosen from a power-law distribution for various values of the power-law exponent γ, while setting all $\alpha_i = 1$. The agreement with the predictions of Eq. (22) is very good so long as the predicted exponent of growth of T with N is greater than 1, that is for $\gamma > 2$. This includes the region $2 < \gamma < 3$, in which the mean fixation time, T, grows more slowly with N than in the usual situation where $T \propto N^2$. That is, the mean time to fixation may be reduced without recourse to any special network structure, merely by allowing heterogeneity in the response of speakers to the utterances of their interlocutors.

For $\gamma \leq 2$, Eq. (22) predicts $T \propto N$. As we approach this region, we find the theoretical predictions break down. This can be seen in the lowest set of data in Fig. 3. This is not unexpected, if we consider the approximations made to derive our estimates of the mean fixation time. We have assumed that there is a short relaxation period after which the dynamics can be well described by considering only the collective variable ξ (see Blythe (2010) for details). Our calculated fixation times are only for this second stage. Typically the initial relaxation happens in a time of order N. We see that if the calculated fixation time is of a similar time scale, the initial relaxation can no longer be ignored.

So, in summary, choosing an extreme distribution for β_i of the type described by (21) can reduce the growth of T with population size, the slowest growth (and hence the shortest fixation times) being when γ approaches 2 when $T \propto N^\nu$ tends to $\nu = 1$.

The complementary situation is to set the β_i to be all equal, while the α_i vary. In this situation, some speakers give more attention to others' utterances, and some less, but the identity of their interlocutor is not taken into account. However in this

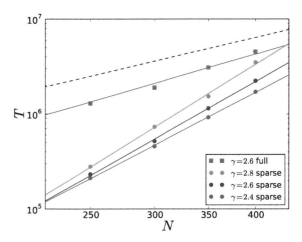

Fig. 4 Numerical results for mean fixation time for α_i distributed according to inverted power law distributions, with values of γ given in the legend. Top line (squares, red online) is for a fully connected graph with $\gamma = 2.6$. Lower lines (circles) are for a sparse graph with mean degree 10 and $\gamma = 2.8, 2.6, 2.4$ from top to bottom. Lines are fitted functions of the form $T = aN^\xi$. Dashed line is aN^2 for comparison.

situation the method we used when the α_i were all equal does not apply, and we cannot obtain any simple analytic results. However, by examining Eq. (17), we see that it is the smallest values of α_i which contribute most to r. In fact we find that $r \sim \langle 1/\alpha \rangle / N^2$. This result is similar to that found in Masuda et al. (2010) and Baxter (2011) where different agents in the network could change state with different rates: this is one way to interpret variation of the α parameter in the present work. Consider a power-law distribution of values, such that $P(\alpha) \propto \alpha^{-\gamma}$. The moment $\langle 1/\alpha \rangle$ is independent of γ in this case, so we would expect to find $T \propto N^2$. Indeed this is exactly what is observed through numerical simulations, with the mean fixation time growing as N^2, exactly as in the standard case, regardless of the value of γ.

Considering the fact that the smallest α values make the largest contribution, we can also consider an 'inverted' power-law distribution, $P(1/\alpha) \propto (1/\alpha)^{-\gamma}$, that is $P(\alpha) \propto \alpha^{+\gamma}$ with an imposed upper bound instead of a lower bound. In this case we do see mean fixation times changing with γ, but rather than fixation being sped up, it is slowed down. We find $T \propto N^\nu$, with ν approaching the baseline value of 2 when $\gamma = 3$, and increasing as γ decreases, as shown in Fig. 4. Here we do find a difference relative to other cases we investigated, in that the density of the graph also has an effect on the exponent ν: it grows more quickly with decreasing γ on a sparse network than a fully connected network.

Returning to heterogeneous β_i values, let us investigate the effect of correlations between the β_i values of neighboring speakers. To do this, we place the speakers on a random sparse network, in which each speaker has approximately the same number of neighbors. As before, power law distributed β values are chosen, but now the largest value is assigned first to a randomly chosen speaker. The next largest values are then assigned to this speaker's neighbours, and so on. These correlations only slightly affect the scaling of mean fixation time with population size N, with T growing as N^ν with exponent ν similar to that found in the uncorrelated case described above for the same γ. This is confirmed by simulation results, Fig. 5. Simulations

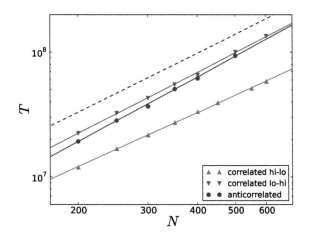

Fig. 5 Numerical results for mean fixation time for correlated β_i. Speakers are located on a sparse network and β_i values distributed according to a power law with exponent $\gamma = 2.4$, and correlated (see text) from highest to lowest (triangles), from lowest to highest (inverted triangles) and anti-correlated (circles). Lines are fitted functions of the form $T = aN^\nu$, with $\nu = 1.44, 1.63, 1.72$ respectively. For comparison the black dashed line has $\nu = 1.57$ which is the slope expected for uncorrelated β_i.

in which β values are assigned from lowest to highest, and also anticorrelations, in which the lowest β values are located on the neighbors of the highest value show similar results. In each case the growth of T with N was similar, though the overall prefactor was different to that found in the uncorrelated case. Results are plotted in Fig. 5, compare with Fig. 7.

Finally we consider the effect of inhomogeneity in the initial conditions. After randomly assigning power law distributed β_i values (without correlations), the speakers with the largest β_i's have their initial grammar value $x_i(0)$ set to 1, while the remainder were set to 0, such that the overall mean grammar was x_0. In this case T scales with N exactly as found above. The mean fixation time is affected by initial conditions through the center-of-mass parameter $\xi(0)$ which appears in Eq. (13). This affects the prefactor but not the scaling of T with N. The value of $\xi(0)$ differed from the homogeneous case value of x_0, as expected, as evidenced by a much greater probability of fixation to state 1.

These last numerical investigations thus support the value of the simpler cases for which we made analytic predictions. They give a good indication of the general conditions for finding fixation times shorter than the standard $T \propto N^2$.

6.2 Asymmetry Depends on Speakers Degree

We now consider the more general situation in which a speaker's influence may be related to his or her position in the network. This might occur, for example, if a popular speaker (i.e., one with many neighbors) is given more weight by her interlocutors, e.g. as in Fagyal et al. (2010). Alternatively, speakers might divide their attention between all of their interlocutors.

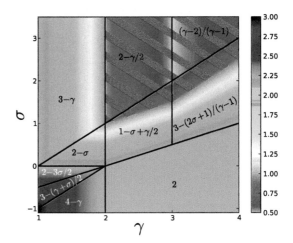

Fig. 6 (Color online.) Scaling of mean time to reach fixation with population size. Shading represents the value of exponent v where $T \propto N^v$. The labels give expressions for v in each region, with black lines marking boundaries between regions. The diagonal hatches cover the region in which the approximations used are not expected to be accurate.

We gain suppose that α_i is independent of i, say $\alpha = 1$, and now suppose that β_i is proportional to k_i raised to some power,

$$\beta_j = A k_j^\sigma \qquad (23)$$

for some constants A and σ. This allows us to explore a range of both positive and negative correlations between influence and degree.

We follow the same procedure as in Sec. 6.1, but now the moments of β are not independent of the network degree moments. In fact, everything can be written in terms of moments of the degree distribution μ_s. Expanding Eq. 12 for r in powers of $k_i^{\sigma+1}/N$ we find

$$r \approx \frac{4}{N^2 \mu_1 \mu_\sigma^2} \left\{ \frac{\mu_{\sigma+1} \mu_{2\sigma+1}}{[2\mu_{\sigma+1} + \mu_1 A^{-1}]} - \frac{\mu_{3\sigma+2} \mu_1 A^{-1}}{N[2\mu_{\sigma+1} + \mu_1 A^{-1}]^2} - \frac{2\mu_{4\sigma+3} \mu_1 A^{-1}}{N^2 [2\mu_{\sigma+1} + \mu_1 A^{-1}]^3} - \ldots \right\}. \qquad (24)$$

For conventional degree distributions, all the moments tend to N-independent values as N becomes large, and we have $r \sim 1/N^2$ as usual.

Suppose however that the degree distribution obeys a power law. In different regions of the $\gamma - \sigma$ plane different moments appearing in (24) diverge with N. By carefully examining the ratios between subsequent terms in the series, we find that the first term always dominates. This gives

$$r \approx \frac{4\mu_{\sigma+1} \mu_{2\sigma+1}}{N^2 \mu_1 \mu_\sigma^2 [2\mu_{\sigma+1} + \mu_1 A^{-1}]}. \qquad (25)$$

The scaling with respect to N depends on whether any or which combination of the moments $\mu_1, \mu_\sigma, \mu_{\sigma+1}, \mu_{2\sigma+1}$ diverge with N. This divides the $\sigma-\gamma$ plane into a number of regions, as seen in Fig. 6. The mean fixation time grows with population

Fig. 7 Mean fixation time as a function of population size with β_i depending on speaker degree, as described in Sect. 6.2. Results are for a random network whose degree distribution obeys a power law $p(k) \propto k^{-\gamma}$. The interaction weights depend on degree through $\beta_i \sim k_i^\sigma$. Markers are average fixation times for 5000 numerical runs. Solid lines are expected scaling as given by Eq. (25) and Fig. 6, dashed lines are best fit curves of the form $T = AN^\zeta$.

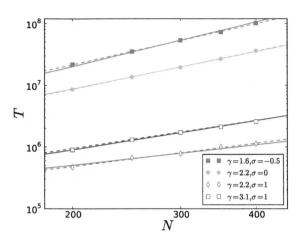

size in general as $T \propto N^\nu$, and we give expressions for ν in the various regions in Fig. 6. We see that in a large area, $\nu = 2$ as in the standard case. For $\gamma < 3$ and $\sigma < 0$ the mean time to fixation may grow faster than N^2. On the other hand, for $\sigma > 0$ and above the line $\sigma = \gamma - 1$, T may grow more slowly than N^2, with the slowest growth rate $T \propto N^{1/2}$ being achieved when $\gamma = 3$ for $\sigma \geq 2$ (though, as we will see, our approximations start to break down when $\nu < 1$).

In Fig. 7 we present simulation results for various representative locations in the γ–σ plane (see Fig. 6). The numerical results are in excellent agreement with the ν values predicted by Eq. (25) for values both smaller and larger than the baseline value of 2. As before, the agreement fails when the predicted value of ν is 1 or less. This occurs in the region marked with diagonal hatching in Fig. 6.

7 Conclusions

In this Chapter, we have investigated how social network and heterogeneous social interactions in a model of language change affect the behaviour of the model, particularly the mean time to reach a state of fixation (all speakers using a common conventional variant). Remarkably, when social interactions are symmetric, the details of the network structure have no effect on the mean time to fixation, or the probability that a particular variant fixes. In contrast, the mean time to fixation can be dramatically affected by the presence of large disparities in the influence of different speakers, for example when the H_{ij} are constructed to be drawn from a power-law distribution. This is the case even when the social network is rather homogeneous, and the H_{ij} entries are separated as properties of individual speakers (e.g.. $H_{ij} = \beta_i$, independent of j). A wide variety of scaling relationships between the mean fixation time and network size are possible when node influence and degree (a measure of 'popularity') co-vary. In different cases fixation may be accelerated or decelerated relative to the baseline case of uniform influence, depending on how influence and degree are correlated. The derived analytical predictions hold when the initial relaxation to the quasi-stationary state is of a much shorter time scale than the overall time to reach fixation. When the second period becomes close to order N^1, this is no longer true, and the analytical results are no longer valid.

In more general terms, the results obtained here underline the importance of developing numerical models from first principles, based on a careful consideration of the phenomena occurring in the system under study. In this case, the distinction of replicator selection from interactor selection emerged naturally from consideration of this model (Baxter et al. 2009), something which might otherwise have been missed. Here we have outlined in general terms the effects of neutral and weighted interactor selection. These represent the complexities of social interactions: group identity, social class, gender, age, etc. and it remains to apply the methods developed to specific linguistic problems. In many such systems, replicator selection is of course also likely to be present. Nevertheless, effects of both types of selection ought to occur simultaneously. It remains for future investigations to delineate these situations and the interactions between the different forms of selection.

Acknowledgements. Many of the results presented here have previously been presented in Baxter et al. (2008) and Baxter et al. (2012). Figures 3–7 are reproduced from Baxter et al. (2012) with permission of the Authors.

Appendix: Derivation of Fokker-Planck Equation

Here we give a brief outline of the derivation of the Fokker-Planck Equation (4), using a standard method that can be found in, for example Risken (1989) and Gardiner (2004). A similar derivation can be found in Baxter et al. (2006).

The time derivative of an evolving probability distribution can be related to derivates of the state variables using the Kramers-Moyal expansion

$$\frac{\partial}{\partial t}P(\mathbf{x},t) = -\sum_{i=1}^{N} \frac{\partial}{\partial x_i}\{\alpha_i(\mathbf{x})P(\mathbf{x},t)\} + \frac{1}{2}\sum_{i=1}^{N}\sum_{j=1}^{N} \frac{\partial^2}{\partial x_i \partial x_j}\{\alpha_{ij}(\mathbf{x})P(\mathbf{x},t)\} + \ldots \quad (26)$$

where the α functions are called the jump moments and are defined as

$$\alpha_i(\mathbf{x}) = \lim_{\delta t \to 0} \frac{\langle \delta x_i(t) \rangle}{\delta t} \quad (27)$$

$$\alpha_{ij}(\mathbf{x}) = \lim_{\delta t \to 0} \frac{\langle \delta x_i(t) \delta x_j(t) \rangle}{\delta t}, \quad (28)$$

where $\delta x_i(t) \equiv x_i(t+\delta t) - x_i(t)$, and angled brackets represent ensemble averages. The ellipsis indicates that this series continues indefinitely, to higher and higher order derivatives. The Fokker-Planck equation is nothing more than the truncation of this expansion at the level of the second derivatives. Derivation of the

Fokker-Planck equation is then a matter of finding the jump moments. In the present case, the jump moments can be found from Eq. (3). The change in $x_i(t)$ in a single time increment, given that i has already been chosen to interact with j, is

$$\delta x_i = \frac{\lambda}{1+\lambda(1+H_{ij})} \left[\frac{n_i}{w} - x_i + H_{ij}\left(\frac{n_j}{w} - x_{iv}\right)\right]. \tag{29}$$

The expected change $\langle \delta x_i \rangle$ can then be found using the fact that n_i is a multinomial sample of w trials with probability x_i, and taking into account that the probability that i interacts with j at time t is G_{ij}. We must then sum over all possible j, giving

$$\langle \delta x_i \rangle = \lambda^2 \sum_{j \neq i} \frac{\lambda}{1+\lambda(1+\lambda h_{ij})} G_{ij} h_{ij}(x_j - x_i), \tag{30}$$

where we have used the rescaled variable $h_{ij} = \lambda H_{ij}$. Identifying δt with λ^2 then gives us the first term in the Fokker-Planck Equation (4).

For the second jump moment, we must use the variance of the multinomial distribution

$$\langle n_i n_j \rangle - \langle n_i \rangle \langle n_j \rangle = \begin{cases} wx_i(1-x_i) & i = j \\ 0 & i \neq j \end{cases} \tag{31}$$

and following a similar procedure, one finds that

$$\langle \delta x_i \delta x_j \rangle = \sum_i \sum_j G_{ij} \frac{\lambda^2}{w} x_i(1-x_i) \tag{32}$$

if $i = j$ and $\langle \delta x_i \delta x_j \rangle = 0$ otherwise. Again, using $\delta t = \lambda^2$ we immediately find the required jump moment, giving the second term in (4).

References

Baxter, G.J.: A voter model with time dependent flip rates. J. Stat. Mech. 2011, P09005 (2011)
Baxter, G.J., Blythe, R.A., Croft, W., McKane, A.J.: Utterance selection model of language change. Phys. Rev. E 73, 046118 (2006)
Baxter, G.J., Blythe, R.A., Croft, W., McKane, A.J.: Modeling language change: An evaluation of Trudgill's theory for the emergence of New Zealand English. Language Variation and Change 21, 257–296 (2009)
Baxter, G.J., Blythe, R.A., McKane, A.J.: Fixation and consensus times on a network: a unified approach. Phys. Rev. Lett. 101, 258701 (2008)
Baxter, G.J., Blythe, R.A., McKane, A.J.: Fast fixation with a generic network structure. Phys. Rev. E. 86, 031142 (2012)
Blythe, R.A.: Ordering in voter models on networks: exact reduction to a single-coordinate diffusion. J. Phys. A: Math. Theor. 43(38), 385003 (2010)
Blythe, R.A., Croft, W.: S-curves and the mechanism of propagation in language change. Language 88, 269 (2012)
Bybee, J.L.: Phonology and language use. Cambridge University Press, Cambridge (2001)

Castellano, C., Fortunato, S., Loreto, V.: Statistical physics of social dynamics. Rev. Mod. Phys. 81, 591–646 (2009)

Croft, W.: Explaining Language Change: An Evolutionary Approach. Long-man Linguistics Library. Pearson Education, Harlow (2000)

Crow, J.F., Kimura, M.: An Introduction to Population Genetics Theory. Harper and Row, New York (1970)

Dall'Asta, L., Baronchelli, A., Barrat, A., Loreto, V.: Non-equilibrium dynamics of language games on complex networks. Phys. Rev. E 74, 036105 (2006)

Fagyal, Z., Swarup, S., Escobar, A.M., Gasser, L., Lakkaraju, K.: Centers and peripheries: Network roles in language change. Lingua 120, 2061–2079 (2010)

Gardiner, C.W.: Handbook of Stochastic Methods, 3rd edn. Springer, Berlin (2004)

Gordon, E., Campbell, L., Hey, J., MacLagan, M., Sudbury, A., Trudgill, P.: New Zealand English: its Origins and Evolution. Cambridge University Press, Cambridge (2004)

Hawkins, J.A.: Efficiency and Complexity in Grammars. Oxford University Press, Oxford (2004)

Hubbell, S.: The Unified Neutral Theory of Biodiversity, Princeton (2001)

Hull, D.L.: Science as a Process: An Evolutionary Account of the Social and Conceptual Development of Science. University of Chicago Press, Chicago (1988)

Hull, D.L.: Science and selection: essays on biological evolution and the philosophy of science. Cambridge University Press, Cambridge (2001)

Kandler, A., Steele, J.: Ecological Models of Language Competition. Biol. Theor. 3, 164–173 (2008)

Kirby, S.: Natural Language From Artificial Life. Artif. Life 8, 185 (2002)

Labov, W.: Principles of Linguistic Change: Social Factors. Blackwell, Oxford (2001)

LePage, R.B., Tabouret-Keller, A.: Acts of Identity. Cambridge University Press, Cambridge (1985)

Masuda, N., Gibert, N., Redner, S.: Heterogeneous voter models. Phys. Rev. E 82, 010103(R) (2010)

McMahon, A.M.S.: Understanding Language Change. Cambridge University Press (1994)

Niyogi, P., Berwick, R.C.: Evolutionary consequences of language learning. Linguistics Philos. 20, 697 (1997)

Nowak, M.A., Komarova, N.L., Niyogi, P.: Evolution of universal grammar. Science 291, 114 (2001)

Ohala, J.: The origin of sound patterns in vocal tract constraints. In: MacNeilage, P.F. (ed.) The Production of Speech, pp. 189–216. Springer, New York (1983)

Pierrehumbert, J.: Phonetic diversity, statistical learning, and acquisition of phonology. Language and Speech 45, 115–154 (2003)

Risken, H.: The Fokker-Planck Equation. Study. Springer, Berlin (1989)

Trudgill, P.: New-dialect formation: the inevitability of colonial Englishes. Edinburgh University Press, Edinburgh (2004)

Vogt, P.: Modeling Interactions between Language Evolution and Demography. Human Biology 81, Article 7 (2009)

Emergence of Dominant Opinions in Presence of Rigid Individuals

Suman Kalyan Maity and Animesh Mukherjee

Abstract. In this chapter, we study the dynamics of the so-called naming game as an opinion formation model with a focus on how the presence of a set of rigid minorities can result in the emergence of a dominant opinion in the system. These rigid minorities are "speaker-only", i.e., they only "speak" and never "listen" thus strongly affecting the course of a social agreement process. We show that for a moderate α (fraction of rigid minorities), the agreement dynamics results in an emergence of a dominant opinion. We extensively study the property of such dominant opinions and observe that the dominance is not the characteristic property of only the "speaker-only" opinions; other opinions under certain circumstances can also become dominant. However, with increasing α, the chances of a "speaker-only" opinion becoming dominant increases. We also find early invented opinions possess higher chances of becoming dominant. We embed this model on various static interaction topologies and real-world time-varying face-to-face interaction data. Importantly, for a reasonably static societal structure the presence of rigid minorities influences the emergence of a dominant opinion to a much larger extent than in case where the societal structure is very dynamic.

1 Introduction

Our social behavior is to a large extent determined by the society we live in. The social interactions among different individuals in a society shape/reshape or may determine how an individual will adopt new ideas or opinions. A group of individuals that strongly advocate compelling points of view can influence public sentiments

---

Suman Kalyan Maity · Animesh Mukherjee
Dept. of Computer Science and Engineering,
Indian Institute of Technology,
Kharagpur, India - 721302
e-mail: {sumankalyan.maity,animeshm}@cse.iitkgp.ernet.in

and opinions on a particular issue. This phenomenon is evident in multi-party election campaigns where politicians belonging to multiple parties try to influence the public and try to align their opinion toward that of the politician's party. Each party here represents a political ideology and is therefore representation of a rigid opinion. Each politician bearing the rigid political ideology of a particular party tries to win the votes of other individuals in the population. Each group of the rigid individuals convey the opinion of the party. Thus, the rigid group of individuals who never change their opinions play a vital role in opinion formation and may influence the agreement process. Such type of competition is also valid for describing the language competition among variety of languages where a very few survive while a majority die competing with others (MARIAN and SPIVEY 2003; Patriarca and Leppänen 2004; Stauffer et al. 2007; COSTA et al. 2003).

In this chapter, we focus on the popular naming game framework (NG) (Baronchelli et al. 2006) as a model of opinion formation to study how social dominance emerge in the presence of rigid individuals and how such dominance could influence societies to align toward the dominant opinion through negotiation. The naming game is a simple multi-agent model that employs local communications which leads to the emergence of a shared communication scheme/common opinion in a population of agents. The system evolves through pairwise interactions among agents that necessarily capture the generic and essential features of an agreement process. This model was conceived to explore the role of self-organization in the evolution of languages (Steels 1995; Steels 1996). Steels (1995) primarily focused on the formation of vocabularies, i.e., a set of mappings between words and their meanings (for physical objects). In this context, each agent develops its own vocabulary in a random and private fashion. Agents are forced to align their vocabularies, through successive conversation, in order to obtain the benefit of cooperating through communication. Thus, a globally shared vocabulary emerges as a result of local adjustments of individual word-meaning associations. The communication evolves through successive conversations. It is worth pointing out that these conversations are particular cases of language games, which are used to describe linguistic behavior but can also describe non-linguistic behavior, such as pointing. As a practical example of this model, Talking Heads experiment in Steels (1999) was carried out where robotic agents develop their vocabulary observing objects through digital cameras, assigning them randomly chosen names and sharing these names in pairwise interactions. This model has also acquired attention in the field of semiotic dynamics (Golder and Huberman 2006; Cattuto et al. 2007) that studies evolution of languages through invention of new words, grammatical constructions and more specifically, through adoption of new meanings for different words. For instance, the proliferation of new generation of web-tools enabling human web-users to self-organize a system of tags in such a way to ensure a shared classification of information about different arguments, for example, del.icio.us or www.flickr.com has took place (Cattuto et al. 2007).

The minimal naming game (NG) consists of a population of N agents observing a single object in the environment (may be a discussion on a particular topic) and opining for that through pairwise interactions, in order to reach a global agreement. We consider discrete opinions and the all the opinions to be distinct. We also assume the opinions to be uncorrelated. The agents have at their disposal an internal inventory, in which they can store an unlimited number of different words or opinions. At the beginning, all the individuals have empty inventories. At each time step, two individuals are chosen - the "speaker" chosen randomly and the "hearer" also chosen randomly but from the neighborhood of the "speaker". The speaker voices to the hearer a possible opinion for the object under consideration; if the speaker does not have one, i.e., his inventory is empty, he invents a brand new opinion which is completely different from other opinions present before it. In case where he already has many opinions stored in his inventory, he chooses one of them randomly. The hearer's move is deterministic: if she possesses the opinion pronounced by the speaker, the interaction is a "success", and both speaker and hearer retain that opinion as the right one, removing all other competing opinions/words from their inventories; otherwise, the new opinion is included in the inventory of the hearer, without any cancellation of opinions in which case the interaction is termed as a "failure" (see Fig. 1).

We recast this model to incorporate a small set of rigid individuals and investigate the effect of their presence on the overall dynamical properties of the system. One of them could be the unwillingness to listen. This type of rigid individuals try to speak a lot and never listen to the others. Therefore, these rigid individuals take part in the interaction only as "speaker", and never as hearer. Since, an adoption of an opinion is only possible in the role of a hearer, these rigid agents never undergo any change in their opinion. The interactions between pairs of rigid individuals are therefore forbidden as they can never influence one another. Consequently, these rigid/"speaker-only" agents do not allow the population to reach a global consensus except the case where there is only one "speaker-only" agent in the population and rest of the population adopts the opinion of this particular agent in order to reach final agreement. However, in general, stable polarized/multi-opinion states are observed in the system at long times; such multi-opinion states have also been reported by Baronchelli et al. (2007) however, in a different context of modeling trust among agents. Examples of such forms of rigidity are found in languages quite frequently; for instance, some languages are very rigid in their word order (e.g., Irish, English, Persian) as opposed to certain others (e.g., Turkish, Russian) that are very flexible (Odlin 1989).

The rest of the chapter is organized as follows. In section 2, we discuss the state-of-the-art. In section 3, we describe the model in detail. Section 4 is devoted for the discussion and analysis of interesting insights that we obtain from the model in presence of a set of "speaker-only" agents for different social structures. In section 4, we draw conclusions and point to future direction of this research.

Fig. 1 Agent's interaction rules in NG on a topic, say "Who is the best soccer player?". The speaker selects the opinion highlighted. If the hearer does not possess that opinion she includes it in her inventory and the interaction is a "failure" (top). Otherwise both agents erase their inventories only keeping the winning opinion and interaction is a "success" (bottom).

2 Related Work

Opinion dynamics models involving committed individuals have been studied previously in Mobilia et al. (2007), Mobilia (2003), Galam and Jacobs (2007), Galam (2010), Biswas and Sen (2009), Yildiz et al. (2011), Lu et al. (2009), Xie et al. (2011), Xie et al. (2012). Mobilia et al. (2007) studied how the presence of zealots (rigid individuals) affected the distribution of opinions in the case of the voter model. Similarly, Mobilia et al. (2007) and Yildiz et al. (2011) studied the properties of steady-state opinion distribution for the voter model with stubborn agents, but additionally considered the optimal placement of stubborn agents so as to maximally affect the steady-state opinion on the network. The effect of rigid/committed individuals has also been studied on the binary naming game (Lu et al. 2009; Xie et al. 2011; Xie et al. 2012) where they introduce a set of individuals committed to a single opinion and study how majority opinion get rapidly reversed by the presence of a small fraction of them. In our model, we consider every "speaker-only" agent to possess one different opinion each and try to investigate the competition dynamics in the population finally, leading to the emergence of a single dominant opinion which is held by a majority of the agents. However, when the fraction of "speaker-only" agents crosses a critical value in the population, no single dominant opinion emerges, instead almost equal-sized multi-opinion clusters are formed. We further investigate the specific properties of the opinion that manifests as the most dominant one. In particular, we observe that those opinions that are invented early and, in most cases, by one of the "speaker-only" agents get elected as the dominant one. Nevertheless, for a reasonably low fraction of "speaker-only" agents an early invented opinion from the "non-speaker-only" group may also emerge as the most dominant one - we investigate in details the properties of these opinions in order to precisely reason for such a non-intuitive phenomena.

3 The Model Description

In this section, we discuss the model in detail. The population consists of N agents and out of them αN no. of agents are "speaker-only" agents. In each iteration of the game at $t = 1, 2, \ldots$ following happens :

1. An agent is randomly selected from the population and act as "speaker".
2. Another agent is chosen among the rest of the population except the "speaker-only" agents and is designated as "hearer"
3. If there is no name for the object/topic the "speaker" invents one brand new name otherwise he selects from list of already existing names/words and conveys it to the "hearer". Please note that every agent has an inventory/memory to keep the names.
4. The hearer then searches for the topic name in her inventory. If the search is successful, then the game is called "successful game" and both the agents delete all other competing names from their inventory.
5. On an unsuccessful search, the "hearer" learns the name (adds the name/word into the already existing name-inventory) and the game is termed as "failure" interaction.

The emergent properties that are of interest in the naming game are the sum of memory sizes of all agents at a particular time step t denoted by $N_w(t)$, the number of unique words/opinions/names in the system at t ($N_d(t)$) and the time needed to reach the global consensus (t_{conv}).

4 Results and Discussion

In this section, we shall try to elaborate the impact of having a fraction of "speaker-only" agents (α) in the population embedded on various types of social graphs and point to possible explanations of our findings.

4.1 The Mean-Field Case

The mean-field case corresponds to a fully connected network in which all agents are in mutual contact. Thus, every individual can, in principle, talk to every other individual. On this topology, we try to investigate the microscopic activity pattern of the game dynamics driven by the parameter α where α is the fraction of "speaker-only" agents chosen uniformly at random from the population. In Figs. 2(a) and (b), we have shown the time evolution of $N_w(t)$ and $N_d(t)$ where we observe that the system does not reach consensus and we have αN number of opinions left even after 5×10^7 games. The acquired state is the steady state which does not change as we have already reached the lower bound on the number of opinions with small fluctuations of N_w. Therefore, it is apparent that the system breaks down into clusters of opinions. Now, if we find the typical cluster sizes, we observe that there is usually one cluster which is extremely large compared to the other clusters. In other words, the distribution of the frequency of opinions (describing the number of agents who have possessed the opinion in their inventories) show skewness for lower values of α (see Figs. 2(c) and (d) describing the relation between the frequency of opinions($freq_r$) vs the rank (r)) pointing to the emergence of a dominant opinion (the opinion that is present in most of the agent's inventories) in the system. However, for a large enough α, the frequency of winner decreases allowing an increase in the size of the other clusters. To find the cut-off α for which this phenomenon happens, we observe the dependence of the frequency of the dominant opinion (f_w) on the "speaker-only" fraction α and find a "mirrored" S-shaped curve - the first part shows a linear decrease, then an abrupt fall and then again a steady linear dependence (see Figs. 2(e) and (f)). The cut-off value of α decreases with increasing N.

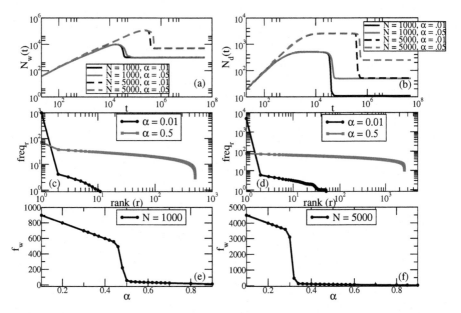

Fig. 2 Time evolution of (**a**) $N_w(t)$ and (**b**) $N_d(t)$ for $N = 1000$, 5000 and $\alpha = 0.01$, 0.5. Frequency of the opinions ranked in order of decreasing frequency ($freq_r$) present in the system after 5×10^7 games for (**c**) $N = 1000$ (**d**) $N = 5000$. Relation between the frequency of the dominant opinion (f_w) and α for (**e**) $N = 1000$ (**f**) $N = 5000$. The curves are averaged over 100 simulation runs.

In particular, we shall attempt to understand the reasons for the high frequency of the dominant opinion. Therefore, the natural question that arises is : Why does a sheer majority in the population agree to this opinion? Do the "speaker-only" opinions (opinions invented by "speaker-only" agents) always emerge as winners? The answers to these questions can be found out from Fig. 3 where in Fig. 3(a), we show the winning probability of the "speaker-only" opinions (W_{sp}) and its variation with α. We refer to the winning probability as the fraction of simulations in which a "speaker-only" opinion is elected as the most dominant one. We observe that for a small enough α, the chance of winning of a "speaker-only" opinion is more than 50% and as we increase N the required value of α to guarantee more than 50% winning chance is even lowered. Therefore, for a moderate α, "the speaker-only" opinions suppress the chances of other opinions' survival in the system and create a sheer monopoly for themselves.

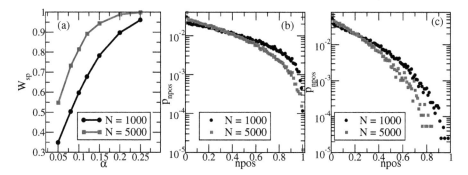

Fig. 3 (a) Relation between the winning probability of "speaker-only" opinions (W_{sp}) and α after 5×10^7 games for different N. Probability that a given opinion becomes the dominating one (p_{npos}) is plotted as a function of the normalized invention position ($npos$) when b) "speaker-only" opinions c) other opinions become the winner for $\alpha = 0.1$. The curves are averaged over 10^5 simulation runs.

Furthermore, we shall observe the effect of the creation time of the opinions on the phenomenon of emergence of dominance in the system. It turns out that the early invented opinions are in advantageous position in this context irrespective of whether they are invented by a "speaker-only" agent or by any other agent in the system. In Figs. 3(b) and (c), we plot the probability for an opinion to become the dominating one as a function of its normalized creation position. This means that each opinion is identified by its creation order: the first invented opinion is labeled as 1, the second as 2 and so on. To normalize the labels, they are then divided by the label of the last invented opinion. Clearly, the early invented opinions have higher chances of becoming dominant compared to the others invented late in time. To explore the property of the dominant opinion further, we delve deeper into the dynamics to identify the characteristic properties of the competing opinions invented before the dominant opinion. From Table 1, it is clear that the dominant opinion has

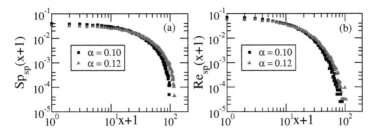

Fig. 4 Distribution of the number of "speaker-only" opinions before the winner when (**a**) "speaker-only" opinion ($Sp_{sp}(x)$) (**b**) "non-speaker-only" opinion ($Re_{sp}(x)$) is the winner for $N = 1000$. The curves are averaged over 10^5 simulation runs.

a fewer number of competitors facilitating its emergence as a winner. For an opinion to become winner, there has to be less number of competing opinions invented before it. In addition, from among "speaker-only" and "non-speaker-only" opinions, the proportion of "speaker-only" ones invented before the winner is significantly lower since they have a higher potential to quickly modify the opinion of the others. The presence of the "speaker-only" opinions are even much less in case a winner is a "non-speaker-only" opinion. The number of such opinions are approximately half as compared to the case when a "speaker-only" opinion emerges as winner. Now, if we observe the typical distribution of "speaker-only" opinions invented before the wining opinion, we find exponential-like distribution for both the cases (see Figs. 4(a) and (b)). Therefore, fewer the number of "speaker-only" opinions invented before an opinion, more is its chance to become the dominant or the wining opinion. This quantity is even lesser for the case when the dominant opinion has been invented from the "non-speaker-only" group as is also evident from the figure.

Table 1 Number of different types of opinions invented before the winner after 5×10^7 games, averaged over 10^5 simulation runs. "spo" refers to "speaker-only" opinions.

N	α	Winner	Earlier opinions (spo)	Earlier opinions (Rest)
1000	0.10	Spo	20.6	151.4
		Rest	12.1	96.4
	0.12	Spo	25.2	149.1
		Rest	14.3	92.7
5000	0.10	Spo	86.1	647.2
		Rest	47.3	385.8
	0.12	Spo	105.9	641.4
		Rest	55.9	369.7

Although we observe that the presence of fewer competitors invented before the invention of a particular opinion increases its chance of becoming dominant, an important question is that why exactly one of them emerge as a winner? It seems that

Table 2 Percentages of successes for the winning opinion compared to other opinions invented before it till dominance time, averaged over 10^5 simulation runs

N	α		Winner	Earlier opinions
1000	0.10	Spo	88.9	5.36
		Rest	93.1	3.07
	0.12	Spo	89.0	6.2
		Rest	93.5	3.42
5000	0.10	Spo	91.0	5.8
		Rest	94.34	3.38
	0.12	Spo	90.67	6.64
		Rest	94.77	3.46

the winning opinion takes part in a significantly larger fraction of successful interactions compared to the other competing opinions created before it till the dominance time which helps in propelling its dominance in the system. We define dominance time as the time when the most frequent opinion in the system reaches frequency $\sim (1-\alpha)N$ and next most frequent opinion has frequency less than 10% of the population. This phenomenon is supported by Table 2. The competition seems to be even less pronounced if the winner has to be from the rest of the population as compared to the case where a "speaker-only" opinion is the winner.

4.2 Scale-Free Networks

Social networks are far from being fully-connected or homogeneous. Most of them show a large skew in the distribution of node-degrees resulting in the so-called scale-free networks. In this section, we shall study the effect of α on the Barabási & Albert (BA) network (Barabasi and Albert 1999) which follows a scale-free degree distribution. Since low degree nodes form a vast majority in such networks, any randomly chosen node is, with high probability, a low degree node. The neighbors of this low degree node, however, should be high degree nodes (since the high degree nodes tend to be connected to almost all low-degree nodes by virtue of their high degree). Therefore, in this case we adopt two strategies while selecting the α fraction of "speaker-only" agents - (a) we select uniformly at random and (b) we select preferentially based on degree so that the high-degree hubs are more probable of being selected as a "speaker-only" participant.

Similar to the mean-field case, we observe a single dominant opinion in the system with majority of agents aligning to this opinion. Nevertheless for higher α, this dominance of a single opinion does not persist in the system (see Figs. 5(a), (b), (e) and (f)). The decomposition of the system into multiple similar size clusters occurs for a lesser value of α as compared to the mean-field case. In fact, this decomposition is even more pronounced for the case where the "speaker-only" population is chosen preferentially. A similar mirrored S-shaped dependence of the frequency of

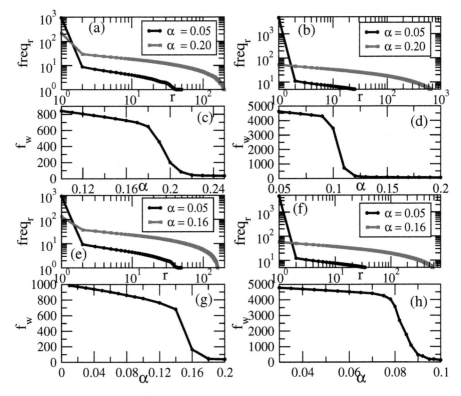

Fig. 5 $freq_r$ vs r after 5×10^7 games for "speaker-only" agents distributed randomly with (**a**) $N = 1000$, (**b**) $N = 5000$ and preferentially with (**e**) $N = 1000$, (**f**) $N = 5000$ respectively. f_w vs α for (**c**) $N = 1000$, (**d**) $N = 5000$ when "speaker-only" agents are selected randomly and (**g**) $N = 1000$, (**h**) $N = 5000$ when "speaker-only" agents are selected preferentially. The curves are averaged over 100 simulation runs for 10 network realizations each.

the winning opinion on α is again observed. However, the sharp transition occurs at a much lower α here and more so in case of preferential selection of the "speaker-only" agents (see Figs. 5 (c), (d), (g) and (h)). This is the consequence of the fact that social networks are sparse (with agents mostly interested in the local neighborhood) and therefore more vulnerable to such decomposition than the mean-field scenario. Further, selecting preferentially the hubs as the "speaker-only" agents help manifold in propelling their opinions very fast in the population mostly because of their disproportionately high connectivity. Next we study the relationship between the winning probability of the "speaker-only" opinions and α. The winning probability of the "speaker-only" opinion increases as we increase α. The typical α for which the winning probability becomes 50% is much lower compared to the mean-field case (see Figs. 6(a) and (b)).

The creation time of an opinion plays an important role in deciding the dominance of the opinion. For the case when a "speaker-only" opinion becomes winner,

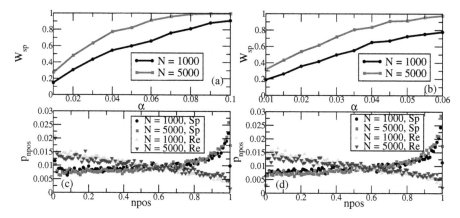

Fig. 6 W_{sp} vs α after 5×10^7 games when the "speaker-only" agents are selected (**a**) randomly (**b**) preferentially from the population. p_{npos} vs $npos$ when "speaker-only" opinions (Sp) and other opinions (Re) become the winner with "speaker-only" agents (**c**) randomly and (**d**) preferentially selected respectively for $\alpha = 0.03$. The curves are averaged over 10^4 simulation runs for 10 network realizations each.

irrespective of the way the "speaker-only" agents are selected (random/preferential) the lately invented opinions seem to mostly emerge as the winner which is possibly due to the inherent skewed degree distribution of agents (see Figs. 6(c) and (d)). Note that this result is markedly in contrast with those observed for the mean-field case. It is actually the late inventors in the "speaker-only" population who seem to be in advantageous position because by being late they are able to align their local neighborhood to their opinion in the last stages (i.e, there are no further scopes of opinion switch) and the larger this neighborhood, the higher is the chance that the lately invented "speaker-only" opinion is the winner. Further, a "speaker-only" agent has mostly "non-speaker-only" agents in his local neighborhood which makes it easy for the "speaker-only" agent to align them to his opinion and the later this takes place, the higher is the chance that this "speaker-only" opinion would emerge as the winner. However, the early invented opinions seem to be more probable winners in case this opinion has been initiated by a "non-speaker-only" member of the population (see Figs. 6(c) and (d)). We observe that the number of "speaker-only" opinions invented before the dominant opinion is always less in case this opinion is from the "non-speaker-only" group as in the mean-field scenario (see Table 3). In addition, the number of "non-speaker-only" opinions invented before the winner is larger compared to the "speaker-only" opinion pool. We also analyze the proportion of successes of the winner in comparison to the opinions created before the invention of the winner till the dominance time and observe that a huge majority of the successful interactions are witnessed by the winner (see Table 4). Although the mean-field trends are preserved here, one important point of difference is that the proportion of successes witnessed by the competitors of the winner is much larger here indicating a tougher competition.

Table 3 Number of different types of opinions invented before the winner, averaged over 10^4 simulation runs for 10 network realizations each when the "speaker-only" agents are randomly selected. The numbers in parentheses correspond to the case of preferential selection of "speaker-only" agents.

N	α	Winner	Earlier opinions (spo)	Earlier opinions (Rest)
1000	0.01	Spo	3.9 (3.9)	318.9 (320.8)
		Rest	3.1 (3.1)	240.1 (240.5)
	0.03	Spo	12.8 (13.0)	315.5 (316.6)
		Rest	9.2 (9.2)	234.6 (234.2)
5000	0.01	Spo	22.6 (22.4)	1670.3 (1654.7)
		Rest	16.0 (16.0)	1245.3 (1247.0)
	0.03	Spo	69.4 (70.4)	1643.6 (1649.4)
		Rest	48.3 (49.0)	1248.8 (1228.8)

Table 4 Percentage of successes for the winner compared to other opinions invented before it till the dominance time, averaged over 10^4 simulation runs for 10 network realizations each when the "speaker-only" agents are randomly selected. The numbers in parentheses correspond to the case of preferential selection of "speaker-only" agents.

N	α	Winner	Earlier opinions
1000	.01	Spo 72.97 (72.68)	2.27 (2.50)
		Rest 81.22 (81.41)	1.63 (1.76)
	.03	Spo 91.57 (95.34)	2.49 (1.64)
		Rest 97.43 (98.65)	0.84 (0.59)
5000	.01	Spo 90.01 (89.92)	1.51 (1.69)
		Rest 95.68 (95.86)	0.65 (0.68)
	.03	Spo 92.1 (93.3)	2.53 (2.48)
		Rest 98.03 (99.05)	0.55 (0.43)

5 Time-Varying Networks

In all the above discussions, we have considered static graphs where the link structure is known a priori and the game is played on top of the same. However, real-world social networks show dynamicity. Links appear and disappear over time. Friendship relations change with due course of time. Hence, it could be interesting to study the effect of rigid individuals embedded on such time-varying social networks.

5.1 Dataset Description

For the purpose of the investigation of the naming game dynamics on time-varying networks, we consider two specific real-world face-to-face contact datasets and present our results on each of them. Both the datasets are obtained from SocioPatterns Collaboration (http://www.sociopatterns.org/datasets/). The data collection infrastructure uses active RFID devices embedded in conference badges to detect and store face-to-face proximity relations of persons wearing the badges. These devices can detect face-to-face proximity (1–1.5 m) of individuals wearing the badge with a temporal resolution of 20 sec. The first dataset comprises face-to-face interaction data of visitors of the Science Gallery in Dublin, Ireland during the spring of 2009 (Isella et al. 2011). This dataset consists of time-varying snapshots of interactions at 20 s time interval for 69 consecutive days. We investigate the time-varying snapshots of one such representative day (22nd day). This network consists of 240 science gallery visitors. For future invocation of this dataset, we shall refer to it as SG22. The other data consists of the conference attendees of ACM Hypertext 2009 held in ISI Foundation in Turin, Italy. The dataset contains the dynamical network of face-to-face proximity of 113 conference attendees over about 2.5 days. For future invocation of this dataset, we shall refer to it as HT.

5.2 The Model Adaptation in the Time-Varying Setting

The naming game on time-varying network has already been studied by Maity et al. (2012) where they play the game in complete synchronization with real time, i.e., at each time step $t = 1, 2, \ldots$ (the elementary unit of time being second), a game is played among those agents that are alive at that particular instant of time (those agents having degree at least one) in the network. In this setting, at each time instant, the network snapshot of the agents at that particular time instant is considered. Therefore, this essentially boils down to having a series of network snapshots (one per second) and one game being played on each network snapshot. Please note that, as the RFID device can only consider an interaction if it stays for 20 seconds, the network of agents essentially change after 20 second. This study reports that behavior of the emergent properties of the system for the time-varying case is markedly in contrast with that of the static counterparts. Motivated by the above work, we investigate how the presence of rigid individuals in the population shapes the agreement process on the above real-world dataset. We adopt the same game playing strategy as earlier mentioned with a set of rigid individuals marked before the game starts.

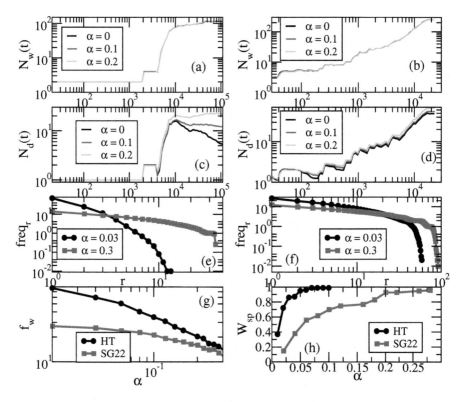

Fig. 7 Time evolution of $N_w(t)$ for (**a**) HT and (**b**) SG22 data, $N_d(t)$ for (**c**) HT and (**d**) SG22 data. $freq_r$ vs r for (**e**) HT and (**f**) SG22 data. (**g**) f_w vs α for HT and SG22 data. (**h**) W_{sp} vs α for HT and SG22 data. The curves are averaged over 10^3 simulation runs.

5.3 Results and Discussion

The evolution of $N_w(t)$ and $N_d(t)$ on the time-varying graph of HT and SG22 data (see Figs. 7(a), (b), (c) and (d)) show a drastically different behavior from the case where these quantities are measured on the static networks. A global consensus usually does not take place on such networks because of their inherent community structure and the openness of the system, i.e., the agents coming in and going out of the system leading to late-stage failures in the system which hinders the consensus (see Maity et al. (2012) for further details). Therefore, several opinion clusters get already formed with one becoming the dominant one. The presence of rigid individuals (chosen randomly from the population prior to the beginning of game) breaks such large opinion clusters into several smaller clusters. To analyze this phenomenon further, we observe the frequency of each opinion and find that the dominance of an opinion is not as pronounced in case of time-varying networks (see Figs. 7(e) and (f)) as in static social networks. As one increases α, the size of the clusters tend to become more and more uniform indicating that the frequency of the winner is almost as close as the others. Further, we investigate the relationship between the frequency of the winner in the system and α indicating a drastically different behavior as compared to the static networks (see Fig. 7(g)). In particular, we do not observe the mirrored S-shaped curve and associated transition; instead, there is a steep drop from the very beginning indicating that these networks already have a strong community structure leading to such multi-opinion states and injecting "speaker-only" agents in the system makes it even more fragmented. Figure 7(h) shows the relation between the winning probability of the "speaker-only" opinions and the "speaker-only" fraction α. As in case of static networks, here also the 50% winning probability is achieved at a very low value of α.

We further analyze the effect of the presence of "speaker-only" agents on the game interactions. We calculate the occurrence frequency of the number of "speaker-only" agents actively present at different time steps denoted by $size(sp)$ and observe the number of success/failure interactions experienced by $size(sp)$ "speaker-only" agents. The smaller the value of $size(sp)$ the larger is the number of success/failure interactions (success being orders of magnitude higher than failures) as is indicated through Figs. 8(a) and (b). This is due to the fact that in majority of the time steps $size(sp)$ is relatively quite low (see Fig. 8's insets). This indicates that in time-varying networks, it is not only the fraction of "speaker-only" agents that determine the winning opinion but also the number of such agents that are actually actively participating in the social interactions at a time step. The lower is the number of such active participants the higher is the chance that a single dominant opinion emerges.

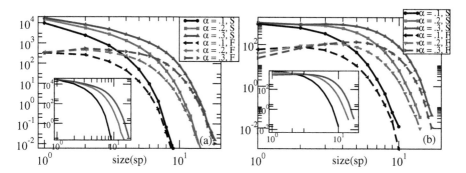

Fig. 8 Number of successes (S) and failures (F) vs $size(sp)$ for (**a**) HT, (**b**) SG22 data. Figures in the insets display the occurrence frequencies of different sizes of active "speaker-only" groups for HT and SG22 data respectively. The curves are averaged over 10^3 simulation runs.

6 Conclusions and Future Works

In conclusion, we have studied the effect of the rigid individuals on the naming game dynamics and how such rigid minorities influence the emergence of a dominant opinion in the system. We observe that the dominance is not the characteristic property of only the "speaker-only" opinions; other opinions from the population may also become dominant. However, with increasing α, the winning probability for a "speaker-only" opinion increases. The dominance index (i.e., f_w) decreases linearly as one increases α and at critical α, it shows an abrupt transition pointing to fragmented state with similar opinion clusters formed around the stubborn opinions. The creation time of an opinion also plays a vital role in the dynamics. Opinions that are invented early in time, possess higher overall chances of becoming the winner except for in the case of "speaker-only" opinions corresponding to static scale-free networks. This observation is quite interesting as this suggest that late-inventors can also produce dominant opinion. This is probably due to the high network heterogeneity in terms of network connectivity. Opinions that are created earlier in time compete among themselves to become the dominant one and the characteristic property of such a dominant opinion is that it takes part in a disproportionately large number of successful interactions (above 80%) compared to its competitors. We have also elucidated the game dynamics on diverse topological structures from homogeneous fully-connected network to heterogeneous scale-free networks and on real world social networks. On static social networks we observe similar results as in case of mean-field, however, for a significantly lower value of α. This indicates that the presence of rigid minorities can strongly affect a society that hardly changes over time. However, if the society is changing fast then such minorities do not seem to have a pronounced effect on the dynamics of opinion formation.

There are quite a few other interesting directions that can be explored in the future. One such direction could be to investigate the effect of introducing a flexibility component of the agents in adopting new opinions (traditionally modeled by a system parameter β that encodes the probability with which the agents update their inventories in case of successful interactions (Baronchelli et al. 2007)) rather than making them fully rigid. Finally, a thorough analytical estimate of the important dynamical quantities and the cut-off α reported only through empirical evidence here is needed to have a deeper understanding of the emergent behavior of the system.

References

Barabasi, A.L., Albert, R.: Emergence of scaling in random networks. Science 286(5439), 509–512 (1999)

Baronchelli, A., Dall'Asta, L., Barrat, A., Loreto, V.: Nonequilibrium phase transition in negotiation dynamics. Physical Review E 76(5), 051102 (2007)

Baronchelli, A., Felici, M., Loreto, V., Caglioti, E., Steels, L.: Sharp transition towards shared vocabularies in multi-agent systems. Journal of Statistical Mechanics: Theory and Experiment 2006(06) (2006)

Biswas, S., Sen, P.: Model of binary opinion dynamics: Coarsening and effect of disorder. Phys. Rev. E 80(2), 027101 (2009)

Cattuto, C., Loreto, V., Pietronero, L.: Semiotic dynamics and collaborative tagging. Proceedings of the National Academy of Sciences 104(5), 1461–1464 (2007)

Costa, A., Colomé, A., Gómez, O., Sebastián-Gallés, N.S.: Another look at cross-language competition in bilingual speech production: Lexical and phonological factors. Bilingualism: Language and Cognition 6(03), 167–179 (2003)

Galam, S.: Public debates driven by incomplete scientific data: The cases of evolution theory, global warming and h1n1 pandemic influenza. Physica A: Statistical Mechanics and its Applications 389, 3619–3631 (2010)

Galam, S., Jacobs, F.: The role of inflexible minorities in the breaking of democratic opinion dynamics. Physica A: Statistical Mechanics and its Applications 381, 366–376 (2007)

Golder, S., Huberman, B.A.: The Structure of Collaborative Tagging Systems. Journal of Information Science 32(2), 198–208 (2006)

Isella, L., Stehlé, J., Barrat, A., Cattuto, C., Pinton, J., Van den Broeck, W.: What's in a crowd? Analysis of face-to-face behavioral networks. Journal of Theoretical Biology 271(1), 166–180 (2011)

Lu, Q., Korniss, G., Szymanski, B.K.: The Naming Game in social networks: community formation and consensus engineering. Journal of Economic Interaction and Coordination 4(2), 221–235 (2009)

Maity, S.K., Manoj, T.V., Mukherjee, A.: Opinion formation in time-varying social networks: The case of the naming game. Physical Review E 86(3), 036110 (2012)

Marian, V., Spivey, M.: Competing activation in bilingual language processing: Within- and between-language competition. Bilingualism: Language and Cognition 6(02), 97–115 (2003)

Mobilia, M.: Does a Single Zealot Affect an Infinite Group of Voters? Physical Review Letters 91(2), 028701 (2003)

Mobilia, M., Petersen, A., Redner, S.: On the role of zealotry in the voter model. Journal of Statistical Mechanics: Theory and Experiment 2007(08), P08029 (2007)

Odlin, T.: Language Transfer: Cross-Linguistic Influence in Language Learning. Cambridge University Press, Cambridge (1989)

Patriarca, M., Leppänen, T.: Modeling language competition. Physica A: Statistical Mechanics and its Applications 338(1-2) (2004); Proceedings of the conference A Nonlinear World: the Real World, 2nd International Conference on Frontier Science, pp. 296–299

Stauffer, D., Castelló, X., Eguíluz, V.M., Miguel, M.S.: Microscopic Abrams-Strogatz model of language competition. Physica A: Statistical Mechanics and its Applications 374(2), 835–842 (2007)

Steels, L.: A self-organizing spatial vocabulary. Artificial Life 2(3), 319–332 (1995)

Steels, L.: Self-organizing vocabularies. In: Langton, C.G., Shimohara, K. (eds.) Artificial Life V, Nara, Japan, pp. 179–184 (1996)

Steels, L.: The Talking Heads Experiment. Words and Meanings, vol. 1. Laboratorium, Antwerpen (1999)

Xie, J., Emenheiser, J., Kirby, M., Sreenivasan, S., Szymanski, B.K., Korniss, G.: Evolution of opinions on social networks in the presence of competing committed groups. PLoS ONE 7(3), 33215 (2012)

Xie, J., Sreenivasan, S., Korniss, G., Zhang, W., Lim, C., Szymanski, B.K.: Social consensus through the influence of committed minorities. Physical Review E 84(1), 011130 (2011)

Yildiz, E., Acemoglu, D., Ozdaglar, A., Saberi, A., Scaglione, A.: Model of binary opinion dynamics: Coarsening and effect of disorder. SSRN eLibrary (2011)

Part V
Resources

Considerations for a Linguistic Network Markup Language

Maik Stührenberg, Nils Diewald, and Rüdiger Gleim

1 Introduction

As the previous chapters have shown, the possible ways of representing linguistic data as a graph are as diverse as the data itself. For the process of graph modeling, the decision as to what information will be represented as nodes and what information as relations is of great importance. In addition, what kind of added value is going to be expected by the representation of the data as a graph and what kinds of scientific questions should be answerable by the model.

In this chapter we discuss the requirements of a data format that is neutral regarding these questions while being applicable to a wide range of linguistic network data. We look especially at existing formats and the applications that process them and propose an extension of one of the afore-discussed formats in order to demonstrate the use of the proposed markup language. We do this by the example of Wiki-Graphs (Mehler 2008).

2 Data Formats

When speaking of a digital "format", one usually refers to a specific method to represent a specific kind of information – either in memory, as a file, or as a database entry. For example there are commonly known formats to represent formatted text, such as HTML, RTF, LaTeX or Microsoft DOC. Given a piece of formatted text which is stored in those formats, the content of each file will be quite different (for

---

Maik Stührenberg · Nils Diewald
IDS Mannheim
e-mail: {stuehrenberg,diewald}@ids-mannheim.de

Rüdiger Gleim
Goethe-University Frankfurt
e-mail: gleim@em.uni-frankfurt.de

example, when viewed in raw text or binary view) – even if they are rendered equally by an application capable of processing different formats. Even though the formats represent the same piece of information (based on a model reflecting paragraphs, headlines and alike), the *serialization into a file* of a specific format can be very different. This example illustrates that when thinking of standard formats to represent a specific kind of information such as graph structures, one should distinguish the model from the form instances.

2.1 Data Models

The task to accurately model and describe linguistic information may result in complex data models, although the established tools to analyze linguistic structures usually work on trees or directed graphs at best. So why invest time and energy into corpora of linguistic data based on complex data models in the first place? The question is intentionally provocative but nonetheless addresses the observation that data models of corpora used in computational linguistic analyses tend to be more complex than is actually needed by the algorithms being run on them.

A corpus which is richly annotated offers more degrees of freedom in terms of possible analyses. A simple tree or graph view of a more complex data structure can easily be extracted for a specific algorithm. Instead, using a simple data model from the beginning narrows down the spectrum of possible applications for research (and may lead to information loss and biased results). Therefore we argue for an adequate and possibly complex data model of a graph which allows the quick extraction of views for specific tasks, rather than extracting incoherent different views from raw data.

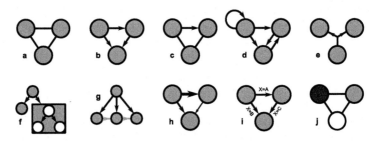

Fig. 1 Network visualizations of **a**) simple graph, **b**) directed graph, **c**) mixed graph, **d**) multigraph, **e**) hypergraph, **f**) hierarchical graph, **g**) ordered graph, **h**) weighted graph, **i**) property graph, and **j**) k-partite graph

Formally, a *graph* is an ordered pair $G = (V, E)$ which consists of a set of vertices V^1 and a set of edges $E \subseteq V \times V$. A graph is commonly understood to be undirected, and to disallow self loops (reflexive arcs) and multiple edges. To emphasize these properties such a graph is also referred to as a *simple graph* (see Fig. 1a). A *directed graph* defines edges as ordered pairs of vertices (see Fig. 1b). Graphs which explicitly allow for directed and undirected edges are called *mixed graphs* (see Fig. 1c). In contrast to a simple graph, a *multigraph* allows self loops and multiple edges (see Fig. 1d). A *hyperedge* generalizes the notion of an edge by putting more than two nodes into a relation. A graph supporting hyperedges is called *hypergraph* (see Fig. 1e). When modeling data collections such as Wikipedia pages, it can be convenient to establish a hierarchical structure; a page is considered to be a node in a graph but contains hierarchically subordinate graphs comprising the sections and page internal links. A graph which allows nodes to contain subordinate graphs is called a *hierarchical graph* (see Fig. 1f). Usually the order of incident edges of a node is not defined – however, if the order is relevant and defined (e. g. in syntax trees), such graphs are referred to as *ordered graphs* (see Fig. 1g). In a *weighted graph* a function is defined which assigns each edge a specific numeric value (a weight, see Fig. 1h). In computer science the notion of a *property graph* is used to state that nodes and edges can be freely annotated with key-value pairs (see Fig. 1i). In this sense a property graph is a generalization of a weighted graph. The nodes of a *k-partite graph* can be (or already are) partitioned into k sub sets such that *all* edges connect nodes of *different* partitions. When talking about linguistic networks, we are refering to linguistic data being modeled following these principles.

2.2 Data Structures

Graph structures are commonly used in computational processing. Depending on the nature of the graph and the types of planned processing, graphs can be represented by different abstract data structures.

2.2.1 Adjacency Matrices

An adjacency matrix can represent a wide range of different types of graphs. A graph comprising n nodes is represented by an $n \times n$ matrix. An edge between two nodes is represented by a cell value in the matrix, which can represent a boolean value in case of a simple graph (meaning two nodes are either connected or not), a weight in case of weighted graphs, or the number of connections between two nodes in the case of a multigraph. As the diagonal values in the matrix denote self loops, this diagonal is empty for the simple graph. In case of an undirected graph, the matrix is symmetric and can be reduced to $\frac{n^2-n}{2}$ cells for the simple graph.

---

[1] In this paper we use the terms "vertice" and "node" synonymously.

The adjacency matrix given in Fig. 2 represents a graph consisting of the three nodes A, B, and C and the directed edges (A,B) and (B,C).

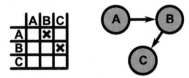

Fig. 2 Adjacency matrix and corresponding graph visualization

Simple directed or undirected graphs can be efficiently represented for processing as bit matrices, however, due to the square space complexity of this form, adjacency matrices may be inefficient for large or sparsely connected graphs, especially with complex edge weight. On the other hand, some graph algorithms rely on the representation of the graph as a matrix for efficient processing.

2.2.2 Adjacency Lists

Adjacency lists are an alternative data structure for representing graphs. They only store existing edges between two nodes, making this form space-efficient for even sparse graphs. In a directed graph for each node all adjacent nodes are listed, possibly with accompanied values for weight (in case of a weighted graph) or the number of connections to the node (in case of a multigraph). In an undirected graph this may lead to redundancy, so it is common to only represent the edge in the list of one of two adjacent nodes. Edge lists may be seen as a special case of adjacency lists, where each pair of nodes is listed separately.

The graph consisting of the three nodes A, B and C and the directed edges (A,B), (A,C) and (B,C) can be represented in an adjacency list as the one shown in Fig. 3.

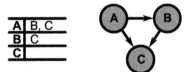

Fig. 3 Adjacency list and corresponding graph visualization

2.3 Data Serialization

In order to store adjacency matrices in files, a character separated value (CSV) representation is typically used (see Listing 1). This simple format has the advantage to be easily written and parsed. It is also natively supported by a wide range of applications (cf. Sect. 4).

Listing 1 Adjacency Matrix of Figure 2 on the preceding page as CSV

```
1   ,A,B,C
2   A,0,1,0
3   B,0,0,1
4   C,0,0,0
```

This serialization of the abstract data structure however does not contain further information regarding the types of nodes and edges, the directionality of the graph, etc. The concrete data structure may contain this information and simple serialization formats may be applicable to resemble these structures. JSON (Crockford 2006) – a popular character based data serialization format – can be used to serialize the graph represented as an adjacency list in Fig. 3 (see Listing 2).

Listing 2 Adjacency list as JSON string

```
1   {
2     "A": ["B", "C"],
3     "B": ["C"],
4     "C": []
5   }
```

Most standardized serialization formats for data structures are limited to base data types (such as strings, integers or booleans) and base data structures (like lists or key-value pairs). As these base types are commonly understood by various programming languages, serialized data structures can be used as an exchange format between different systems that process the same data in the same manner. Using JSON, the same graph can be enriched with further information upon the nodes and edges (see Listing 3).

Listing 3 Adjacency list as JSON string with additional information

```
1   {
2     "A": {
3       "name": "Alice",
4       "knows": ["B", "C"]
5     },
6     "B": {
7       "name": "Bob",
8       "knows": ["C"]
9     },
10    "C": {
11      "name": "Carol"
12    }
13  }
```

Computationally, serializing and deserializing data structures is in most cases faster and less memory consuming than parsing more sophisticated data formats and building new internal data structures based on the information parsed. Other serialization formats like CBOR (Bormann and Hoffman 2013) even create binary streams instead of character based representations, being more compact and faster to deserialize.

On the other hand, more sophisticated data formats may provide additional information independent of certain processing types, meaning they do not aim to just represent concrete data structures. For example, a graph data format providing information about the positions of nodes in a visual layout, the thickness of edges depending on a relational weight, or the color of nodes in addition to the network data, can be used in computer vision applications, while a system to compute distance measures on the graph data can parse the document ignoring all visual information. In another example described later in this chapter, a wiki can be seen as a large complex network, consisting of three component graphs: a social network of authors, the hyperlinked network of documents, and a network of lexical entities (Mehler 2008). Research questions (and therefore algorithms) may treat these levels of networks separately, while others combined. The data formats described in the following sections are able to distinguish between these different types of processing, using either adjacency matrices or adjacency lists as their foundation.

3 Existing Formats

In this section we give a survey of methods and formats to serialize graph structures into files. We start by examining the expressiveness of the underlying data models and draw a connection from simple adjacency matrices to hierarchical hypergraphs, describing the development of a selection of graph representation formats afterwards. As the required expressiveness for modeling linguistic networks is not fixed and depends highly on the data in question, the selection prefers more expressiveness to less expressiveness. Note that this selection focuses on formats that emphasize an application-independent modeling of graphs, leaving out formats that also more or less explicitly represent graphs, but do so for a specific application or domain, e. g. the specific formats of the tools described in Sect. 4, or the Lexical Markup Framework (LMF, ISO 24613:2008), the recent version of the Guidelines of the Text Encoding Initiative (TEI P5 2.7.0), or the OWL 2 Web Ontology Language (W3C Web Ontology Working Group 2012).

One of the main benefits of XML-based markup languages compared to proprietary serialization formats is the possibility to define it via a formally adequate document grammar (or schema) using one of the three main grammar formalisms: DTD (Document Type Definition, defined in the XML specification Bray et al. 2008), XML Schema (XSD, Gao et al. 2012; Peterson et al. 2012), or RELAX NG (ISO/IEC 19757-2:2003).[2] In addition, there are a large number of XML-supporting technologies available, ranging from XML-accompanying specifications such as XSLT (for transformations, Clark 1999; Kay 2007), the traversal language XPath (Clark and DeRose 1999; Berglund et al. 2010) or XQuery (Boag et al. 2010; Robie et al. 2013), up to the mature XML support in general purpose programming languages by means of software libraries. Therefore, formally (and technically)

---

[2] Regarding a discussion about the formal expressiveness of these formats, refer to Murata et al. (2005) and Stührenberg and Wurm (2010).

defined XML-based markup languages are application-independent, which is the reason why they are the preferred way of storing graph-structured information.

3.1 GML

The Graph Modeling Language (GML, Himsolt 1997) is kind of a veteran compared to other languages describing graphs. It has been designed as an interchange format for Graphlet, a (now discontinued) toolkit for graph algorithms and graph editing. The general structure of a GML document is defined in Backus-Naur-Form, structuring all data as lists of key-value pairs. Values can be literals as well as lists, which allows for a tree-like hierarchical structure. Himsolt (1997) encourages the definition of custom attributes as long as they do not collide with the reserved keys for representing graphs and their basic properties. An advantage of this pragmatic approach is that GML can be extended ad hoc to any given representation needs. On the other hand, there is no means to validate either the core GML or any extension to it since there is no schema involved.

Formally speaking, GML is designed to represent directed or undirected attributed graphs. The recursive definition of the document structure in terms of lists of key-value pairs would allow for hierarchical graphs, but it is not explicitly defined. Listing 4 gives an example of how two nodes representing a Wikipedia user called *Alice* and an article about *Graph Theory* can be expressed using GML. The two nodes are connected by a directed edge to denote the authorship of *Alice*. The example also shows how custom attributes can be used.

Listing 4 GML

```
1  graph [
2    directed 1
3    label "Wikipedia Graph"
4    node [
5      id 1
6      label "Alice"
7      mycomment "A registered User"
8      mycomplexattribute [
9        contributions 5
10       role "editor"
11     ]
12   ]
13   node [
14     id 2
15     label "Graph Theory"
16   ]
17   edge [
18     source 1
19     target 2
20     label "Authorship"
21   ]
22 ]
```

3.2 XGMML

Due to the simplicity of the Graphlet Toolkit syntax, GML quickly gained in popularity. But since it is a proprietary format, it cannot benefit from the range of tools and libraries supporting a common standard. In order to fill this gap, Punin and Krishnamoorthy (2001) proposed the eXtensible Graph Markup and Modelling Language (XGMML). This has been designed explicitly as a successor to GML, thus keeping its vocabulary to describe graphs, but based on XML. As an extension, XGMML explicitly introduces hierarchical graphs (see Fig. 1f) but is still limited to binary edges.

Figure 4 shows a graph containing three nodes, *A*, *B* and *C*, representing interlinked articles of a Wikipedia graph. In this example we assume that the article being represented by node *C* holds an inner structure in terms of sections. Thus "article"-node *C* contains a subgraph with a "section"-node *D*. Listing 5 demonstrates how this graph can be represented using XGMML. It also shows how simple as well as complex custom attributes can be expressed. Note that in this and in the following listings some attributes are left out for the sake of brevity.

Listing 5 XGMML

```
1  <?xml version="1.0" encoding="UTF-8"?>
2  <graph xmlns="http://www.cs.rpi.edu/XGMML"
3    xmlns:xsi="http://www.w3.org/2001/XMLSchema-instance"
4    xsi:schemaLocation="http://www.cs.rpi.edu/XGMML xgmml.xsd"
        directed="1" id="0" label="Graph1">
5    <node id="1" label="A">
6      <att name="title" value="Linguistics"/>
7      <att name="namespace">
8        <att name="namespaceID" "0"/>
9        <att name="namespaceKey" "article"/>
10     </att>
11   </node>
12   <node id="2" label="B"/>
13   <node id="3" label="C">
14     <att>
15       <graph directed="1" id="4" label="Graph2">
16         <node id="5" label="D">
17           <att name="title" value="History"/>
18         </node>
19       </graph>
20     </att>
21   </node>
22   <edge source="1" target="2" label="wikiLink"/>
23   <edge source="2" target="3" label="wikiLink"/>
24 </graph>
```

Metadata can be stored as an att element, that may occur as a child of the elements graph, node and edge. Since the att element itself can be nested recursively, the serialization of structured metadata is possible as well. Apart from that, there are no means to store annotation.

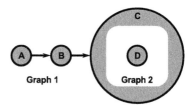

Fig. 4 Hierarchical Graph Structure as example for graph representation

It is uncertain if there is still development going on, since the XGMML web site[3] has not been updated since 2001. The latest specification of XGMML is dated from 2001, too. While Punin and Krishnamoorthy (2001) mentions an XML DTD for validating XGMML instances, there is also an XSD file available on the web site.

3.3 GraphXML

GraphXML (Herman and Marshall 2001) is a graph format which specifically addresses applications for graph layout and visualization. This is reflected by various attributes of graphs and edges such as isPlanar, isAcyclic and preferredLayout.[4] Compared to the formats already presented, GraphXML does not add to the expressiveness – but offers some alternative concepts to describe graphs. One noticeable difference is the way the hierarchy of graphs is modeled. Similar to XGMML, GraphXML uses XLink (DeRose et al. 2001) to refer to subordinate graphs rather than placing them as child nodes within the document object model (DOM) tree. This approach allows to refer to external graph representations.

GraphXML offers three mechanisms to annotate graph elements. Where as the name attribute allows to assign unique identifiers to graphs, nodes and edges, labels can be used to annotate components with arbitrary string literals. If more complex data is to be annotated, this can be done either inline by data elements or by means of external references (dataref). The specification does not add any constraints on the contents of data elements thus allowing arbitrary XML code. On the one hand, this gives the adopter of this format the freedom to add any kind of data. Specifying the structure of the data is neither required nor possible. On the

---

[3] http://www.cs.rpi.edu/research/groups/pb/punin/public_html/XGMML

[4] Note that these attributes, even though they are in part formal properties of graphs, are primarily used by applications to layout and visualize graphs. For example acyclic graphs can be layouted in a way to point out hierarchies. The information that a graph is planar can be used to compute a corresponding layout where no edges cross each other.

other hand this arbitrariness impedes sustainability because of the lack of a proper specification.

The code in Listing 6 shows the difference of representing hierarchical graphs. It represents the graph structure shown in Fig. 4 which has already been used as an example for XGMML. It also adopts the annotation examples of the graphs shown earlier to illustrate the use of the three variants to annotate graph elements.

Listing 6 Code example of GraphXML

```
1  <?xml version="1.0"?>
2  <!DOCTYPE GraphXML SYSTEM "file:GraphXML.dtd">
3  <GraphXML>
4    <graph id="Graph1">
5      <node name="A">
6        <label>Linguistics</label>
7        <data>
8          <namespace>
9            <namespaceID>0</namespaceID>
10           <namespaceKey value="article"/>
11         </namespace>
12       </data>
13       <dataref>
14         <ref xlink:role="WebLink"
15              xlink:href="http://en.wikipedia.org/wiki/
                            Linguistics"/>
16       </dataref>
17     </node>
18     <node name="B"/>
19     <node isMetanode="true" name="C" xlink:href="#Graph2"/>
20     <edge source="A" target="B"/>
21     <edge source="B" target="C"/>
22   </graph>
23   <graph id="Graph2">
24     <node name="D">
25       <label>History</label>
26     </node>
27   </graph>
28 </GraphXML>
```

GraphXML is defined by an XML DTD. The DTD file itself is no longer available at the GraphXML web site[5] and GraphXML's development came to an end years ago.

---

[5] http://projects.cwi.nl/InfoVisu/GraphXML/ – it
seems that a copy of the DTD is still available from
http://wiki.ikmemergent.net/files/hypergraph/GraphXML.dtd,
however, it is uncertain if this is the correct file and the latest version.

3.4 GraphML

The development of the Graph Markup Language (GraphML, Brandes et al. 2004) was initiated during the *Graph Drawing Symposium* in 2000. Today it is considered to be the most popular format for graph representations, at least in the scientific community. Like XGMML and GraphXML it is based on XML and defined using XML Schema (in fact, it uses four modular XML schema files which – like XGMML and GraphXML – import XLink for linking between graphs).[6] Regarding the formal expressiveness, GraphML extends the capability of the former approaches by introducing hyperedges and nested graphs.

The code given in Listing 7 depicts the representation of a hierarchical graph structure as shown in Fig. 5. It extends the example of a Wikipedia graph which has already been used to illustrate hierarchical graphs (see Fig. 4) by including a hyperedge.

Listing 7 Hyperedge in GraphML

```
<?xml version="1.0" encoding="UTF-8"?>
<graphml xmlns="http://graphml.graphdrawing.org/xmlns"
  xmlns:xsi="http://www.w3.org/2001/XMLSchema-instance"
  xsi:schemaLocation="http://graphml.graphdrawing.org/xmlns
  http://graphml.graphdrawing.org/xmlns/1.0/graphml.xsd">
  <graph id="Graph1" edgedefault="directed">
    <node id="A"/>
    <node id="B"/>
    <edge source="A" target="B"/>
    <hyperedge>
      <endpoint node="B"/>
      <endpoint node="C"/>
      <endpoint node="D"/>
    </hyperedge>
    <node id="C">
      <graph id="Graph2" edgedefault="directed">
        <node id="D"/>
      </graph>
    </node>
  </graph>
</graphml>
```

A distinctive feature of GraphML is the concept of *ports*. Each node can contain an arbitrary number of labeled ports. Note that *ports* are a specific construct of GraphML which do not have a direct counterpart in terms of graph theory. Rather, they are introduced to solve specific problems, as for example to simplify the representation of hyperlinks between Wikipedia pages: A hyperlink points from one page to another. In addition, a hyperlink can specify a source and a target section within

---

[6] The schema files consist of the `graphml-structure.xsd`, `graphml-attributes.xsd`, and `graphml-parseinfo.xsd` schemas, plus an additional `graphml.xsd` which redefines some of the before-mentioned elements and types.

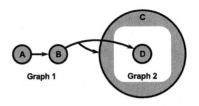

Fig. 5 Hierarchical graph structure including a hyperedge connecting three nodes

the respective pages, similar to HTML. One way to represent such a hyperlink could be to represent the anchors of pages as nodes of a subordinate graph. Hyperedges would then be used to interrelate the the nodes representing the source and target page as well as the source and target section. While this is a clean solution it also bloats the representation. The concept of ports in GraphML allows for a more compact representation: Rather than modeling sections of a page as nodes of a subgraph they can be modeled as ports of a node. Thus an edge connecting two nodes can optionally specify the ports it connects. The code in Listing 8 shows an alternative representation of Fig. 5 defining node D as a port of node C rather than a node in a subordinate graph. That way we can also use a simple binary edge containing the additional information about the target port.

Listing 8 Ports in GraphML

```
<?xml version="1.0" encoding="UTF-8"?>
<graphml xmlns="http://graphml.graphdrawing.org/xmlns"
xmlns:xsi="http://www.w3.org/2001/XMLSchema-instance"
xsi:schemaLocation="http://graphml.graphdrawing.org/xmlns
http://graphml.graphdrawing.org/xmlns/1.0/graphml.xsd">
  <graph id="Graph1" edgedefault="directed">
    <node id="A"/>
    <node id="B"/>
    <node id="C">
      <port name="D"/>
    </node>
    <edge source="A" target="B"/>
    <edge source="B" target="C" targetport="D"/>
  </graph>
</graphml>
```

Another advantage of GraphML is the strict separation of attribute specification and value assignment. Attributes of graphs, nodes and edges are defined once and can then be referred to throughout the document when assigning values. The following code example demonstrates the definition of a node attribute named *namespace* which is used to declare the namespace of a given Wikipedia page. Values need to be of type string and the default value is set to *category*.

Listing 9 Key definition in GraphML

```
...
<key id="k0" for="node"
     attr.name="namespace"
     attr.type="string">
```

```
5    <default>category</default>
6  </key>
7  ...
```

This attribute can now be used to annotate nodes of a graph as the following code illustrates:

Listing 10 Reference to former key definition in GraphML

```
1  ...
2  <node id="A">
3    <data key="k0">article</data>
4  </node>
5  ...
```

The `data` element can be stored underneath the `graphml`, `graph`, `node`, `port`, `edge`, and `hyperedge` elements, making it a ubiquitous place to store annotation. Beyond such simple attribute types as the ones shown in Listings 9 and 10, GraphML also supports complex types as well as extensions thereof.

3.5 GXL

The Graph eXchange Language (GXL) has been proposed by Winter et al. (2002). It benefits greatly from the experiences made with previous proposals and concepts to represent graph structures. Initially GXL evolved from the GRAph eXchange format (GraX, Ebert et al. 1999), the Tuple Attribute Language (TA, Holt 1997), as well as the format of the PROGRES graph rewriting system (Schürr et al. 1999). It aims at bringing together the concepts of GML, XGMML and GraphXML. Formally speaking, GXL supports the representation of attributed, typed, directed and undirected hierarchical hypergraphs. This expressiveness is also provided by GraphML. In extension to that, GXL also provides ordered hyperedges and explicit typing of elements, which is implemented separately from attributing graph elements. The representation of hyperedges is quite similar to GraphML where the elements `hyperedge` and `endpoint` are used rather than `rel` and `relend`.[7]

Listing 11 shows the serialization of the graph as shown in Fig. 5.

Listing 11 Code example of GXL

```
1  <?xml version="1.0" encoding="UTF-8"?>
2  <!DOCTYPE gxl SYSTEM "http://www.gupro.de/GXL/gxl-1.0.dtd">
3  <gxl>
4    <graph id="Graph1" edgemode="directed" hypergraph="true">
5      <node id="A"/>
6      <node id="B"/>
7      <edge id="EdgeAB" from="A" to="B"/>
8      <rel id="RelBCD">
```

---

[7] See Brandes et al. (2005) for a comparison of both GraphML and GXL as well as the transformation of instances from and to each other.

```
 9      <relend dir="in" target="B"/>
10      <relend dir="out" target="C"/>
11      <relend dir="out" target="D"/>
12    </rel>
13    <node id="C">
14      <graph id="Graph3">
15        <node id="D"/>
16      </graph>
17    </node>
18  </graph>
19 </gxl>
```

GXL offers an extensive attribute model to annotate graph elements. The `attr` element is used to express named attributes, typed literals can be specified using `string`, `int`, `bool`, `float`, `enum` or `locator` elements. Furthermore GXL distinguishes four different kinds of containers, namely sets (`set`), lists (`seq`), tuples (`tup`), and multisets (an unordered set in which multiplicity is significant, called `bag`). This distinction is quite noticeable – however GXL does not allow to type or name these containers beyond the ancestor `attribute` element. Listing 12 shows the use of attributes in GXL.

Listing 12 Attribute code example of GXL

```
 1  ...
 2 <node id="A">
 3    <attr name="title">
 4      <string>Linguistics</string>
 5    </attr>
 6    <attr name="ratings">
 7      <set>
 8        <tup>
 9          <string>language</string>
10          <int>4</int>
11        </tup>
12        <tup>
13          <string>innovative</string>
14          <int>5</int>
15        </tup>
16      </set>
17    </attr>
18 </node>
19 ...
```

GXL is defined by an XML DTD and an XSD schema file which both can be obtained from GXL web site.[8] Technically speaking, the XSD schema is not valid, because it uses a wrong (preliminary) XML Namespace for the root element, the wrong `xs:NMTOKEN` data type and some obsolete attributes for default values. However, these mistakes can easily be corrected.

---

[8] http://www.gupro.de/GXL/

Although there is a section regarding possible future updates to GXL at its web site (including an alpha version of a forthcoming GXL 1.1 DTD), the latest changes originate in 2002, leaving GXL's future development uncertain.

3.6 GrAF

The Graph Annotation Format (GrAF) is a so-called pivot format as part of the international standard ISO 24612:2012, the Linguistic Annotation Framework (LAF). LAF was developed for multiple annotated language corpora and has its roots in the Corpus Encoding Standard (CES, Ide 1998; Ide et al. 1996) which started as a TEI P3-based (P3) SGML application and later on evolved into the XML-based XCES (Ide et al. 2000). Since LAF's primary design goal is to store linguistic annotations, there are different data types to be stored, namely the metadata (divided into resource header and document header) and the standoff annotation files which are stored in a graph-based format, which makes GrAF an interesting candidate for a linguistic network markup language.

LAF's data model consists of three parts: (1) regions that are defined by anchors referencing locations in the primary data (that is, the data to be annotated); (2) a directed graph structure, consisting of nodes, edges and links to the before-mentioned regions; and (3) an annotation structure comprising a directed graph referencing regions or other annotations. The nodes and edges in this graph can be annotated via feature structures (attribute-value-graphs). Language resources annotated according to LAF confirm to the following architecture:

- At least one primary data file. Primary data may be in any format; if markup is part of textual primary data it is treated as part of the data stream (in case of XML-based markup, XPath expressions may be supported as well).
- Annotation documents containing linguistic annotation associated to the nodes and edges of the graph structure.
- One or more documents defining regions as base segmentation of the primary data. The concept of base segmentation is derived from Ide et al. (2000). The referencing and segmentation mechanism used is application-dependent, although each anchor can be described by *n*-tuples containing sets of spatial or temporal offsets.
- A set of header files providing metadata for both the primary data and each annotation document.

This data model can be serialized in any given markup language. To assure the exchange of data between different LAF-based markup languages, the pivot format GrAF should be used. GrAF's serialization of this data model is defined by a normative RNG schema.[9] Listing 13 shows a small GrAF instance.

---

[9] Located at http://www.xces.org/ns/GrAF/1.0/graf-standoff.rng; there are additional schema files for the resource and document header files and informative XSD versions of the schemas.

Listing 13 GrAF example

```xml
<?xml version="1.0" encoding="UTF-8"?>
<graph xmlns:xsi="http://www.w3.org/2001/XMLSchema-instance"
  xsi:schemaLocation="http://www.xces.org/ns/GrAF/1.0/
  http://www.xces.org/ns/GrAF/1.0/graf-standoff.xsd"
  xmlns="http://www.xces.org/ns/GrAF/1.0/" xml:id="graph_1">
  <graphHeader>
    <labelsDecl>
      <labelUsage label="sentence" occurs="1"/>
    </labelsDecl>
  </graphHeader>
  <anchor xml:id="a1" value="1"/>
  <anchor xml:id="a2" value="4"/>
  <anchor xml:id="a3" value="5"/>
  <anchor xml:id="a4" value="10"/>
  <region xml:id="r1" anchors="a1 a2"/>
  <region xml:id="r2" anchors="a3 a4"/>
  <node xml:id="n1">
    <link targets="r1"/>
  </node>
  <node xml:id="n2">
    <link targets="r2"/>
  </node>
  <edge xml:id="e1" from="n1" to="n2"/>
  <a xml:id="anno1" ref="e1" label="sentence">
    <fs>
      <f name="s" fVal="#e1"/>
    </fs>
  </a>
</graph>
```

The a element in Line 24 contains the annotation of a given node or edge. In this case the edge identified by *e1* (Line 23) from the node *n1* (Line 17) to the node *n1* (Line 20) is annotated with a feature structure containing a simple feature (Line 26) named *s* (for sentence). The order of the nodes determines the direction of an edge. Annotations may use labels which should be defined in the labelsDecl element of the graph header – in this example we use the more human-readable label "sentence" instead of the feature *s*. The markup for the feature structure is compliant to the one defined by the TEI (P5 2.7.0) and the International Standard ISO 24610-1:2006.[10] This is one of the most distinctive features that separates GrAF from the other serialization formats discussed above, since GrAF allows for structured annotation and not only attributes.

---

[10] There is an additional and more concise form of the f element for simple features, that consists of an empty f element and a name and value attribute. Since this concise form is not compliant to the International Standard ISO 24610-1:2006, we use the longer form in the example.

The most prominent application of GrAF is the Manually Annotated Sub-Corpus (MASC) project which provides a number of freely available corpus files, including a corpus header, document headers, primary data files, segmentation files, and annotation files (with annotation headers) for a number of annotation levels.[11]

3.7 Summary

The development of the different graph formats discussed in this section has shown a natural evolution, in terms of underlying serialization (non-XML vs. XML-based) and in terms of expressiveness. While most of the formats are quite similar regarding their annotation inventory (each of them has a `node` and `edge` element for example), GrAF differs in providing a specific place to store metadata (the graph header) and handling of annotation as well as a more complex segmentation mechanism (with a separate definition of anchors and regions). The reason for this is GrAF's (and LAF's) background of linguistic annotation.

From a technical point of view, it is interesting that XGMML, GraphXML and GraphML use XLink (DeRose et al. 2001) which was supposed to be added as linking mechanism in a *Second Generation Web* where specific document information is encoded in XML (Bosak and Bray 1999). XLink never gained enough support in browsers and although the specification itself has been updated recently (DeRose et al. 2010), its future in the Web is uncertain.

Table 1 shows the differences in terms of formal expressiveness of the different graph serialization formats.

Table 1 Expressiveness of Graph Formats

	directed	attributed	typed	hierarchical	hyperedges	ordered	multiple graphs
GML	x	x	(x)	(x)	–	–	–
XGMML	x	x	(x)	x	–	–	–
GraphXML	x	x	(x)	x	–	–	x
GraphML	x	x	(x)	x	x	–	x
GXL	x	x	x	x	x	x	x
GrAF	x	x	(x)	–	–	x	–

4 Network Tools

An important research purpose when it comes to encoding linguistic data as a graph and storing it in a serialized format is making this information accessible to network science tools. These tools mainly help to analyze the data, describe the information in algorithmic terms to make it possible to compare different linguistic corpora and

---

[11] See http://www.anc.org/MASC/MASC_Structure.html for further details.

different linguistic structures, or even compare linguistic networks with networks induced from other sources (e. g. by the average degree or the graph density).

Beside the mathematical analysis of this data, sometimes there is also a significant benefit in visualizing these graphs. This is especially true for data coming from the humanities like linguistic data, as scientists working in this field sometimes prefer the visual demonstration and exploration of measures like density to pure numerical representations.

The following list is just a small fraction of the wide variety of tools for analyzing and visualizing network data, but can be seen as a list of major representatives.

Cytoscape[12] (Smoot et al. 2011) is a network analysis tool with a focus on bioinformatics. It comes with similar visualizations as *yEd* (see below; by using the *yFiles visualizers*), but with extensive possibilities for network analysis (including measure calculations, but also distribution charts). In addition to that, *Cytoscape* is easily extensible, providing a plugin framework and an application store with user provided plugins (not limited to bioinformatics; Saito et al. 2012). The software is distributed as open source under the *GNU Lesser General Public License* and supports 14 different data formats in the tested version 3.0.2.

Gephi[13] is a graph editing and visualization tool with a focus on real-time visualization and interactive manipulation of even large graphs (more than 20,000 nodes; Bastian et al. 2009). The *Gephi marketplace*[14] provides plugins for additional features, and community driven support (for training and development) can be announced as a service on the Gephi web site.

The software itself is available as open source with a dual license of the *Common Development and Distribtion License* (CDDL) v1.0 and the *GNU General Public License* (GPL) v3. The current version 0.8.2 Beta supports 12 different graph formats, and can be connected to various relational database management systems for importing edge and node lists.

Network Workbench (NWB)[15] is a network analysis, modeling and visualization tool, and a community web platform for network analysis (NWB Team 2006). [16] Its main purpose is the access to well documented implementations of network analysis and visualization algorithms as well as network data to compare networks throughout different disciplines. NWB is open source and available under the *Apache License* v2.0. Version 1.0.0 supports 13 different network formats.

yEd[17] is mainly a graph editing and visualization tool, with just limited support for graph analyses (e. g. the calculation of centrality measures). Other than the former mentioned tools, *yEd* is closed source and distributed under a commercial license, however it can be downloaded and used freely. *yEd* is based on the *yFiles*

---

[12] http://cytoscape.org/
[13] http://gephi.org/
[14] https://marketplace.gephi.org/
[15] http://nwb.cns.iu.edu/
[16] The community portal of Network Workbench at http://nwb.cns.iu.edu/ seems to be no longer accessible (as of 06/01/2015).
[17] http://www.yworks.com/en/products_yed_about.html

Table 2 Graph formats supported by different tools

	Cytoscape	Gephi	NWB	yEd
Version	3.0.2	0.8.2b	1.0.0	3.11.1
n supported Formats	14	12	13	7
GML	X	X	–	X
XGMML	X	–	X	–
GraphXML	–	–	–	–
GraphML	X	X	X	X
GXL	–	–	–	–
GraF	–	–	–	–
CSV	X	X	X	–

Java API[18] which in contrast cannot be used beyond an evaluation phase. It supports seven different graph formats in the tested version 3.11.1.

From the existing formats discussed in Sect. 3, GraphML has the best support among these tools, followed by GML and its XML translation, XGMML[19] (see Table 2). The simple character separated value format (CSV), as discussed in Sect. 2.3, is widely adopted as well. Most formats, however, can be transformed into each other quite easily (e. g. in case of XML formats using XSLT stylesheets), but in some cases, as shown in Table 1 regarding the different levels of expressiveness of the represented graph models, this may come with information loss.

The choice, which tool to choose for the analysis of network data, however, has more (and probably more important) aspects than supported data formats. Some tools focus on certain types of networks (e. g. biological networks, social networks) and therefore provide different feature sets to support different types and sizes of networks. As with formats, sustainability is a major aspect, where it is important, if a software will run on various operating systems[20], if it is still under maintenance, if the license allows a possible community driven continuation of maintenance, and maybe if the software is designed for extensibility, providing an open API (Application Programming Interface) to allow to add custom features. The stability and documentation of the implemented algorithm is also an important aspect when it comes to future reproducibility of research results. Obviously, the performance of the software and the quality of the visualizations play a major role as well.

---

[18] http://www.yworks.com/de/products_yfiles_about.html

[19] *yEd* supports an XML variant of GML as well, called XGML, but the format differs to XGMML, see http://docs.yworks.com/yfiles/doc/developers-guide/xgml.html

[20] All presented tools are available for the major platforms Linux, Windows and OS X. Other tools are limited in this aspect, e. g. *Pajek* (http://pajek.imfm.si/doku.php?id=pajek) natively runs on Windows only.

In addition to the aforementioned tools, mathematical software not primarily focusing on graph analysis like *Matlab*[21], *R*[22], or *Wolfram Mathematica*[23], also support certain graph formats. Graph databases like the popular *Neo4j*[24] provide import handlers for different graph formats. There are also numerous libraries focusing on either the analysis or visualization of graphs. Regarding the latter it is worth to mention the amount of upcoming JavaScript libraries for visualizing network data embedded on the web, using modern techniques like SVG, HTML5 Canvas and WebGL. These libraries mainly use graph data serialized in JSON, as introduced in Sect. 2.3.

5 Proposal for a Linguistic Network Markup Language

In the preceding sections we introduced several existing graph formats as candidates for a Linguistic Network Markup Language. We also discussed the requirements of such a candidate regarding the expressiveness of the graph model to cover most characteristics of linguistic networks as presented in previous chapters. But beside the graph model, a Linguistic Network Markup Language needs to provide a rich toolkit for annotating linguistic phenomena. As the aforementioned graph formats are of a general purpose nature, we have to discuss their capabilities in this aspect a little further and possibly turn the perspective upside down, looking at linguistic markup languages and how we can extend these to represent network information such as the aforementioned WikiGraphs. However, since GrAF (cf. Sect. 3.6) is designed to serialize only single graphs (e. g. those of a single article) without any support for hyperedges or hierarchical graphs, it is not expressive enough for more complex examples; we therefore stick to the general purpose formats as a starting point.

There are two options to define a new linguistic network markup language: the extension of an already established markup language (candidates have been discussed in Sect. 3), or the creation of a new markup language that mimics familiar languages but is defined by a markup grammar on its own. Both options have advantages and disadvantages. The extension of a markup language implies that the document grammar is freely available and can be modified easily. In addition, it is recommended that the newly introduced elements and attributes belong to a different XML Namespace (Bray et al. 2009) to make it easy to distinguish the information items defined in the host language from the others. XML Namespaces use a unique URI reference that serves together with the local name of the elements and attributes of a given markup language as an expanded name, similar to package names in programming languages such as Java. Therefore, two elements with the same local name (e. g. title) can be treated differently (once as the title of a book, the second time as academic title). However, XML Namespaces introduce some pitfalls to the processing of XML instances and must be supported by the document grammar formalism used to define the markup language in question (while

---

[21] http://www.mathworks.de/products/matlab/
[22] http://www.r-project.org/
[23] https://www.wolfram.com/mathematica/
[24] http://www.neo4j.org/

XML DTD do not support XML Namespaces, both RNG and XSD do). A prominent example of a markup language that allows the extension in such a way is the TEI. The Guidelines of the Text Encoding Initiative consists of a highly modular markup language that can be used to annotate various works of digital humanities data, including literary texts, transcriptions, plays, and the like. Another example of a meta markup language that makes heavy use of XML Namespaces is XStandoff (Stührenberg and Jettka 2009; Stührenberg 2013).

A related aspect is the grammar formalism itself. DTD, XSD and RNG are different in terms of their formal expressiveness. Candidate markup languages that easily allow the extension of their markup inventory are especially those that use a recent XML document grammar formalism. However, if one adapts a present markup language, tools that are already available for the original markup language (such as XSLT stylesheets) have to be adapted as well. Otherwise they are not aware of the new markup items. In addition, markup languages that use a not so expressive document grammar formalism or that undergo no active development anymore, are no suitable candidates as a starting point. Apart from that it is sometimes not possible to extend a markup language without removing already present markup items. In this case it is more convenient to start over with a new markup language.

Crucial aspects are the handling of metadata and annotations, since we want to define a markup language for describing linguistic data in networks. All three parts of an instance, the graph (including nodes and edges), the annotation and the metadata, can be stored in a single instance, or – as it is in GrAF – the metadata can be stored separately.

Another interesting aspect is the inclusion of reference points to taxonomies or ontologies of linguistic annotation concepts. With the international standard ISO 12620:2009 (ISO 12620:2009) and its implementation ISOcat[25] there is a reasonable resource that can be used. The current version of the TEI document grammar includes a schema file that defines the two attributes `datcat` and `valueDatcat` which can easily be used to refer to ISOcat data categories and values (for further information see Section 18.3, "Other Atomic Feature Values" in TEIP5 2.7.0). Since the markup inventory (that is, the elements, attributes, and types) of the XML-based markup languages already discussed is very similar, it is easy to start over with a new markup language that takes advantages of the current versions of document grammar formalisms (such as XML Schema 1.1; Gao et al. 2012) and international standards available while still mimicking already-familiar annotation formats.

On the other hand, support of an already established markup language by analysis and visualization tools may be an added value which can be a decisive factor. GraphML (see Sect. 3.4) is widely adopted by analysis and visualization tools (see Table 2), while regarding its formal expressiveness, only GXL is stronger (see Table 1; but less adopted, see Table 2). We therefore decided to extend GraphML.

---

[25] See http://www.isocat.org for further details. Although ISOcat resembles an ontology, it is not one per design since agreement on relation and modeling strategies has been considered stronger at the level of an individual application. By the time of writing, ISOcat's future development is uncertain, however, similar data category registries supporting the ISO standard will be available in the near future.

5.1 Extending GraphML by Redefinition

GraphML is defined by a set of XML Schema files and supports XML Namespaces. As an extension mechanism, GraphML supposes a set of pre-defined attribute groups and the globally defined `data-extension-type`, applicable as a container for user-defined extensions. This can be achieved by using XSD's redefinition mechanism (Walmsley 2012, pp. 448-459; Brandes et al. 2004, Section 4). Before that, we have to streamline GraphML's schema files, since they define a large number of global types that are only used once. In addition, there is a `unique` constraint underneath the declaration of the `data` element, that should ensure uniqueness of the `key` attributes of `data` child elements. However, the content model of the `data` element does not allow any child elements at all. We have addressed this issue as well.

If we want to extend GraphML's `data` and `desc` elements (which both use the `data-extension.type` as base type), we can create a new XML Schema containing the redefinition of the `data-extension.type` (similar to the `graphml.xsd` which is part of the official schema files). For example, if we want to model a graph of wiki articles written by one or more authors as a complex linguistic network (Mehler 2008), we could redefine the `data-extension.type` as shown in Listing 14. Note, that we have to alter the `graphml.xsd` file itself (or duplicate it), since it redefines elements and attributes from the other schemas and if we want to redefine the `data-extension.type` (defined in `graphml-structure.xsd`) we have to include every single redefinition of the `graphml.xsd` schema as well to maintain compatibility.[26]

Listing 14 Redefining GraphML's `data-extension.type` for describing Wiki-Graphs

```
1  <?xml version="1.0" encoding="UTF-8"?>
2  <xs:schema xmlns:xs="http://www.w3.org/2001/XMLSchema"
3    targetNamespace="http://graphml.graphdrawing.org/xmlns"
4    xmlns:graphml="http://graphml.graphdrawing.org/xmlns"
5    elementFormDefault="qualified">
6
7    <xs:import namespace="http://www.w3.org/XML/1998/namespace"/>
8
9    <xs:redefine schemaLocation="graphml-structure.xsd">
10     <xs:complexType name="data-extension.type" mixed="true">
11       <xs:complexContent>
12         <xs:extension base="graphml:data-extension.type">
13           <xs:choice minOccurs="0" maxOccurs="unbounded">
```

---

[26] A redefinition of a type (or element, group, etc.) not directly defined in the referenced schema is not allowed by the standard: "The definitions within the `redefine` element itself are restricted to be redefinitions of components from the redefined schema document, *in terms of themselves*." (Thompson et al. 2004, Sect. 4.2.2 and Schema Representation Constraint: Redefinition Constraints and Semantics 4.1) The popular XSLT and XSD 1.1 schema processor *Saxon* throws a warning regarding this behavior, and although the current version 9.5 does not enforce this rule, future versions will do.

```
14        <xs:element name="author" minOccurs="0"
15                    maxOccurs="unbounded">
16          <xs:complexType>
17            <xs:sequence>
18              <xs:element name="name" type="xs:string"/>
19            </xs:sequence>
20            <xs:attribute ref="xml:id"/>
21          </xs:complexType>
22        </xs:element>
23        <xs:element name="article">
24          <xs:complexType>
25            <xs:sequence>
26              <xs:element name="title" type="xs:string"
                     minOccurs="0"/>
27            </xs:sequence>
28            <xs:attribute ref="xml:id"/>
29            <xs:attribute name="topic" type="xs:string"/>
30          </xs:complexType>
31        </xs:element>
32      </xs:choice>
33     </xs:extension>
34    </xs:complexContent>
35   </xs:complexType>
36  </xs:redefine>
37
38  <!-- further refinements of graphml.xsd -->
39 </xs:schema>
```

In the example instance shown in Listing 15, we use the redefined `complexType` to store structured information about authors working on wiki articles (articles not shown in the graph).

Listing 15 Extended GraphML instance for describing relations between authors of wiki articles

```
1 <?xml version="1.0" encoding="UTF-8"?>
2 <graphml xmlns="http://graphml.graphdrawing.org/xmlns"
      xmlns:xsi="http://www.w3.org/2001/XMLSchema-instance"
3     xsi:schemaLocation="http://graphml.graphdrawing.org/xmlns
          GraphML-extended.xsd">
4  <key id="key1" for="graph" attr.name="author"/>
5  <graph id="AuthorGraph" edgedefault="undirected">
6    <node id="Author1">
7      <data key="key1">
8        <author>
9          <name>Alice</name>
10       </author>
11     </data>
12   </node>
13   <node id="Author2">
14     <data key="key1">
```

```
15      <author>
16        <name>Bob</name>
17      </author>
18    </data>
19  </node>
20  <edge source="Author1" target="Author2"/>
21 </graph>
22 </graphml>
```

XML Schema's redefinition feature is sometimes criticized because of its complexity, possible side-effects, and inconsistent implementation among XSD processors (Walmsley 2012, p. 447). Therefore, it was marked as deprecated in the current version 1.1 (Gao et al. 2012; Peterson et al. 2012) and replaced with the new `override` feature. However, compared to the latter, `redefine` is still supported by a larger number of tools. In addition, it allows us to maintain backwards compatibility with GraphML and the tools supporting it, therefore we will use it throughout the remainder of this chapter as the method to extend this format. A future version of the proposed format should make use XSD 1.1's `override` feature.

5.2 Extending GraphML by XML Namespaces

If we want to use an already established markup language for structuring subtrees underneath the `data` element, it is the canonical way to use XML Namespaces. We have two options for such an extension: (1) We redefine the `data-extension.type` complexType in the `graphml.xsd` schema file to include the corresponding root element of the foreign markup language, or (2) we use element wildcards (Gao et al. 2012, Section 3.10).

The XML Schema shown in Listing 16 uses the first option and extends the type by adding an element sequence containing the `fs` element of the feature structure format defined in the TEI and ISO 24610-1:2006. We have chosen this approach to be compliant with the International Standard which is the base of a large number of accompanying standards for the annotation of linguistic features.

Listing 16 Redefining GraphML's `data-extension.type` by including TEI FS

```
1  <?xml version="1.0" encoding="UTF-8"?>
2  <xs:schema xmlns:xs="http://www.w3.org/2001/XMLSchema"
3    targetNamespace="http://graphml.graphdrawing.org/xmlns"
4    xmlns:graphml="http://graphml.graphdrawing.org/xmlns"
        xmlns:tei="http://www.tei-c.org/ns/1.0"
5    elementFormDefault="qualified">
6
7    <xs:import namespace="http://www.tei-c.org/ns/1.0"
        schemaLocation="../tei_fs.xsd"/>
8    <xs:redefine schemaLocation="graphml-structure.xsd">
9      <xs:complexType name="data-extension.type" mixed="true">
10       <xs:complexContent>
11         <xs:extension base="graphml:data-extension.type">
12           <xs:sequence>
```

Considerations for a Linguistic Network Markup Language

```
13          <xs:element ref="tei:fs" minOccurs="0"
14                      maxOccurs="unbounded"/>
15        </xs:sequence>
16      </xs:extension>
17    </xs:complexContent>
18   </xs:complexType>
19  </xs:redefine>
20
21  <!-- further refinements of graphml.xsd -->
22
23 </xs:schema>
```

An example using the extended schema can be seen in Listing 17.

Listing 17 Instance using TEI-FS-extended GraphML

```
1  <?xml version="1.0" encoding="UTF-8"?>
2  <graphml xmlns="http://graphml.graphdrawing.org/xmlns"
       xmlns:xlink="http://www.w3.org/1999/xlink"
3    xmlns:tei="http://www.tei-c.org/ns/1.0"
4    xmlns:xsi="http://www.w3.org/2001/XMLSchema-instance"
5    xsi:schemaLocation="http://graphml.graphdrawing.org/xmlns
       GraphML-TEI-FS.xsd">
6   <key id="key1" for="graph" attr.name="color"
7        attr.type="string">
8    <default>
9     <tei:fs>
10      <tei:f name="title">
11       <tei:string>Linguistics</tei:string>
12      </tei:f>
13     </tei:fs>
14    </default>
15   </key>
16   <graph id="Graph1" edgedefault="directed">
17    <node id="A">
18     <data key="key1">
19      <tei:fs>
20       <tei:f name="title">
21        <tei:string>Linguistics Online</tei:string>
22       </tei:f>
23      </tei:fs>
24     </data>
25    </node>
26    <node id="B">
27     <data key="key1"/>
28    </node>
29    <edge source="A" target="B"/>
30   </graph>
31  </graphml>
```

To realize the second option, we only have to insert the xs:any wildcard as a child element of the redefined data-extension.type complexType (as shown in Listing 18).

Listing 18 Redefining GraphML's data-extension.type with the xs:any wildcard

```xml
<?xml version="1.0" encoding="UTF-8"?>
<xs:schema xmlns:xs="http://www.w3.org/2001/XMLSchema"
  targetNamespace="http://graphml.graphdrawing.org/xmlns"
  xmlns:graphml="http://graphml.graphdrawing.org/xmlns"
      xmlns:dcr="http://www.isocat.org/ns/dcr"
  elementFormDefault="qualified">

  <xs:import namespace="http://www.w3.org/XML/1998/namespace"/>

  <xs:import namespace="http://www.isocat.org/ns/dcr"
      schemaLocation="dcr.xsd"/>

  <xs:redefine schemaLocation="graphml-structure.xsd">
    <xs:attributeGroup name="common.extra.attrib">
      <xs:attributeGroup ref="common.extra.attrib"/>
      <xs:attribute ref="dcr:datcat"/>
      <xs:attribute ref="dcr:valueDatcat"/>
    </xs:attributeGroup>
    <xs:complexType name="data-extension.type" mixed="true">
      <xs:complexContent>
        <xs:extension base="data-extension.type">
          <xs:choice minOccurs="0" maxOccurs="unbounded">
            <xs:sequence>
              <xs:any namespace="##other" maxOccurs="unbounded"
                  processContents="lax" minOccurs="0" />
            </xs:sequence>
          </xs:choice>
        </xs:extension>
      </xs:complexContent>
    </xs:complexType>
  </xs:redefine>

  <!-- further refinements of graphml.xsd -->

</xs:schema>
```

Note that we use the value "lax" for the processContents attribute of the xs:any element which allows for optional validation of embedded markup (if a schema location is provided). A benefit of this solution is that elements of any other markup language (as long as it is XML-based and defined via an XML schema) can be used as a virtual root element of an element subtree underneath GraphML's default and data elements, allowing for a shallow or deeply structured annotation of nodes and edges of linguistic networks and graphs.

We have imported the `dcr.xsd` XSD file from TEI's P5 as well, to include the two `datcat` and `valueDatcat` ISOcat attributes in the `common.extra.attrib` attribute group. As a result, we are able to add these two attributes that store a reference to an ISOcat category and value as a means to provide a semantic concept to a (linguistic) annotation.

5.3 Example Instance

For a brief demonstration of the outlined GraphML based format for describing linguistic networks, we will prepare a small instance of a Wiki-Graph, roughly following Mehler (2008).[27] A Wiki-Graph consists of three graph components: (1) an *agent graph*, with agents as vertices and collaborations as arcs, (2) a *document graph*, with articles as vertices and *intertextual relations* as arcs, and (3) a *word graph*, with words as vertices and word associations as arcs (Mehler 2008, p. 641 f.; see Fig. 6).

A possible serialization of an example instance of such a complex graph is shown in Listing 19.

Listing 19 Example demonstrating Wiki-Graphs

```
1  <?xml version="1.0" encoding="UTF-8"?>
2  <graphml xmlns="http://graphml.graphdrawing.org/xmlns"
        xmlns:xsi="http://www.w3.org/2001/XMLSchema-instance"
        xmlns:dcr="http://www.isocat.org/ns/dcr"
3    xmlns:tei="http://www.tei-c.org/ns/1.0"
4    xsi:schemaLocation="http://graphml.graphdrawing.org/xmlns
        GraphML-wildcard_verbose-dcr.xsd">
5    <key id="pT" for="node" attr.name="pageType"
6        attr.type="string">
7      <default>
8        <tei:fs>
9          <tei:f name="pageType">
10           <tei:string>article</tei:string>
11         </tei:f>
12       </tei:fs>
13     </default>
14   </key>
15   <key id="pos" for="node"/>
16   <key id="documentStructureIntro" for="node">
17     <default>
18       <tei:fs>
19         <tei:f name="Part">
20           <tei:string>Introduction</tei:string>
21         </tei:f>
22       </tei:fs>
23     </default>
24   </key>
25   <graph id="WikiGraph" edgedefault="directed">
```

---

[27] The presented model is simplified for demonstration purposes and not related to the expressiveness of the format.

```
26      <node id="Articles">
27        <graph id="ArticleGraph" edgedefault="directed">
28          <node id="Page1"/>
29          <node id="Page2">
30            <graph id="Page2SectionGraph" edgedefault="directed">
31              <node id="Page2Section1"/>
32              <node id="Page2Section2"/>
33              <edge source="Page2Section1" target="Page2Section2"/>
34            </graph>
35          </node>
36          <hyperedge>
37            <endpoint node="Page1"/>
38            <endpoint node="Page2"/>
39            <endpoint node="Page2Section1"/>
40          </hyperedge>
41        </graph>
42      </node>
43      <node id="Authors">
44        <graph id="AuthorGraph" edgedefault="undirected">
45          <node id="Author1"/>
46          <node id="Author2"/>
47          <edge source="Author1" target="Author2"/>
48        </graph>
49      </node>
50      <hyperedge>
51        <endpoint node="Author1"/>
52        <endpoint node="Page1"/>
53      </hyperedge>
54      <hyperedge>
55        <endpoint node="Author2"/>
56        <endpoint node="Page2"/>
57        <endpoint node="Page2Section2"/>
58      </hyperedge>
59    </graph>
60  </graphml>
```

Note that we neither include the full authors' graph (underneath the node identified by "Authors") nor the text internal structure of a wiki page due to space constraints. For the former, the graph serialization shown in Listing 15 can be used.

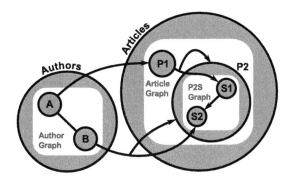

Fig. 6 Visualization of a Wiki-Graph

6 Conclusion

Since linguistic analysis is increasingly advancing in terms of heterogeneity of data, adequate formats to describe this data are required. Graphs are able to model a wide range of heterogeneous structures without narrowing the spectrum of possible research questions and without enforcing information loss. We have discussed different data models and formats that may be taken into account when considering a markup language for linguistic networks. Among these, we have chosen GraphML as an initial point, since it is reasonably expressive and supported by a large number of existing tools for network analysis and visualization. A proper new version of GraphML could include more aggressive cleaning of the underlying schema, resulting in a loss of backwards-compatibility and tool support. Depending on reactions from the GraphML community, however, such a development could be worthwhile. For the time being, the extensions of GraphML by both redefinition and XML Namespaces discussed in this chapter allow for additional annotations for complex linguistic networks such as Wiki-Graphs.

Acknowledgements. The foundation of this chapter was a working group meeting at the "Modeling Linguistic Networks: from Language Structures to Communication Processes" conference, held in December 2012 at the Goethe-University in Frankfurt am Main. Further participants included Jérôme Kunegis and Vincent Esche. We also want to thank Sven Banisch for his support in evaluating network analysis and visualization tools.

References

Bastian, M., Heymann, S., Jacomy, M.: Gephi: An Open Source Software for Exploring and Manipulating Networks. In: International AAAI Conference on Weblogs and Social Media (2009)

Berglund, A., Boag, S., Chamberlin, D., Fernández, M.F., Kay, M., Robie, J., Siméon, J.: XML Path Language (XPath). Version 2.0, 2nd edn. W3C Recommendation. World Wide Web Consortium, W3C (2010)

Boag, S., Chamberlin, D., Fernández, M.F., Florescu, D., Robie, J., Siméon, J.: XQuery 1.0: An XML Query Language, 2nd edn. W3C Recommendation. World Wide Web Consortium, W3C (2010)

Bormann, C., Hoffman, P.: Concise Binary Object Representation (CBOR). Specification. Internet Engineering Task Force (2013)

Bosak, J., Bray, T.: XML and the Second-Generation Web. Scientific American, 89–93 (1999)

Brandes, U., Eiglsperger, M., Lerner, J.: GraphML Primer (2004), http://graphml.graphdrawing.org/primer/graphml-primer.html

Brandes, U., Lerner, J., Pich, C.: GXL to GraphML and Vice Versa with XSLT. Electronic Notes in Theoretical Computer Science 127(1), 113–125 (2005), doi:10.1016/j.entcs.2004.12.037

Bray, T., Hollander, D., Layman, A., Tobin, R., Thompson, H.S.: Namespaces in XML 1.0, 3rd edn. W3C Recommendation. World Wide Web Consortium, W3C (2009)

Bray, T., Paoli, J., Sperberg-McQueen, C.M., Maler, E., Yergeau, F.: Extensible Markup Language (XML) 1.0, 5th edn. W3C Recommendation. World Wide Web Consortium, W3C (2008)

Burnard, L., Bauman, S. (eds.): TEI P5: Guidelines for Electronic Text Encoding and Interchange. Text Encoding Initiative Consortium, Charlottesville (2014), Version 2.7.0. Last updated on 16th September 2014, revision 13036

Clark, J.: XSL Transformations (XSLT) Version 1.0. W3C Recommendation. World Wide Web Consortium, W3C (1999)

Clark, J., DeRose, S.J.: XML Path Language (XPath). Version 1.0. W3C Recommendation. World Wide Web Consortium, W3C (1999)

Crockford, D.: The application/json Media Type for JavaScript Object Notation (JSON). Specification. Internet Engineering Task Force (2006)

DeRose, S.J., Maler, E., Orchard, D.: XML Linking Language (XLink) Version 1.0. W3C Recommendation. World Wide Web Consortium, W3C (2001)

DeRose, S.J., Maler, E., Orchard, D., Walsh, N.: XML Linking Language (XLink) Version 1.1. W3C Recommendation. World Wide Web Consortium, W3C (2010)

Ebert, J., Kullbach, B., Winter, A.: GraX – An Interchange Format for Reengineering Tools. In: Sixth Working Conference on Reverse Engineering, pp. 89–98. IEEE Computer Society, Los Alamitos (1999)

Gao, S.(S.)., Sperberg-McQueen, C.M., Thompson, H.S.: W3C XML Schema Definition Language (XSD) 1.1 Part 1: Structures. W3C Recommendation. World Wide Web Consortium, W3C (2012)

Herman, I., Marshall, M.S.: GraphXML – An XML-Based Graph Description Format. In: Marks, J. (ed.) GD 2000. LNCS, vol. 1984, pp. 52–62. Springer, Heidelberg (2001)

Himsolt, M.: GML: A portable Graph File Format. Tech. rep. Universitaät Passau (1997)

Holt, R.: An Introduction to TA: The Tuple Attribute Language (1997), http://plg.uwaterloo.ca/~holt/papers/ta-intro.html (updated 2002)

Ide, N.M.: Corpus Encoding Standard: SGML Guidelines for Encoding Linguistic Corpora. In: Proceedings of the First International Language Resources and Evaluation (LREC 1998), pp. 463–470. European Language Resources Association (ELRA), Granada (1998)

Ide, N.M., Bonhomme, P., Romary, L.: XCES: An XML-based Encoding Standard for Linguistic Corpora. In: Proceedings of the Second International Language Resources and Evaluation (LREC 2000), pp. 825–830. European Language Resources Association (ELRA), Athen (2000)

Ide, N.M., Priest-Dorman, G., Véronis, J.: Corpus Encoding Standard (CES). Tech. rep. Expert Advisory Group on Language Engineering Standards, EAGLES (1996)

ISO 12620:2009. Terminology and other language and content resources — Specification of data categories and management of a Data Category Registry for language resources. International Standard. International Organization for Standardization, Geneva

ISO 24610-1:2006. Language Resource Management — Feature Structures – Part 1: Feature Structure Representation. International Standard. International Organization for Standardization, Geneva

ISO 24612:2012. Language Resource Management — Linguistic annotation framework (LAF). International Standard. International Organization for Standardization, Geneva

ISO 24613:2008. Language Resource Management — Lexical markup framework (LMF). International Standard. International Organization for Standardization, Geneva

ISO/IEC 19757-2:2003. Information technology — Document Schema Definition Language (DSDL) – Part 2: Regular-grammar-based validation – RELAX NG (ISO/IEC 19757-2). International Standard. International Organization for Standardization, Geneva

Kay, M.: XSL Transformations (XSLT) Version 2.0. W3C Recommendation. World Wide Web Consortium, W3C (2007)

Mehler, A.: Structural Similarities of Complex Networks: A Computational Model by Example of Wiki Graphs. Applied Artificial Intelligence 22(7&8), 619–683 (2008), doi:10.1080/08839510802164085

Murata, M., Lee, D., Mani, M., Kawaguchi, K.: Taxonomy of XML Schema Languages Using Formal Language Theory. ACM Transactions on Internet Technology 5(4), 660–704 (2005)

NWB Team. Network Workbench Tool. Tech. rep. Northeastern University and University of Michigan (2006)

Peterson, D., Gao, S.(S.)., Malhotra, A., Sperberg-McQueen, C.M., Thompson, H.S.: W3C XML Schema Definition Language (XSD) 1.1 Part 2: Datatypes. W3C Recommendation. World Wide Web Consortium, W3C (2012)

Punin, J., Krishnamoorthy, M.: XGMML (eXtensible Graph Markup and Modeling Language). XGMML 1.0 Draft Specification. Dept. of Computer Science RPI, NY (2001)

Robie, J., Chamberlin, D., Dyck, M., Snelson, J.: XQuery 3.0: An XML Query Language. W3C Candidate Recommendation. World Wide Web Consortium, W3C (2013)

Saito, R., Smoot, M.E., Ono, K., Ruscheinski, J., Wang, P.-L., Lotia, S., Pico, A.R., Bader, G.D., Ideker, T.: A travel guide to Cytoscape plugins. Nature Methods 9, 1069–1076 (2012)

Schürr, A., Winter, A.J., Zündorf, A.: PROGRES: Language and Environment. In: Kreowski, H.-J., Ehrig, H., Engels, G., Rozenberg, G. (eds.) Handbook on Graph Grammars: Applications, Languages and Tools, vol. 2, pp. 487–550. World Scientific, Singapore (1999)

Smoot, M., Ono, K., Ruscheinski, J., Wang, P.-L., Ideker, T.: Cytoscape 2.8: new features for data integration and network visualization. Bioinformatics 27(3), 431–432 (2011)

Sperberg-McQueen, C.M., Burnard, L. (eds.): TEI P3: Guidelines for Electronic Text Encoding and Interchange. Published for the TEI Consortium by Humanities Computing Unit, University of Oxford, Oxford (1994)

Stührenberg, M.: What, when, where? Spatial and temporal annotations with XStandoff. In: Proceedings of Balisage: The Markup Conference, Montréal. Balisage Series on Markup Technologies, vol. 10 (2013), doi:10.4242/BalisageVol10.Stuhrenberg01

Stührenberg, M., Jettka, D.: A toolkit for multi-dimensional markup: The development of SGF to XStandoff. In: Proceedings of Balis-age: The Markup Conference, Montréal. Balisage Series on Markup Technologies, vol. 3 (2009), doi:10.4242/BalisageVol3.Stuhrenberg01

Stührenberg, M., Wurm, C.: Refining the Taxonomy of XML Schema Languages. A new Approach for Categorizing XML Schema Languages in Terms of Processing Complexity. In: Proceedings of Balisage: The Markup Conference. Vol. 5. Balisage Series on Markup Technologies, Montréal (2010), doi:10.4242/BalisageVol5.Stuhrenberg01

Thompson, H.S., Beech, D., Maloney, M., Mendelsohn, N.: XML Schema Part 1: Structures Second Edition. W3C Recommendation. World Wide Web Consortium, W3C (2004)

W3C Web Ontology Working Group, OWL 2 Web Ontology Language. Document Overview, 2nd edn. W3C Recommendation. World Wide Web Consortium, W3C (2012)

Walmsley, P.: Definitive XML Schema, 2nd edn. The Charles F. Gold-farb Definitive XML Series. Prentice Hall PTR, Upper Saddle River (2012)

Winter, A.J., Kullbach, B., Riediger, V.: An Overview of the GXL Graph Exchange Language. In: Diehl, S. (ed.) Dagstuhl Seminar 2001. LNCS, vol. 2269, pp. 324–336. Springer, Heidelberg (2002)

Linguistic Networks – An Online Platform for Deriving Collocation Networks from Natural Language Texts

Alexander Mehler and Rüdiger Gleim

1 Introduction

This section describes the *Linguistic Networks System* (LNS).[1] Its primary goal is to allow users for exploring texts from a network-oriented perspective. One aim is to let researchers – especially from the area of historical semantics (Jussen et al. 2007) – reveal particularities of the underlying texts that are hardly accessible otherwise. This relates to research questions of the following sort: *How is a given lexical unit, sentence or historical text connected to other linguistic instances in a usage-based manner? In which lexical context is it typically used and what are the lexical contexts of the context-forming units when explored in a recursive manner? Does this usage-based network reveal unexpected relations that can be related to historical, social-semantic processes?* Tackling such questions requires that the views provided by LNS go beyond collocation tables (as they are still predominant in historical semantics and corpus linguistics – cf. Evert (2008)). Networks may be an alternative in this sense: ideally, users can grasp lexical characteristics of texts visually by detecting clusters, extracting long-distance relations etc. Or they make them input to text classification and, more generally, to text mining (Masucci and Rodgers 2006; Amancio et al. 2008; Amancio et al. 2012). Networks may also be used to approximate the expressiveness of collocation networks derived from syntactic analyses (Ferrer i Cancho et al. 2004; Liu 2008; Liu and Hu 2008; Abramov and Mehler 2011) since the latter are computationally still very expensive – all the

---

Alexander Mehler · Rüdiger Gleim
Goethe-University Frankfurt
e-mail: {mehler,gleim}@em.uni-frankfurt.de

[1] See http://linguistic-networks.net. The LNS has been developed as part of the research project of the same name in the framework of the Priority Program "*Wechselwirkungen zwischen Natur- und Geisteswissenschaften*" of the *German Federal Ministry of Education* (BMBF).

more in the case low-resource languages for which parsers are out of reach. Further, lexical networks as computed by LNS may be taken to derive null models that serve for computing significance thresholds of graph invariants derived from linguistic networks (Masucci and Rodgers 2009; Amancio et al. 2013). These is are some tasks to be supported by LNS.

Table 1 gives an overview of different types of approaches to lexical-semantic networks as currently discussed in the literature together with exemplary references. In the cases 2–10, the collocation networks computed by LNS may be used as reference points (null models) to make the respective network models comparable (e.g., by the degree the latter depart from the former seen as null models, cf. Deyne et al. 2015). Alternatively, LNS may be used as a starting point to approximate lexical networks that are hard to observe empirically (e.g., association networks by example of expert languages) or prohibitively complex to compute (e.g., syntactic networks or historical lexico-semantic networks). In this way, LNS addresses two different communities and their tasks: visualizing collocation statistics in computational (historical) semantics for humanities scholars and inducing complex lexical networks for experts in network theory. From a linguistic point of view, a network-related variant of the contextual similarity hypothesis of Miller and Charles (1991) can be seen to underly this approach: since the similarities of contextual representations of words do not form equivalence classes when analyzed in a recursive manner, a lexical network is a natural candidate to represent the graph of these (syntagmatic or paradigmatic) relations. Note also that other than in the case of simple co-occurrence networks, we deal with weighted collocation networks where the words' edges are weighted by some collocation measure that assesses frequencies of co-occurrences in relation to some expected value (Evert 2008).

Generally speaking, to get from an input text to its quantitative analysis is a complex task – especially for humanities scholars: starting from model selection over linguistic preprocessing, network induction and edge weighting it traverses a huge parameter space. The motivation of developing LNS is to enable researchers from various disciplines to study their texts from a network perspective by bypassing this complexity. LNS supports the latter workflow and guides the user throughout the process of network induction. Once networks have been extracted the user can query and browse the resulting network data. This is especially needed for humanities scholars, who ultimately aim at qualitative network analyses instead of quantitative ones. However, since LNS stores network data in GraphML, it can be used to start quantitative analyses too.

In this paper, we outline the workflow served by LNS, its underlying software architecture and technical implementation. We also outline the parameter space that is currently traversable by LNS. Finally, we conclude with a prospect on future work.

Table 1 Approaches to lexical-semantic networks

No. Approach	Vertices	Edges	Weighting	Reference
1. co-occurrence network	word form ∨ syntactic word ∨ lemma	co-occurrence	syntagmatic association	Ferrer i Cancho and Solé (2001)
2. nth order collocation network	word form ∨ syntactic word ∨ lemma	similarity	syntagmatic / paradigmatic association	Bordag et al. (2003)
3. syntactic word network	head/modifier ∨ lemma	dependency	degree of dependency	Ferrer i Cancho et al. (2004)
4. syntactic word network	word form ∨ syntactic word ∨ lemma	co-constituency	degree of embedding	Corominas-Murtra et al. (2007)
5. concept network	concept (e.g. *synset*)	sense relation	sense-relational neighborhood	Steyvers and Tenenbaum (2005)
6. thesaurus network	lemma	thesaurus relation	lexicographic relation	Jesus Holanda et al. (2004)
7. association network	prime ∧ response word	priming	cognitive association	Borge-Holthoefer and Arenas (2010) and Deyne et al. (2013)
8. tag network	tag	co-tagging	co-usage	Cattuto et al. (2009)
9. encyclopedic network	article, category, …	encyclopedic relation	ontological distance	Mehler (2011)
10. *mixture models*	heterogeneous	heterogeneous	multiple	Gravino et al. (2012)

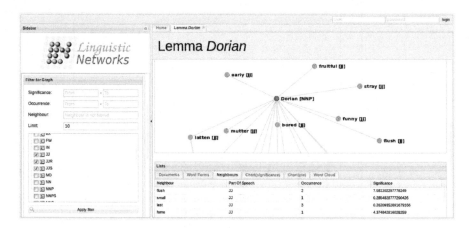

Fig. 1 Screenshot of the Linguistic Networks WebApp showing query results of "Dorian" based on Oscar Wilde's novel "Dorian Gray"

Fig. 2 The workflow of LN

2 On the Parameter Space of LN

Fig. 1 shows the interface of LNS. A sidebar to the left allows for selecting the working language (currently, English, German or Latin – see Fig. 4), the corpus underlying the analysis and the parameter setting of network induction. Query results (which serve to traverse the output networks) are shown in a tab panel. The sidebar holds additional panels depending on the active view in the center. It allows, for example, filtering the network by the number of nodes to be displayed, the degree of collocation (as a lower bound) and for projecting onto selected parts of speech. In this way, the results of different lexical networks can be compared with each other according to multiple settings and views.

As illustrated in Fig. 2, LNS manifests a typical NLP pipeline. Note that evaluation and quantitative network analysis occur outside of LNS (e.g., by means of software like Gephi, see Bastian et al. 2009). From the user's perspective the initial step is to select an input text (corpus) together with the collocation model that underlies edge generation. Model selection (of a collocation measure and its parameters – see below) determines the number of edges and the way they are weighted. Further, by specifying the type of constituents (wordforms or lexemes) of the nodes to be generated, the user determines indirectly the order (i.e., number of nodes) of the network. By sending the text input together with the selected collocation model to the server, text preprocessing and graph induction are triggered. To this end, the texts are automatically preprocessed and converted into TEI P5[2] (TEI Consortium 2007) – see Mehler et al. (2015) for the most current description of this procedure by example of Latin texts. The meaning of the preprocessing step is twofold: Firstly, by extracting the logical document structure of a text and the sequence of its tokens, the counting frames and the units of collocation statistics are determined (i.e., the entities that generate the nodes). Secondly, by lemmatizing and PoS-tagging the tokens, information is gained that allows for applying filters during network induction and subsequent query processes. As soon as the computation is completed, the input texts and the derived networks are accessible for querying. The user can browse through different views of the data, query for specific nodes of selected types and filter their contexts. Currently, the system includes views of the preprocessed text using the TEI format, a graph view of the resulting network and two additional views (distribution and word cloud) of the frequency distribution of the text's lexical constituents. Once the user has selected a sub-graph of interest, she can download it as a GraphML file for further analysis. GraphML (GraphML Project Group 2014) is a standard format of machine-based graph representation. It can be imported by Gephi (Bastian et al. 2009) to perform quantitative analyses of the networks generated by LNS. In this way, LNS bridges between the original documents and state-of-the-art software for network analysis using the standard approach of collocation statistics – without the need of any programming expertise on the side of the user. There are currently only few web-based systems like LN that make lexical networks in this way accessible to the research community. A related example is ConText (Diesner 2014) which differs from LN because of its more general network perspective and its integration of ML components, while LN focuses more on linguistic networks of modern as well as ancient languages.

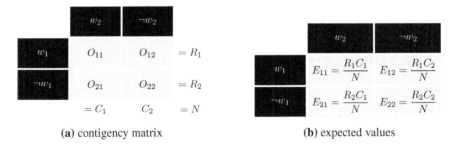

(a) contingency matrix   (b) expected values

Fig. 3 Computation of the contingency matrix (left) and the matrix of expected values (right) (Evert 2008)

The data model of LNS is general enough to support the level of texts, sentences and tokens so that networks on these three levels can, in principle, be computed and visualized. However, only lexical networks are currently supported – on the level of lemmata and wordforms. The network induction supports a set of collocation

---

[2] TEI is a *de facto* standard for text representation.

measures to compute weighted edges. Parameterizing such a measure for network induction includes the specification of its counting frame. This frame can be of type *textual* (where sentences are selected as part of the logical document structure) or of type *surface structural* (if skip-bigrams have to be explored). In the latter case, the user needs to specify the size of the frame in terms of the number of left and right neighbors around focal nodes. Intuitively, the degree of collocation of two words w_1, w_2 should be rated higher if they co-occur more often in instances of the selected frame than expected by chance. This intuition is reflected by collocation measures (Evert 2008). The (absolute) number of observations of co-occurrences of pairs of words is recorded in a contingency matrix (see figure 3a). In combination with the respective expected values (see figure 3b) and the total number of co-occurrence frames, the edges can be weighted. Table 2 lists the collocation measures that are currently supported by LNS whose implementation allows for extensions of this list. See Church and Hanks (1990), Evert (2008), Heyer et al. (2006), Pecina (2010) and Rieger (1989) for more information about these and related measures.

The parameter space being implemented by LNS is summarized in Fig. 4. Using large PoS tagsets (e.g., the Brown tagset in the case of English or (a subset of) the STTS in the case of German), a huge set of alternative views can be laid over the same output network. In this way, purely nominal networks or networks of personal names can be computed as well as networks including only adjectives, adverbs, nouns or verbs, or networks that focus on function words or pronouns. This gives access to text representation models as recently used in text classification (Stamatatos 2011), network motif detection (Biemann et al. 2015) or in syntactic network analysis (Chen and Liu 2015). Note that as shown in Fig. 4, the choice of the type of lexical network (wordform or lemma) does not restrict the subsequent query process. That is, for example, lemma networks can be queried by means of wordforms. This query component qualifies LNS for its usage in historical semantics, which, so far, is restricted to concordances in the form of *Key Word in Context*-lists.

3 The Software Architecture of LN

LNS' architecture is mainly driven by the requirement to combine generic, extensible representations with efficient means to browse and query them. The aim is to find a balance between performance and generality: (over-)optimization to a specific task likely restricts extensibility. On the other hand, keeping every aspect generic induces additional, time-consuming overhead. In this section, we outline the modular approach chosen to implement LNS in the light of this trade-off.

Fig. 5 depicts LNS' architecture: boxes represent modules implementing specific functionalities. Breaking down the overall architecture into modules helps to abstract from implementation details and, if necessary, to replace a given solution by another one. The *LN Data Core* provides means to specify classes of objects, their attributes and interrelations as well as mechanisms to build and query them. This module serves as a layer that abstracts from the storage engine representing and holding the data. Generally speaking, by allowing the design of one module to

Table 2 List of collocation measures implemented in Linguistic Networks

CHI Square	$\sum_{ij} \frac{(O_{ij}-E_{ij})^2}{E_{ij}}$
Binomial	$\begin{cases} 0 & \text{if } \frac{O_{11}+1}{E_{11}} \leq 2.5 \\ \frac{E_{11}-O_{11}*\log(E_{11})+\log(O_{11}!)}{\log(N)} & \text{if } 2.5 < \frac{O_{11}+1}{E_{11}} \leq 10 \\ \frac{O_{11}*(\log(O_{11})-\log(E_{11})-1)}{\log(N)} & \text{if } \frac{O_{11}+1}{E_{11}} > 10 \end{cases}$
Local Mutual Information	$O_{11} * log_2(\frac{O_{11}}{E_{11}})$
Log Likelihood	$\sum_{ij} O_{ij} \log_2(\frac{O_{ij}}{E_{ij}})$
Log Odds Ratio	$\log_2(\frac{(O_{11}+\frac{1}{2})*(O_{22}+\frac{1}{2})}{(O_{12}+\frac{1}{2})*(O_{21}+\frac{1}{2})})$
Mutual Information	$\log_2(\frac{O_{11}^R}{E_{11}})$
Mutual Information ^{R}	$\log_2(\frac{O_{11}}{E_{11}})$
Rieger's α (a correlation coefficient operating on certain observed and expected values)	
Simple Log Likelihood	$2(O_{11} \cdot \log_2\left(\frac{O_{11}}{E_{11}}\right) - (O_{11} - E_{11}))$
T Score	$\frac{O_{11}-E_{11}}{\sqrt{O_{11}}}$
Z Score	$\frac{O_{11}-E_{11}}{\sqrt{E_{11}}}$

influence the one of other modules makes it difficult to replace and extend them. Therefore, we use this abstraction layer.

Entities, attributes and relations are specified in terms of *Contracts*. *Builders*, *Queries* and *Cursors* are built on top of contract specifications. The *LN Data Core* is implemented in a generic manner so that storing and retrieving data is kept abstract. A further component is given by *Data formats*. Clients connecting to the *LN Server* can differ by required data formats: the *LN WebApp* relies on *JSON*, a format which is common to JavaScript-based web applications. However, when acting as a web service, XML formats are used instead. To keep the way data is returned to clients flexible, the *LN Data Core* provides the *Format API*. It allows for rendering data objects into different formats.

The current release of LNS uses Neo4j (Neo Technology 2014) a freely available[3] graph database known for its performance and robustness. All aspects of data modeling and retrieval left abstract in *LN Data Core* are implemented in the *LN Data Neo* module. This module contains extensions to pick search strategies (index vs. graph-based) for answering queries. Though Neo4j offers a query language for graph databases called *Cypher*, we do not use it because it collides with the requirement to keep a generic design.

---

[3] Neo4j is dual-licensed as GPLv3 and AGPLv3 as well as commercial.

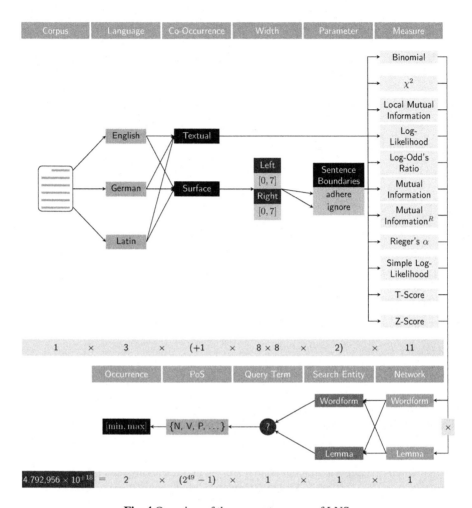

Fig. 4 Overview of the parameter space of LNS

The number of networks to be hosted by LNS should not be limited by the capacity of a single DB instance. Even if a DB can hold dozens of large networks, it complicates maintenance and backup. Therefore, we store networks derived from corpora in separate Neo4j DBs. That is, LNS works on multiple databases. In cases like this, it is common practice to use connection pools. Since opening and closing a DB takes time, these operations should be avoided whenever possible. A connection pool offers a solution for this task. If the server receives a request for a new corpus, it checks whether the pool contains a fitting connection. If available, as in the case of frequently queried databases, this connection is used. Otherwise, a new connection is created and added to the pool. Since the number of connections kept in a pool is limited by server memory, connections which have not been used for a longer time

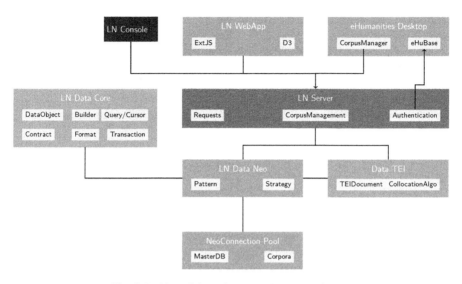

Fig. 5 Outline of the software architecture of LNS

are closed. Note that Neo4j does not provide mechanisms for connection pooling so that we developed the *NeoConnection Pool* module for this task.

The *Data TEI* module completes the server components for data modeling and representation. It provides all means to import, represent and browse TEI P5 documents. It also provides functions for inducing collocation networks including implementations of the collocation measures of Table 2.

The *LN Server* is implemented as a *RESTfull Web Service* based on *Jersey*[4]. Its primary function is to receive requests from clients, to use the underlying data modules and to return output according to the requested formats. LNS distinguishes between public and private corpora. This requires a session management by the *LN Server*: *who is currently logged in and which corpora are visible to whom?* LNS does not maintain its own user database but uses the one of the *eHumanities Desktop* (Gleim et al. 2012) for authentication. Thus, any user of the *eHumanities Desktop* granted access to LNS, can log in, upload texts and create networks. In addition, the *eHumanities Desktop* allows for directly sending documents from its *Corpus Manager* to LNS. The Linguistic Networks web application (*LN WebApp*) is implemented based on the *ExtJS* (Sencha 2014) JavaScript Framework and *D3* (Bostock 2014) for visualization. Finally, the *LN Console* is a maintenance tool which can be used, amongst others, to directly upload large corpora into the system.

Having all modules together, how is a document imported into the system and what steps are involved to generate a network thereof? Typically, users login via the *LN WebApp*. If authentication is successful, the user can upload documents as plain text. Then, the source language, counting frame and collocation measure are selected. Users can additionally decide to make the generated networks visible for

---

[4] https://jersey.java.net

other users. By activating the upload button, the computation is triggered and run asynchronously in the background. On the server, the uploaded texts are preprocessed and converted into TEI. The TEI document is parsed by the *Data TEI* module and imported into a newly created Neo4j Graph database via *LN Data Neo* and *LN Data Core*. Next, the network structure is extracted by the *Data TEI* module and inserted into a new graph database. Finally, access permissions and meta data are stored in the *MasterDB* to make the networks ready for qualitative analysis.

4 Summary

We presented the LNS web application. It enables users to upload texts, derive linguistic networks thereof and explore their visualizations. LNS focuses on offering views on linguistic networks derived per input text (corpus). Currently, the system does not support automatic comparisons of search results and networks. The former are still better done with the help of the *Historical Semantics Corpus Management System* (www.comphistsem.org) of the *eHumanities Desktop*. Thus, future work focuses on integrating graph similarity measures for labeled graphs (Schenker et al. 2005). Sentence and text networks are a second area of extension as are visualizations of multilevel linguistic networks. A special task comes from historical semantics that requires editable networks for eliminating preprocessing errors, clustering semantically related units and, more generally, for curating the edition of the networks. This requirement shows that whatever is provided by a tool like LNS from the area of computational humanities, it has to allow for manual corrections and additions by humanities scholars who expect error rates currently out of reach for automatic text analysis.

Acknowledgements. LNS was developed in a research project funded by the German Federal Ministry of Education (BMBF). The first releases of LNS have been implemented by Markus Lux, Christian Menßen and Uli Waltinger based on MySQL. The second release based on Neo4j has been implemented by Christian Peil, Benjamin Bronder and Pavel Balazki. Recent additions have been made by members of the Text Technology Lab at Frankfurt University.

References

Abramov, O., Mehler, A.: Automatic Language Classification by Means of Syntactic Dependency Networks. Journal of Quantitative Linguistics 18(4), 291–336 (2011)

Amancio, D.R., Altmann, E.G., Rybski, D., Oliveira Jr., O.N., da F. Costa, L.: Probing the Statistical Properties of Unknown Texts: Application to the Voynich Manuscript. PLoS ONE 8(7), e67310 (2013), doi:10.1371/journal.pone.0067310

Amancio, D.R., Antiqueira, L., Pardo, T.A.S., da Fontoura Costa, L., Oliveira, O.N., das Graças Volpe Nunes, M.: Complex Networks Analysis of Manual and Machine Translations. International Journal of Modern Physics C 19(4), 583–598 (2008)

Amancio, D.R., Oliveira Jr., O.N., da Fontoura Costa, L.: Identification of literary movements using complex networks to represent texts. New Journal of Physics 14, 043029 (2012)

Bastian, M., Heymann, S., Jacomy, M.: Gephi: An Open Source Software for Exploring and Manipulating Networks. In: Proceedings of the 3rd International AAAI Conference on Weblogs and Social Media (2009)

Biemann, C., Krumov, L., Roos, S., Weihe, K.: Network Motifs Are a Powerful Tool for Semantic Distinction. In: Mehler, A., Lücking, A., Banisch, S., Blanchard, P., Job, B. (eds.) Towards a Theoretical Framework for Analyzing Complex Linguistic Networks. Springer, Berlin (2015)

Bordag, S., Heyer, G., Quasthoff, U.: Small worlds of concepts and other principles of semantic search. In: Böhme, T., Heyer, G., Unger, H. (eds.) IICS 2003. LNCS, vol. 2877, pp. 10–19. Springer, Heidelberg (2003)

Borge-Holthoefer, J., Arenas, A.: Semantic Networks: Structure and Dynamics. Entropy 12(5), 1264–1302 (2010), doi:10.3390/e12051264

Bostock, M.: D3 Data-Driven Documents (2014), http://d3js.org

Cattuto, C., Barrat, A., Baldassarri, A., Schehr, G., Loreto, V.: Collective dynamics of social annotation. PNAS 106(26), 10511–10515 (2009)

Chen, X., Liu, H.: Function Nodes in the Chinese Syntactic Networks. In: Mehler, A., Lücking, A., Banisch, S., Blanchard, P., Job, B. (eds.) Towards a Theoretical Framework for Analyzing Complex Linguistic Networks. Springer, Berlin (2015)

Church, K.W., Hanks, P.: Word Association Norms, Mutual Information, and Lexicography. Computational Linguistics 16(1), 22–29 (1990)

Corominas-Murtra, B., Valverde, S., Solé, R.V.: Language Networks: their structure, function and evolution. ArXiv e-prints (2007)

de Deyne, S., Navarro, D., Storms, G.: Associative strength and semantic activation in the mental lexicon: evidence from continued word associations. In: Proceedings of the 35th Annual Meeting of the Cognitive Science Society, pp. 2142–2147 (2013)

de Deyne, S., Verheyen, S., Storms, G.: Structure and Organization of the Mental Lexicon: a Network Approach Derived from Syntactic Dependency Relations and Word Associations. In: Mehler, A., Lücking, A., Banisch, S., Blanchard, P., Job, B. (eds.) Towards a Theoretical Framework for Analyzing Complex Linguistic Networks. Springer, Berlin (2015)

Diesner, J.: ConText: Software for the Integrated Analysis of Text Data and Network Data. Paper presented at the Social and Semantic Networks in Communication Research. Preconference at Conference of International Communication Association (ICA), Seattle, WA (2014)

Evert, S.: Corpora and collocations. In: Lüdeling, A., Kytö, M. (eds.) Corpus Linguistics. An International Handbook of the Science of Language and Society, pp. 1212–1248. Mouton de Gruyter, Berlin (2008)

Ferrer i Cancho, R., Solé, R.V.: The Small-World of Human Language. Proceedings of the Royal Society of London. Series B, Biological Sciences 268(1482), 2261–2265 (2001)

Ferrer i Cancho, R., Solé, R.V., Köhler, R.: Patterns in Syntactic Dependency-Networks. Physical Review E 69(5), 051915 (2004)

Gleim, R., Mehler, A., Ernst, A.: SOA implementation of the eHumanities Desktop. In: Proceedings of the Workshop on Service-oriented Architectures (SOAs) for the Humanities: Solutions and Impacts, Digital Humanities 2012, Hamburg, Germany (2012)

GraphML Project Group (ed.): The GraphML File Format (2014), http://graphml.graphdrawing.org

Gravino, P., Servedio, V.D.P., Barrat, A., Loreto, V.: Complex Structures and Semantics in Free Word Association. Advances in Complex Systems 15(3-4) (2012)

Heyer, G., Quasthoff, U., Wittig, T.: Text Mining: Wissensrohstoff Text. W3L, Herdecke (2006)

de Jesus Holanda, A., Pisa, I.T., Kinouchi, O., Martinez, A.S., Ruiz, E.E.S.: Thesaurus as a complex network. Physica A: Statistical Mechanics and its Applications 344(3-4), 530–536 (2004), doi:10.1016/j.physa.2004.06.025

Jussen, B., Mehler, A., Ernst, A.: A Corpus Management System for Historical Semantics. Sprache und Datenverarbeitung. International Journal for Language Data Processing 31(1-2), 81–89 (2007)

Liu, H.: The complexity of Chinese syntactic dependency networks. Physica A: Statistical Mechanics and its Applications 387(12), 3048–3058 (2008), doi:10.1016/j.physa.2008.01.069

Liu, H., Hu, F.: What role does syntax play in a language network? Europhysics Letters 83, 18002–18008 (2008)

Masucci, A.P., Rodgers, G.J.: Differences Between Normal and Shuffled Texts: Structural Properties of Weighted Networks. Advances in Complex Systems 12(1), 113–129 (2009)

Masucci, A., Rodgers, G.: Network properties of written human language. Physical Review E 74, 1–8 (2006)

Mehler, A.: Social Ontologies as Generalized Nearly Acyclic Directed Graphs: A Quantitative Graph Model of Social Ontologies by Example of Wikipedia. In: Dehmer, M., Emmert-Streib, F., Mehler, A. (eds.) Towards an Information Theory of Complex Networks: Statistical Methods and Applications, pp. 259–319. Birkhäuser, Boston (2011)

Mehler, A., vor der Brück, T., Gleim, R., Geelhaar, T.: Towards a Network Model of the Coreness of Texts: An Experiment in Classifying Latin Texts using the TTLab Latin Tagger. In: Biemann, C., Mehler, A. (eds.) Text Mining: From Ontology Learning to Automated Text Processing Applications. Theory and Applications of Natural Language Processing, pp. 87–112. Springer, Berlin (2015)

Miller, G.A., Charles, W.G.: Contextual Correlates of Semantic Similarity. Language and Cognitive Processes 6(1), 1–28 (1991)

Neo Technology (ed.): Neo4j (2014), http://neo4j.com

Pecina, P.: Lexical association measures and collocation extraction. In: Rayson, P., Piao, S., Sharoff, S., Evert, S., Moiron, B.V. (eds.) Language Resources and Evaluation, vol. 44(1), pp. 137–158 (2010)

Rieger, B.B.: Unscharfe Semantik: Die empirische Analyse, quantitative Beschreibung, formale Repräsentation und prozedurale Modellierung vager Wortbedeutungen in Texten. Peter Lang, Frankfurt a. M. (1989)

Schenker, A., Bunke, H., Last, M., Kandel, A.: Graph-Theoretic Techniques for Web Content Mining. World Scientific, New Jersey (2005)

Sencha (ed.): ExtJS (2014), http://www.sencha.com/products/extjs/

Stamatatos, E.: Plagiarism Detection Using Stopword N-grams. Journal of the American Society for Information Science and Technology 62(12), 2512–2527 (2011), doi:10.1002/asi.21630

Steyvers, M., Tenenbaum, J.: The large-scale structure of semantic networks: Statistical analyses and a model of semantic growth. Cognitive Science 29(1), 41–78 (2005)

TEI Consortium (ed.): TEI P5: Guidelines for Electronic Text Encoding and Interchange (2007), http://www.tei-c.org/Guidelines/P5/

Author Index

Čech, Radek   167

Araújo, Tanya   107

Banisch, Sven   V, 107
Baxter, Gareth J.   257
Beckage, Nicole M.   3
Biemann, Chris   83
Blanchard, Philippe   V

Chen, Xinying   187
Colunga, Eliana   3

De Deyne, Simon   47
Diewald, Nils   299

Eguíluz, Víctor M.   133

Ferrer-i-Cancho, Ramon   203
Frank-Job, Barbara   V

Gleim, Rüdiger   299, 331
Goldstein, Rutherford   29
Gong, Tao   237

Hernández-García, Emilio   133

Johnson, Elizabeth   29

Kalampokis, Alkiviadis   133
Krumov, Lachezar   83

Lücking, Andy   V
Liu, Haitao   167, 187

Mačutek, Ján   167
Maity, Suman Kalyan   279
Masucci, A. Paolo   133
Mehler, Alexander   V, 331
Mukherjee, Animesh   279

Roos, Stefanie   83

Shuai, Lan   237
Stührenberg, Maik   299
Storms, Gert   47

Verheyen, Steven   47
Vitevitch, Michael S.   29

Weihe, Karsten   83

Zweig, Katharina Anna   153

Printed by Printforce, the Netherlands